KU-255-039

Plant Gene Research

Basic Knowledge and Application

Edited by
E. S. Dennis, Canberra
B. Hohn, Basel
Th. Hohn, Basel (Managing Editor)
P. J. King, Basel
J. Schell, Köln
D. P. S. Verma, Montreal

Springer-Verlag Wien New York

Genetic Flux in Plants

Edited by B. Hohn
and E. S. Dennis

Springer-Verlag Wien New York

Dr. Barbara Hohn
Friedrich Miescher-Institut, Basel

Dr. Elisabeth S. Dennis
CSIRO Division of Plant Industry, Canberra

With 40 Figures

Printed in Austria

Library of Congress Cataloging in Publication Data

Main entry under title:

Genetic flux in plants.

(Plant gene research)
Includes bibliographies.
1. Plant genetics. I. Hohn, B. (Barbara), 1939– .
II. Dennis, E. S. (Elisabeth S.), 1943– .
III. Series.
QK981.G4354 1985 581.1'5 85-17278

ISBN 0-387-81809-X (U.S.)

ISSN 0175-2073
ISBN 3-211-81809-X Springer-Verlag Wien – New York
ISBN 0-387-81809-X Springer-Verlag New York – Wien

Preface

Genetic material is in flux: this is one of the most exciting recent concepts in molecular biology. This volume of "Plant Gene Research" describes changes that occur in the genetic material of plants. It is worthwhile remembering that the first examples of unstable genomes were described for maize before DNA was known to be the genetic material. Now transposable elements like the ones found in maize have been described in almost all organisms and have become incorporated into our thinking about genome structure.

Flux in the plant genome is not restricted to transposable elements or to nuclear genes. Exchanges of genetic material have been demonstrated within organelle DNA, between organelle DNAs or between organelle and nuclear DNAs. Such exchanges may only occur over evolutionary times or may be a continuing process.

Also the environment alters the plant genome. Stress, either viral, nutritional or tissue-culture induced causes heritable changes in the genome. Infection with the crown gall bacterium *Agrobacterium tumefaciens* results in the transfer of bacterial DNA into the plant genome.

There are many questions to be answered. Since plant genomes can change in response to the environment, does this process make plant more adaptable? Do any of the DNA changes regulate gene expression during development? Surely, the presence of these mechanism provides a basis for selection during evolution. Is it because plants have so much non-genic DNA that these changes can be accomodated? Or is it that these processes do occur in all living organisms but that we can see them more clearly in plants?

Although our concepts of flux in the plant genome are rapidly developing, we hope that the present volume will give a good view of the present "state of the art" and stimulate future thinking and experimentation.

Basel and Canberra, June 1985 B. Hohn, E. S. Dennis

Contents

Section I.
Movement of Genetic Information from the Environment to the Plant

Section II.
Movement of Genetic Information Between the Plant Organelles

Section III.
Movement of Genetic Information Within Plant Organelles

Chapter 6 **Supernumerary DNAs in Plant Mitochondria**
R. R. Sederoff and C. S. Levings III, Raleigh, N. C., U.S.A.

Chapter 7 **Plant Mitochondrial DNA: Unusual Variation on a Common Theme**
Arnold J. Bendich, Seattle, Wash., U.S.A.

Chapter 8 **Repeated Sequences and Genome Change**
R. B. Flavell, Cambridge, U. K.

Chapter 9 **Sequence Variation and Stress**
C. A. Cullis, Norwich, U. K.

Chapter 10 **The Activation of Maize Controlling Elements**
S. L. Dellaporta and P. S. Chomet, Cold Spring Harbor, N. Y., U. S. A.

Section I.
Movement of Genetic Information from the Environment to the Plant

Chapter 1

Viruses

H. Fraenkel-Conrat

Department of Molecular Biology, University of California
Berkeley, CA 94720, U.S.A

Near the end of the past century a completely new type of transmissible plant pathogen was discovered, at first termed *contagium vivum fluidum,* then filterable virus, and finally just virus, the agent that caused tobacco mosaic disease: the tobacco mosaic virus (TMV). Soon many animal and other plant diseases were found to be due to similar agents. What differentiates the viruses from previously known infectious agents, besides their submicroscopic size, was their obligatory parasitic nature: They could only become replicated within the host's cells. Viruses lack the ability to metabolize and thus to produce the energy and the materials needed for their replication. Their dependence on the host cell makes them a cell-biological phenomenon. That many viruses also interact with the host's genome became realized later.

Since its first recognition as representing a new biological phenomenon by Beijerinck (1898), TMV has continued to occupy center stage in subsequent elucidations of the nature of viruses, with particular reference to plant viruses: first to be isolated in pure pseudocrystalline form by Stanley (1935); first to be split into and reconstituted from its two types of components, protein and RNA (Fraenkel-Conrat and Williams, 1955; Fraenkel-Conrat *et al.,* 1957); first demonstration of genomic and infectious RNA (Fraenkel-Conrat, 1956; Fraenkel-Conrat and Singer, 1957; Gierer and Schramm, 1956); first plant viral protein and later RNA to be sequenced (Tsugita *et al.,* 1960; Anderer *et al.,* 1960; Goelet *et al.,* 1982)*.

---

* The literature on TMV, not to mention plant viruses in general, is so enormous that only a few randomly selected citations can be included in this chapter. For more complete documentation the reader is referred to textbooks, monographs, and major handbook series, such as *Virology* (H. Fraenkel-Conrat and P. Kimball, Plenum Publishing Corporation, New York, 1982); Vols. 11, 13 and others of *Comprehensive Virology* (Eds. H. Fraenkel-Conrat and R. R. Wagner, Plenum Publishing Corporation, New York, 1974—1984); *Plant Virology* (R. E. F. Matthews, Academic Press, New York, 1970).

The characterization of TMV was soon followed by those of many other viral plant pathogens. We will briefly describe some of the more important groups of the about 20 that have now been officially classified (Table 1), and then focus our attention on the question to what extent various plant viruses are dependent upon, and interact with, the host's genome.

Table 1. *Most Studied "Plus-strand RNA Plant" Viruses*

Group Name	Prototype	Shape of virions and requirements for infectivity ( )*	Approx. RNA M.W.$\times 10^{-6}$	Approx Protein M.W.$\times 10^{-3}$
		A. Typical		
Tobamoviruses	tobacco mosaic virus (TMV)	300 × 18 nm rods	2.1	17
Tymoviruses	turnip yellow mosaic virus (TYMV)	29 nm isometric	2.1	20
Potexviruses	potato virus X (PVX)	13 × 500 nm filaments	2.3	14
Potyviruses	potato virus Y (PVY)	11 × 750 nm filaments	3.3	33
Tombusviruses	tomato bushy stunt virus (TBSV)	30 nm isometric	1.5	40
Tobraviruses	tobacco rattle virus (TRV)	long and short 22 nm rods (2)	2.4, ~ 1	22
Almoviruses	alfalfa mosaic virus (AMV)	4 lengths, nonhelical (3)	1, 0.7, 0.6, 0.3	24
Bromoviruses	brome mosaic virus (BMV)	26 nm isometric (3)	1.1, 0.75, 0.3	20
Cucumoviruses	cucumber mosaic virus (CMV)	28 nm isometric (3)	1.3, 1.1, 0.8, 0.3	24
		B. Non-Typical		
Comoviruses	cowpea mosaic virus (CpMV) related to animal picornaviridae	28 nm isometric (2)	2.4, 1.4	42, 22
Tobacco necrosis virus (TNV)	translated like RNA phages	28 nm isometric	1.4	23
Tobacco necrosis satellite (STNV): replicated only with TNV		17 nm isometric	0.4	23

* Number of particles carrying different RNAs.

We must begin by stating that there is no strict meaning to the term plant viruses. Very many of these viruses are transmitted by animals, mostly insects, and many are replicated in these animal vectors. Further, many plant viruses show remarkable similarities to members of certain animal virus families, and are thus classified as genera of these families. Examples are plant rhabdoviridae (minus-strand RNA viruses, the type species of which is vesicular stomatitis virus of cattle); plant reoviridae (doublestranded RNA viruses, the type species being human orthoreovirus); cauliflower mosaic virus, a DNA virus which shows marked similarities to human hepatitis B virus, and may become replicated via a reverse transcription step, analogous to the RNA tumor viruses of mammals and birds. For the geminiviruses with two singlestranded DNAs the corresponding animal viruses have not yet been found, but probably will be. All of these important and numerous "plant viruses" I will omit from further discussion.

Finally there are the viroids, infectious small circular RNA molecules (350—450 nucleotides) which again represent a new biological phenomenon, since they are neither messenger nor strictly genomic nucleic acid, yet as transmissible and pathogenic as viruses.

"Typical" plant virions consist of only plus-stranded RNA and one or rarely two coat proteins. Most are isometric, generally icosahedral bodies of about 25—35 nm diameter, composed usually of 180 identical protein molecules; others are rigid rods (e. g., TMV, tobacco rattle virus); others variously long filaments (e. g., potato virus X and Y), all consisting to about 95 % of several thousand helically arranged protein molecules. The virions' particle weights range from 5 to 50×10^6, their RNA representing 30 – 6 %, a total of 1.5 to 3×10^6 dalton. In most plant virus groups the RNA is segmented into two or three components of characteristic molecular weights. These segments may be packaged within the same virion (e. g., brome mosaic and cucumber mosaic viruses, BMV, CMV), or separately encapsidated in virions of different size (e. g., alfalfa mosaic virus and tobacco rattle virus, AMV, TRV). Even viruses with unsegmented RNA, such as TMV, have strains in which a specific RNA segment is separately encapsidated as a short rod (Higgins *et al.,* 1976; Fukuda *et al.,* 1981).

The simplest hypothesis accounting for the mode of operation of such viruses leading to their replication is that they carry genomic mRNAs into the cell which are translated into their coat protein and possibly other virus-specific proteins by the host's ribosomes. Since their RNA must be replicated and RNA replicating enzymes were believed not to exist in cells, virus-specific RNA replicases were surmised to be essential gene products of RNA viruses. However, while genes for such enzymes have definitely been found in simple animal plus-strand RNA viruses (e. g., polio virus), they have not been found in the "typical" plant viruses. The explanation for this has become evident in recent years when it was found that plant cells contain RNA dependent RNA polymerases, as will be discussed in more detail below. Actually the only plant viral gene product that is clearly characterized in terms of function is that for the coat protein. It appears that the short segments of most divided genome viruses are the genes for the coat protein, and that the long single RNAs of other viruses may become intracellularly processed to yield short coat protein mRNAs (Hunter *et al.,* 1976). Surely the purpose of this is the greater ease and higher rate of translatability of short mRNAs which must yield great numbers of coat protein molecules (2130 for a TMV rod), while other gene products may be required in much smaller amounts. The functions of the subgenomic RNAs and/or their translation products (at least two and probably usually 3—6) remains unclear in molecular terms. It appears probable that they interact with the host genome and in some way account for the host specificity and pathogenicity (if any) of the viruses. It appears very likely that the viroids rely entirely on the latter type of action for their pathogenicity (Diener, 1979).

Regarding the mode of translation of "typical" plant viral RNAs, these generally follow the standard rules, starting at initiation codons variously far from the usually capped 5' end (e. g., 9 nucleotides for BMV RNAs [Hall *et al.*, 1972] and 63 nucleotides for TMV RNA [Richards *et al.*, 1977]), and terminating at the nonsense codons, with frequent read-through to yield proteins of the same amino acid sequence, but additional lengths (Pelham, 1978). The coat protein mRNAs are, as stated, usually short and monocistronic. The larger mRNAs may generally code for single proteins, but this may not always be the case, with the read-through phenomenon being one means of two proteins arising from one RNA. Most of these viruses show the oddity that their 3' end is not polyadenylated, but in many instances shows specific t-RNA-like amino acid binding capabilities (e. g., valine for turnip yellow mosaic virus [Yot *et al.*, 1970] and tyrosine for BMV [Shih and Kaesberg, 1973; Dasgupta and Kaesberg, 1977]), while not having sequences which lend themselves to typical clover-leaf structures. The 5'-terminus initially carries the 7-methyl Gppp cap, but in contrast to most mRNAs no adjacent methylated bases (Keith and Fraenkel-Conrat, 1975).

There are two virus groups which in regard to translation differ from this general scheme, cowpea mosaic virus (CMV), also 5'-capped, has 3'-terminal poly A on both strands, and appears to become translated like animal picornaviruses (e. g., polio RNA), i. e., as single large "polyproteins" corresponding to most of the length of the RNAs; the ultimate functional proteins result from specific proteolytic cleavages, as is the case with the picornaviruses (Pelham and Jackson, 1976). Thus one is tempted to hypothesize that this group of viruses is evolutionarily related to such animal viruses. An added feature supporting this hypothesis is that the cowpea mosaic virus group has two coat proteins, compared to one for all "typical" plant viruses and four for polio virus.

Another special group is represented by tobacco necrosis virus (TNV). The RNA of this virus lacks the cap as well as the poly A, resembling RNA phages in these respects, and it also seems to be translated, like these RNAs, as a multicistronic messenger with translation of three protein starting and stopping at three initiation and termination codons on the same single mRNA molecule (Salvato and Fraenkel-Conrat, 1977). Again one is tempted to hypothesize a special evolutionary relationship, in this case to the RNA phages.

TNV has another unusual feature, in that it is frequently associated with another very small (17 nm diameter) so-called satellite virus (STNV). STNV has its own icosahedrally arranged protein coat and an RNA of only

0.3×10^6 dalton that codes for that protein, but little if anything else (Leung *et al.*, 1976). Thus STNV is a parasite of a parasite, depending on TNV for all other viral functions. This dependence is very specific, certain strains of STNV being associated with only specific TNV strains, and diminishing their pathogenicity.

Several other viruses, e. g., CMV, frequently carry small (ranges $0.1-0.5 \times 10^6$ dalton) RNAs in their virion, unrelated to the viral genomic RNAs, which also affect the pathogenicity of the virus to varying degree and in either direction, leading, to more or less severe disease symptoms (Waterworth *et al.*, 1979). These "satellite RNAs", however, do not code for, and become encapsidated by, a coat protein other than that of the virus they are associated with. At present STNV is the only known plant satellite virus. Similar situations, however, exist with animal viruses (the adeno-associated parvovirus) and bacteriophages (the P2-P4 system).

The hypothesis that STNV's dependence on its helper was due to its lacking an RNA polymerase gene which was present in all autonomous RNA viruses lost most of its appeal when it was found that the host, tobacco, contained such an enzyme and that no good evidence for the presence of an RNA polymerase gene in TNV RNA was found. Such enzymes were first detected in healthy Chinese cabbage and cauliflower (Astier-Manifacier and Cornuet, 1971, 1978) and have since been demonstrated in all plants that were studied. The fact that relates these findings to plant virology is the observation that viral infection causes more or less dramatic increases in the amount of these enzymes in plants. The finding that the sedimentation rates and all known enzymological properties of plant RNA polymerases are very similar if not the same when isolated from virus-infected or healthy tobacco and cowpea, makes it appear improbable that the viral genome contributes an "activating" component to the preexisting plant enzyme (Ikegami and Fraenkel-Conrat, 1978; Fraenkel-Conrat 1983; Takanami and Fraenkel-Conrat, 1982).

The stimulation of the production of these enzymes by viruses has facilitated the isolation of pure RNA dependent RNA polymerases from tobacco, cucumber (Takanami and Fraenkel-Conrat, 1982) and cowpea (Dorssers *et al.*, 1982). These are proteins of 100 to 130×10^3 daltons, characteristically different for the different plants, and unaffected by the type of virus used to stimulate their production. This does not preclude that a host (or viral) gene product is also required for the actual replication of viral RNA, the second step of transcribing the minus strand back to a plus strand, as appears to be the case for polio and Q β RNA. Without doubt, viruses are able to influence the transcription rate of host genes or, less probably, the translation rate of host mRNAs. Other viral functions, such as the nature (e. g., local lesion vs. systemic disease) and the severity of the disease, are probably also the result of the viral RNA or its gene products affecting the host's genome. None of these phenomena have as yet been elucidated in molecular terms.

We earlier mentioned viroids as phenomena quite separate and different from viruses (Diener, 1979, 1981; Diener and Raymer, 1967; Gross *et al.*, 1978; Rackwitz *et al.*, 1979). It now appears that this difference may not be so absolute. One among various hypotheses suggests that viroids represent atavistic RNA viruses that have sacrificed their genome mRNA capabilities in favor of developing particularly powerful host genome affecting functions. It has on the other hand been suggested that viroids represent escaped introns and as such can have transposon-like properties (Diener, 1981). That viroids have been observed only in plants suggests that their replication requires RNA dependent RNA polymerases, enzymes which occur in plants (Boege *et al.*, 1982), but seem not to exist in animal cells, although their replication by a host transcriptase is more likely. In any case viroids surely are pathogenic through acting on the host's genome. This pathogenicity can range from nondetectability to lethality as a consequence of small RNA structural changes. The same is true for the symptomology of classical RNA viruses. It therefore appears most likely that these viruses also affect the host largely through acting on its genome expression. This is in contrast to their replication which need not involve the host's genome.

This brief and superficial survey of plant virology has led me to the conclusion that the various agents that I discussed, namely competent viruses, satellite viruses, satellite RNAs, and viroids may turn out to have more in common than is usually realized. Their replication is more or less dependent on their ability to interact with the host's genome, and these interactions may be primarily responsible for the nature and severity of their pathogenicity, but may also be more or less essential for their "survival", most for the viroids, and least for viruses of plants rich in RNA polymerase, such as tobacco mosaic virus. The mode by which viral or nonviral RNAs interact with the host's genome may become illuminated by subsequent chapters of this book.

References

Anderer, F. A., Uhlig, H., Weber, E., Schramm, G., 1960: Primary structure of the protein of tobacco mosaic virus. Nature **186,** 922—925.

Astier-Manifacier, S., Cornuet, P., 1978: Purification et poids moléculaire d'une RNA polymérase RNA = dépendante de *Brassica oleracca* var. *Botrytis.* C. R. Acad. Sci. Paris **287,** 1043—1046.

Astier-Manifacier, S., Cornuet, P., 1971: RNA-dependent RNA polymerase in Chinese cabbage. Biochim. Biophys. Acta **232,** 484—493.

Beijerinck, M. W., 1898: Verhandl. Kon. Akad. Wetenschap. Amsterdam **6,** 1—24.

Boege, F., Rohde, W., Sanger, H. L., 1982: In vitro transcription of viroid RNA into full-length copies by RNA-dependent RNA polymerase from healthy tomato leaf tissue. Bioscience Reports **2,** 185—194.

Dasgupta, R., Kaesberg, P., 1977: Sequence of an oligonucleotide derived from the 3' end of each of the four brome mosaic viral RNAs. Proc. Natl. Acad. Sci., U.S.A. **74,** 4900—4904.

Diener, T. O., 1981: Are viroids escaped introns? Proc. Natl. Acad. Sci., U.S.A. **78,** 5014—5015.

Diener, T. O., 1979: Viroids: structure and function. Science **205,** 859—866.
Diener, T. O., Raymer, W. B., 1967: Potato spindle tuber virus: A plant virus with properties of a free nucleic acid. Science **158,** 378—381.
Dorssers, L., Zabel, P., Van der Meer, J., Van Kammen, A., 1982: Purification of a host-encoded RNA-dependent RNA polymerase from cowpea mosaic virus-infected cowpea leaves. Virology **116,** 236—249.
Fraenkel-Conrat, H., 1956: The role of the nucleic acid in the reconstitution of active tobacco mosaic virus. J. Am. Chem. Soc. **78,** 882.
Fraenkel-Conrat, H., 1979, RNA-dependent RNA polymerases of plants. Trends Biochem. Sci. **4,** 184—186.
Fraenkel-Conrat, H., 1983: RNA-dependent RNA polymerases of plants. Proc. Natl. Acad. Sci., U.S.A. **80,** 422—424.
Fraenkel-Conrat, H., Singer, B., 1957: Virus reconstitution. II. Combination of protein and nucleic acid from different strains. Biochim. Biophys. Acta **24,** 540—548.
Fraenkel-Conrat, H., Singer, B., Williams, R. C., 1957: Infectivity of viral nucleic acid. Biochim. Biophys. Acta **25,** 87—96.
Fraenkel-Conrat, H., Williams, R. C., 1955: Reconstitution of active tobacco mosaic virus from its inactive protein and nucleic acid components. Proc. Natl. Acad. Sci., U.S.A. **41,** 690—698.
Fukuda, M., Meshi, T., Okada, Y., Otsuki, Y., Takebe, I., 1981: Correlation between particle multiplicity and location on virion RNA of the assembly initiation site for viruses of the tobacco mosaic virus group. Proc. Natl. Acad. Sci., U.S.A. **78,** 4231—4235.
Gierer, A., Schramm, G., 1956: Infectivity of RNA from tobacco mosaic virus. Nature **177,** 702—703.
Goelet, P., Lomonossoff, G. P., Butler, P. J. G., Akam, M. E., Gait, M. J., Karn, J., 1982: Nucleotide sequence of tobacco mosaic virus RNA. Proc. Natl. Acad. Sci., U.S.A. **79,** 5818—5822.
Gross, H. J., Domdey, H., Lossow, C., Jank, P., Raba, M., Alberty, H., Sanger, H., 1978: Nucleotide sequence and secondary structure of potato spindle tuber viroid. Nature **273,** 203—208.
Hall, T. C., Shih, D. S., Kaesberg, P., 1972: Enzyme-mediated binding of tyrosine to brome-mosaic-virus ribonucleic acid. Biochem. J. **129,** 969—976.
Higgins, T. J. V., Goodwin, P. B., Whitfeld, P. R., 1976: Occurrence of short particles in beans infected with the cowpea strain of TMV. II. Evidence that short particles contain the cistron for coat-protein. Virology **71,** 486—497.
Hunter, T. R., Hunt, T., Knowland, J., Zimmern, D., 1976: Messenger RNA for the coat protein of tobacco mosaic virus. Nature **260,** 759—764.
Ikegami, M., Fraenkel-Conrat, H., 1978: RNA-dependent RNA polymerase of tobacco plants. Proc. Natl. Acad. Sci., U.S.A. **75,** 2122—2124.
Keith, J., Fraenkel-Conrat, H., 1975: Tobacco mosaic virus RNA carries 5'-terminal triphosphorylated guanosine blocked by 5'-linked 7-methylguanosine. FEBS Letters **57,** 31—33.
Leung, D. W., Gilbert, C. W., Smith, R. E., Sasavage, N. L., Clark, Jr., J. M., 1976: Translation of satellite tobacco necrosis virus ribonucleic acid by an invitro system from wheat germ. Biochemistry **15,** 4943—4950.
Pelham, H. R. B., 1978: Leaky UAG termination codon in tobacco mosaic virus RNA. Nature **272,** 469—471.
Pelham, H. R. B., Jackson, R. J., 1976: An efficient mRNA-dependent translation system from reticulocyte lysates. Eur. J. Biochem. **67,** 247.

Rackwitz, H.-R., Rohde, W., Sänger, H. L., 1979: DNA-dependent RNA polymerase II of plant origin transcribes viroid RNA into full-length copies. Nature **291,** 297—301.

Richards, K., Jonard, G., Guilley, H., Keith, G., 1977: Leader sequence of 71 nucleotides devoid of G in tobacco mosaic virus RNA. Nature (London) **267,** 548—550.

Salvato, M. S., Fraenkel-Conrat, H., 1977: Translation of tobacco necrosis virus and its satellite in a cell-free wheat germ system. Proc. Natl. Acad. Sci., U.S.A. **74,** 2288—2292.

Shih, D. S., Kaesberg, P., 1976: Translation of the RNAs of brome mosaic virus: the monocistronic nature of RNA 1 and RNA 2. J. Mol. Biol. **103,** 77.

Stanley, W. M., 1935: Isolation of a crystalline protein possessing the properties of tobacco mosaic virus. Science **81,** 644—645.

Takanami, Y., Fraenkel-Conrat, H., 1982: Comparative studies on RNA-dependent RNA polymerases in cucumber mosaic virus-infected cucumber, tobacco, and uninfected tobacco. Biochem. **21,** 3161—3167.

Tsugita, A., Gish, D. T., Young, J., Fraenkel-Conrat, H., Knight, C. A., Stanley, W. M., 1960: The complete amino acid sequence of the protein of tobacco mosaic virus. Proc. Natl. Acad. Sci., U.S.A. **46,** 1463—1469.

Waterworth, H. E., Kaper, J. M., Tousignant, M. E., 1979: CARNA 5, the small cucumber mosaic virus-dependent replicating RNA, regulates disease expression. Science **204,** 845—847.

Yot, P., Pinck, M., Haenni, A.-L., Duranton, H. M., Chapeville, F., 1970: Valine-specific tRNA-like structure in turnip yellow mosaic virus RNA. Proc. Natl. Acad. Sci., U.S.A. **67,** 1345—1352.

Chapter 2

DNA Flux Across Genetic Barriers: The Crown Gall Phenomenon

G. Gheysen, P. Dhaese, M. Van Montagu, and J. Schell*

Laboratorium voor Genetica, Rijksuniversiteit Gent, B-9000 Gent, Belgium

* Permanent Address:
Max-Planck-Institut für Züchtungsforschung, D-5000 Köln 30,
Federal Republic of Germany

With 5 Figures

Contents

I. Introduction: *Agrobacterium tumefaciens*, a Natural Instance of Genetic Engineering

A. General Introduction

Crown gall tumors are neoplastic proliferations induced by the soil bacterium *Agrobacterium tumefaciens* on wounded dicotyledonous plants (for recent reviews, see Kahl and Schell, 1982; Caplan *et al.*, 1983; Depicker *et al.*, 1983; Zambryski *et al.*, 1983a). In nature, the infection is often located at or near the junction of the root and the stem, the crown of the plant. Since the turn of the century, plant pathologists have been interested in this malignant transformation not only because of its agricultural consequences, but also for the unusual observation that a bacterium induces plant neoplasia.

It also became a favorite model system to study plant cell growth and differentiation. Braun and his collaborators succeeded in growing sterile tumor tissue on a simple medium containing only sucrose and inorganic salts (White and Braun, 1942; Braun and White, 1943; Braun and Mandle, 1948). They found that, while most plant cells require auxin and cytokinin to grow vigorously *in vitro*, the crown gall cells could proliferate indefinitely on hormone-free medium. Since this tissue was free of the inciting bacteria, Braun concluded that the bacteria permanently change the normal mechanisms controlling growth and differentiation of the plant cells. He introduced the idea that a "tumor-inducing principle" (TIP) present in the oncogenic bacteria is transmitted upon infection to the host cells where it is responsible for the neoplastic transformation.

It took 30 years to convincingly demonstrate by genetic and physical means that the nature of the TIP was DNA (Zaenen *et al.*, 1974; Van Larebeke *et al.*, 1974, 1975; Watson *et al.*, 1975; Chilton *et al.*, 1977; Schell *et al.*, 1979). It is striking that in the same decade that man started to manipulate bacteria to express animal proteins it was found that nature itself had evolved an efficient system allowing a bacterium to transfer DNA to plant cells. In the next section, we describe some of the interesting features of the *Agrobacterium*/crown gall system that led to this conclusion.

B. In Search of the TIP

Several attempts to identify the hypothetical tumor-inducing principle led to success only after the genetic properties of *Agrobacterium* and the biological significance of the crown gall formation to the oncogenic bacteria were better understood.

First, there was the discovery by Morel (1956) and Lioret (1956) that crown gall cells synthesize arginine derivatives that are not present in untransformed tissues of the same plant. The importance of these observations was, however, understood only after it was found that the type of arginine derivative (such as octopine or nopaline) synthesized in the tumor was not dependent on the host plant but rather on the particular *Agrobac-*

terium strain that induced the tumor (Goldmann *et al.*, 1968; Petit *et al.*, 1970). Furthermore, these authors demonstrated that a given bacterial strain is able to grow on either octopine or nopaline, but not on both, and that it can selectively catabolize the compound whose synthesis it has induced in the tumor.

Considering this metabolic relation, the hypothesis was suggested that transfer of genetic information such as DNA from the bacterium to the plant cell might be responsible for this specificity (Petit *et al.*, 1970). The name "opine" was proposed to describe metabolites specifically synthesized by crown gall cells, that can be used by agrobacteria as carbon and/or nitrogen energy sources (Tempé and Schell, 1977; Tempé *et al.*, 1978). More recently, the identification of other opines such as agropine (Firmin and Fenwick, 1978), agrocinopine (Ellis and Murphy, 1981) and succinamopine (Chilton *et al.*, 1984) reinforced the significance of opines to the crown gall/*Agrobacterium* relationship (for more details, see Tempé and Petit, 1982; Tempé *et al.*, 1984). It has been proposed that the inciting bacteria have a competitive advantage over other soil organisms in the rhizosphere of infected plants by their ability to catabolize the opines released by the crown gall. *Agrobacterium* is a sophisticated parasite, forcing the infected plant by a gene transfer mechanism to divert part of its arginine supply into nutrients which the infecting bacterium can specifically metabolize. This type of parasitic interaction has been called genetic colonization (Schell *et al.*, 1979).

The other question was what part of the bacterial genome was responsible for the oncogenic and metabolic characteristics of the agrobacteria. Kerr (1969, 1971) observed that oncogenicity was transferred from a virulent *Agrobacterium* strain to an avirulent one when the two strains were inoculated onto the same plant, which was a strong indication that the virulence genes were carried by an infectious entity.

Soon, it was shown that all virulent strains contain a very large plasmid that was not detected in the avirulent strains under study (Zaenen *et al.*, 1974). The loss of this plasmid which resulted from growing the *Agrobacterium* strain C 58 at 37° C, invariably led to the loss of the tumorigenic capacity. Introduction of this plasmid by conjugation in an avirulent strain was correlated with the acquirement of the virulence trait (Van Larebeke *et al.*, 1975). These plasmid transfers revealed that not only the capacity for tumor induction but also some other characteristics such as opine synthesis and opine catabolism were linked to the Ti plasmid (Ti stands for tumor-inducing) (Bomhoff *et al.*, 1976). These genetic studies pointed out that the Ti plasmid was the most likely candidate for the role of the tumor-inducing principle.

The first physical evidence for the transfer and stable inheritance of prokaryotic DNA in plant cells came from renaturation kinetics analysis: Chilton *et al.*, (1977) demonstrated that particular fragments of the Ti plasmid reassociated significantly faster in the presence of tumor DNA, indicating that these fragments were present in crown gall tissue. At the same time, Southern blot analysis conclusively demonstrated the presence of Ti plasmid DNA in plant tumor cell DNA (De Beuckeleer *et al.*, 1978).

The transferred segment of DNA that is present in crown gall cells and determines their specific tumorous and metabolic characteristics is now called T-DNA, and the part of the plasmid from which it is derived is called the T-region. The next questions arising were: Where is this T-region situated on the Ti plasmid? Is it a continuous fragment or are several separated regions involved? Is the T-DNA integrated in the plant nucleus or in organelle DNA, or is it independently replicating? Is the T-DNA a well-defined DNA sequence, or does it vary in different tumor lines?

C. A More Precise Picture of the T-DNA

The Ti plasmids have been subdivided according to the opines they specify (for details, see Guyon *et al.*, 1980; Hille *et al.*, 1984a; Hooykaas and Schilperoort, 1984; Petit and Tempé, 1985). The best studied types are the nopaline and the octopine Ti plasmids. The restriction endonuclease map of these plasmids has been constructed (Depicker *et al.*, 1980; De Vos *et al.*, 1981) and has been correlated with the functional map obtained by genetic analysis (Koekman *et al.*, 1979; Holsters *et al.*, 1980; De Greve *et al.*, 1981). Heteroduplex mapping and Southern blot analysis indicated four major regions of homology between octopine and nopaline plasmids (Engler *et al.*, 1981). Two of them (regions A and D) were recognized by genetic analysis to be involved in oncogenicity, one (B) corresponds to the replication control region of the plasmid and the other (C) encodes conjugative transfer functions (Holsters *et al.*, 1980; De Greve *et al.*, 1981).

Octopine and nopaline tumor lines were studied intensively by several groups to determine the state of the T-DNA in the crown gall cell. By isolating DNA from purified nuclei, chloroplasts and mitochondria of some tumors it was shown that the T-DNA was present in the nucleus (Chilton *et al.*, 1980; Willmitzer *et al.*, 1980). A more precise picture of the organization and copy number of T-DNA came from Southern blot hybridization analysis (e. g. Lemmers *et al.*, 1980; Thomashow *et al.*, 1980a; De Beuckeleer *et al.*, 1981).

Hybridization of Ti plasmid-specific probes to restriction endonuclease digests of tumor DNA revealed that the sequences comprising the homology region A (the "common" T-DNA) were found in octopine as well as in nopaline tumors induced on different plants. The T-DNA present in several tumors induced on tobacco by the nopaline strains (C 58 and T 37) was found to be very similar, 24 kb in size (Lemmers *et al.*, 1980). Each of the internal T-region fragments from the plasmid was shown to hybridize to a single restriction fragment of identical size in the crown gall tumor. These observations indicate that the T-DNA is colinear with the T-region and usually undergoes no major rearrangements during stabilization in the plant genome. In contrast, the outermost fragments of the T-region hybridized to tumor DNA fragments of sizes different from the corresponding Ti plasmid fragments. These "composite fragments" were thought to consist of T-DNA covalently linked to plant sequences which was later confirmed by molecular cloning (see below III.C.). The exact number and the sizes of these composite fragments varied from tumor to tumor indicating that more than one copy of T-DNA can be integrated and that the T-DNA can be linked to different plant sequences. In addition, in several lines the nopaline T-DNA was present in a tandem configuration (Lemmers *et al.*, 1980; Zambryski *et al.*, 1980, 1982).

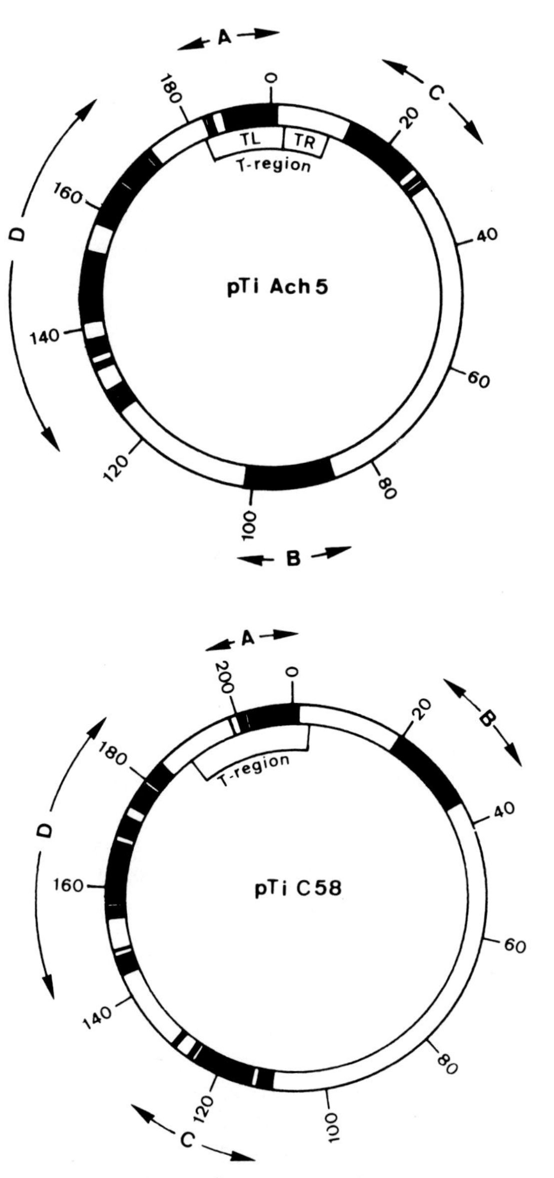

Fig. 1. Circular maps of an octopine Ti plasmid (pTiAch5) and a nopaline Ti plasmid (pTiC58). The homology regions (A, B, C, and D) are indicated as black bars. Region A is part of the T-region, and is shown in more detail in Fig. 2. Region D is the Vir-region. Map coordinates are in kb, starting from the common *Sma*I site within the T-region

In case of the octopine tumor lines, an additional complication was observed: Two non-contiguous segments of the Ti plasmid are present in some transformed lines (Thomashow *et al.*, 1980a; De Beuckeleer *et al.*, 1981). One of these T-DNAs, designated TL (14 kb), is always present and contains the common or "core" T-DNA. The TL-DNA can occur in tandem arrays as was observed for the nopaline T-DNA (Holsters *et al.*, 1983). The other T-DNA segment, designated TR, is about 7 kb in length and originates from a part of the plasmid lying to the right of the TL-region. This TR-DNA is not found in all tumor lines, but when present it can have a different copy number than the TL-DNA in the plant cell.

Although the T-DNA/plant junctions seem to occur preferentially at particular sites of the Ti plasmid, some irregularities have been found. For example, some tumors contain only a truncated form of TL, with a deletion of the right part (De Beuckeleer *et al.*, 1981; Ooms *et al.*, 1982a). Similarly abnormal T-DNA insertions have also been shown to occur in tumors induced by nopaline Ti plasmids (Hepburn *et al.*, 1983a; Van Lijsebettens *et al.*, in preparation).

In the next sections we will primarily focus on discussing some of the recent advances in the understanding at the molecular level of (i) how this bacterial segment of DNA is equipped to be expressed in its eukaryotic host, (ii) how it directs the regulation of plant growth and development, and (iii) how the bacteria might introduce the T-DNA into the plant cell.

II. The T-DNA Is Designed to Be Functional in the Plant Cell

A. T-DNA Gene Structure and Its Expression in Plant Cells

When speculating about the prokaryotic or eukaryotic origin of the T-DNA one must first take into account the fact that these sequences are phenotypically expressed in the plant cell, and that this functional requirement was expected to be reflected in the structure of the transcriptional and translational control signals at the 5′ and 3′ end of the different T-DNA genes. As part of a bacterial plasmid, the T-DNA genes are more readily accessible than the majority of "authentic" plant genes, and therefore might serve as models to identify structural features necessary for gene expression in plants. More specifically, since the T-DNA is integrated in the nuclear plant genome (Chilton *et al.*, 1980; Willmitzer *et al.*, 1980) and its transcription in plant nuclei is inhibited by low concentrations of α-amanitin (Willmitzer *et al.*, 1981), the expectation was that the T-DNA expression signals would resemble those of RNA polymerase II-transcribed eukaryotic genes (Breathnach and Chambon, 1981).

In octopine crown gall tumors of tobacco the TL-DNA is transcribed into eight distinct, well-defined polyadenylated mRNAs (Gelvin *et al.*, 1982; Willmitzer *et al.*, 1982a, 1983), whereas the larger nopaline T-DNA encodes at least thirteen transcripts (Bevan and Chilton, 1982; Willmitzer *et al.*, 1983). Recently, five polyadenylated transcripts specified by the octopine TR-DNA have also been characterized and mapped (Velten *et al.*, 1983; Karcher *et al.*, 1984; Winter *et al.*, 1984). In general, these transcript analyses showed that the T-DNA genes are closely packed (although not overlapping) and that very likely each gene is transcribed independently from its own promoter. The overall efficiency of T-DNA transcription is very low, considerable differences in the steady-state levels exist between different transcripts, and the abundance of a given transcript can vary between tumor lines and with culture conditions (e. g. Karcher *et al.*, 1984).

Table 1. Functional Transcription Initiation and Processing Signals in the T-DNA

	"TATA" box(es)	Position relative to mRNA start	Poly-adenylation signal(s)	Position relative to stop codon	References
	5′ 3′				
Transcript 2 (oct) (a *"tms"* gene)	TATATtT	−31 to −25	tATAAA AATAAt	+ 87 to + 92 + 96 to +101	Klee *et al.*, 1984
Transcript 4 (oct) (the *"tmr"* gene)	aATATAA TATAAAA	−31 to −25 −29 to −23	AATAAA AATAAA	+208 to +213 +276 to +281	Heidekamp *et al.*, 1983; Lichtenstein *et al.*, 1984
Transcript 5 (nop)	TATAgAg	−35 to −29			Dhaese *et al.*, 1983 b
Transcript 7 (oct)	TATATAT	−32 to −26	AATAAA	+119 to +124	Dhaese *et al.*, 1983 a
Octopine synthase	TATtTAA	−32 to −26	AATAtA AATgAA AATAAt	+139 to +144 +143 to +148 +175 to +180	De Greve *et al.*, 1982 a; Dhaese *et al.*, 1983 a
Nopaline synthase	cATAAAT	−27 to −21	AATAAt	+129 to +134	Depicker *et al.*, 1982; Bevan *et al.*, 1983 a
Consensus	TATA$\frac{A}{T}$A$\frac{A}{T}$		AATAAA		Breathnach and Chambon, 1981

The first T-DNA genes whose structure and transcription were studied in detail, were octopine synthase (De Greve *et al.*, 1982 a) and nopaline synthase (Depicker *et al.*, 1982; Bevan *et al.*, 1983 a). DNA sequence determination combined with S 1 nuclease analysis allowed the identification of the coding region, and the determination of the mRNA initiation and poly(A) addition sites at the nucleotide level. Subsequently, this experimental approach was extended to other octopine TL-DNA genes, such as "transcript 7" (Dhaese *et al.*, 1983 a), the *roi* gene or "transcript 4" (Heidekamp *et al.*, 1983; Lichtenstein *et al.*, 1984), and a *shi* gene, "transcript 2" (Klee *et al.*, 1984). Recently, the complete nucleotide sequences of the octopine TL-DNA (Barker *et al.*, 1983; Gielen *et al.*, 1984) and the TR-DNA (Barker *et al.*, 1983) have been elucidated. We have attempted (Table 1) to indicate some common structural characteristics that emerge from a comparison of the 5′-flanking and 3′-noncoding regions of these genes.

First, a TATA or Goldberg-Hogness box can always be identified upstream of the mRNA start. It is not so much the actual structure but rather the distance of this sequence to the transcriptional initiation site (around 30 nucleotides), — as opposed to the "−35" and "−10" promoter regions in bacteria — which classifies this feature as typically eukaryotic. The S1 mapping data have shown that the T-DNA transcripts initiate at unique positions, except for the octopine "transcript 4", which apparently can start at either of two alternative sites (Heidekamp *et al.*, 1983; Lichtenstein *et al.*, 1984) each preceded by a TATA box at the appropriate distance. Interestingly, comparison of the octopine sequence with the homologous nopaline gene sequence reveals that only the first of these two "TATA" boxes is conserved in the pTiT37 T-DNA (Goldberg *et al.*, 1984).

The "TATA" box clearly constitutes only part of the control region of an eukaryotic gene, and its major role might be the positioning of the mRNA start site. Usually, sequences required for efficient *in vivo* transcription or involved in specific regulation of transcription are located at different positions relative to the coding sequence of the gene. We do not know yet if, and how, the expression of T-DNA genes is regulated in tumor cells. Furthermore, this regulation might be different for each individual T-DNA gene. It is thus not surprising that comparison of sequences upstream of the "TATA" box does not reveal regions conserved throughout the T-DNA genes.

The 3′-untranslated region of each of the T-DNA genes contains one or more sequences identical to, or differing in only one base, from the consensus polyadenylation signal AATAAA (Gielen *et al.*, 1984). Multiple sites for polyadenylation have been observed for the octopine synthase gene (Dhaese *et al.*, 1983a) and for the transcript 4 gene (Lichtenstein *et al.*, 1984). The experimentally determined poly(A) addition sites of transcripts 2, 4, 7, octopine synthase, and nopaline synthase are all located within a distance of less than 50 bp downstream from such a hexanucleotide.

Direct characterization of the T-DNA-encoded polypeptide products in tumor cells has been very limited to date. Only the octopine synthase and nopaline synthase enzymes have been purified to some extent, but no amino acid sequence data were reported (Kemp, 1982). Through hybridization selection of polyribosomal mRNA against specific T-DNA fragments and subsequent translation in an eukaryotic cell-free system, a 39,000—40,000 dalton polypeptide was shown to be the product of the octopine synthase gene (Schröder *et al.*, 1981; Murai and Kemp, 1982). Similarly, a 14,000 dalton polypeptide was identified as the possible translation product of "transcript 7" (Schröder and Schröder, 1982). Furthermore, the DNA sequence data have shown that for each of the genes identified by genetical and transcriptional criteria, a corresponding open-reading frame of compatible size can be found, and therefore most probably represents the coding region. Gielen *et al.* (1984) noticed that, for each gene, the first AUG codon encountered from the transcriptional start site is in phase with the open-reading frame representing the coding region, and therefore, most likely serves as translational initiation. One possible exception is the octopine transcript 6a which contains three closely associated in-phase AUG codons, of which only the third is conserved in the homologous nopaline T-DNA counterpart (P. Dhaese, unpublished results).

None of the T-DNA genes appear to contain introns. We do not consider this property as a sound argument against the eukaryotic origin of T-DNA, since several authentic nuclear plant genes also lack intervening sequences. Amongst these are soybean lectin (Goldberg *et al.*, 1983), wheat gliadin (Rafalski *et al.*, 1984), and the maize zeins (Hu *et al.*, 1982; Pedersen *et al.*, 1982).

B. Functional Organization of the T-DNA

To correlate the T-DNA transcripts observed in tumor cells with their function in the establishment of the tumorous phenotype, the T-DNA was mutated by transposon insertions and/or by extensive deletions removing one or more transcripts (Garfinkel *et al.*, 1981; Ooms *et al.*, 1981; Leemans *et al.*, 1982; Joos *et al.*, 1983a; Salomon *et al.*, 1984). Essentially, two classes of mutants can be distinguished.

The first class induces undifferentiated tumors with a wild-type appearance, that fail to produce certain opines. In the nopaline T-DNA the affected genes are nopaline synthase *(nos)* (Holsters *et al.*, 1980; Joos *et al.*, 1983a) or agrocinopine synthase *(acs)* (Joos *et al.*, 1983a). The gene coding for octopine synthase is located in TL *(ocs)* (De Greve *et al.*, 1981; Garfinkel *et al.*, 1981; Ooms *et al.*, 1982b), and TR encodes different enzymes involved in the synthesis of opines: transcripts 1′ and 2′ are necessary for

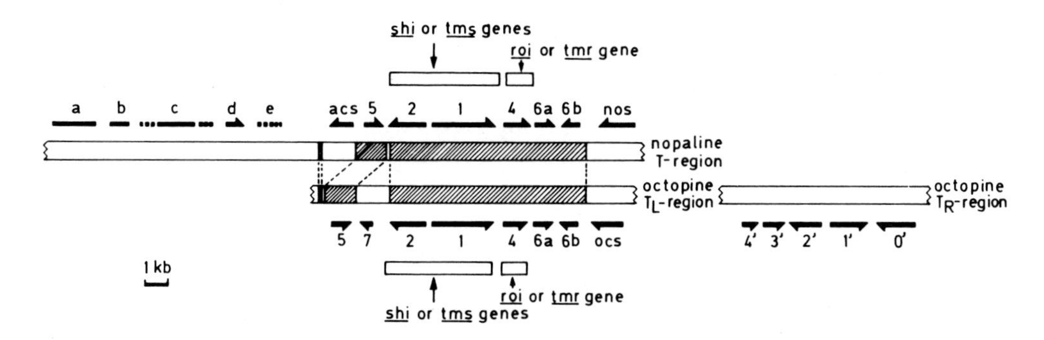

Fig. 2. Functional organization of the T-DNA of nopaline (upper part) and octopine (lower part) Ti plasmids. The arrows indicate the different genes and their direction of transcription in the transformed plant cell. The gene nomenclature follows the original designations according to transcript size (Willmitzer *et al.*, 1982a, 1983; Velten *et al.*, 1983). *Nos,* nopaline synthesis; *acs,* agrocinopine synthesis, *ocs,* octopine synthesis; Roi, root inhibition (also termed *tmr, tumor morphology roots* [Garfinkel *et al.*, 1981]); Shi, shoot inhibition (also termed *tms, tumor morphology shoots* [Garfinkel *et al.*, 1981]). The dashed areas represent regions of homology between the two types of plasmids

the production of mannopine and transcript 0′ is responsible for the conversion of mannopine into agropine (Velten *et al.*, 1983; Salomon *et al.*, 1984; Ellis *et al.*, 1984).

Other mutants, all of them located in the "common" DNA, affect the morphology of the tumor. On *Nicotiana tabacum*, these mutants induce tumors that produce roots or shoots. Since this differentiation is obtained by inactivating one or more transcripts, the corresponding genes must contribute to the tumorous phenotype by actively suppressing normal differentiation. Mutations in gene *4* induce tumors that generate roots (Roi^- phenotype) indicating that gene *4* encodes a function that suppresses root formation in wild-type tumors (Garfinkel *et al.*, 1981; Ooms *et al.*, 1981; Leemans *et al.*, 1982; Joos *et al.*, 1983a). Mutations in genes *1* and *2* induce tumors which produce shoots (Shi^- phenotype) indicating that these genes block the capacity of the plant cells to differentiate into shoots (Garfinkel *et al.*, 1981; Leemans *et al.*, 1982; Joos *et al.*, 1983a).

These phenotypes resemble the effects of imbalances in the levels of plant growth regulators. Skoog and Miller (1957) demonstrated that untransformed tobacco plant tissue cultured *in vitro* grows as callus when auxin and cytokinin is added to the medium in a defined ratio. Changing this ratio results in the development of either roots or shoots. The elimination of transcript 4 results in root formation as if there had been an increase in the auxin content of the tissue or a decrease in its cytokinin concentration. Similarly, the inactivation of genes *1* or *2* induces shoots as if the auxin/cytokinin ratio had decreased. The formation of fast-growing, undifferentiated tumor tissue can be explained by a relatively high concentration of both phytohormones.

Analyses of cytokinin/auxin ratios (Amasino and Miller, 1982; Akiyoshi *et al.*, 1983) in undifferentiated, shoot- and root-producing tumors further supported the idea that genes *1, 2* and *4* are involved in the biosynthesis of phytohormones: Akiyoshi *et al.* (1983) found that the ribosylzeatin/indole acetic acid ratio increases from 0.2 in wild-type tumors to about 20 in shoot-forming tumors, and decreases to 0.004 in root-producing tumors.

Further evidence for the phytohormone-like effects of the T-DNA functions came from the *in vitro* study of these differentiating tumors. Tumors induced on tobacco by mutants in gene *4* grow very poorly *in vitro*, unless cytokinin is supplied (Joos *et al.*, 1983a). Similar tumors on carrot, however, grow vigorously on medium without supplemented cytokinin. Since these carrot tumors produce a considerable amount of roots (in contrast to a limited number or no roots originating from tobacco Roi^- tumors *in vitro*), the cytokinin necessary for tumor growth may be provided by specialized cells in the roots, which are known to be the site of cytokinin synthesis in the plant (for review, see Sembdner *et al.*, 1980). Unlike the tumors induced by Roi^- mutants, tumors induced on tobacco by Shi^- strains grow well *in vitro* without exogenous hormones. In this case, it is presumed that the necessary auxin is provided by the apical meristem in the numerous shoots that arise from the tumors since the apical meristem is known to be the auxin-producing center in normal plants (Sembdner *et al.*, 1980). If these Shi⁻ tumors are transferred *in vitro* to medium where shoots cannot form, then these tissues become auxin-dependent (Binns *et al.*, 1982).

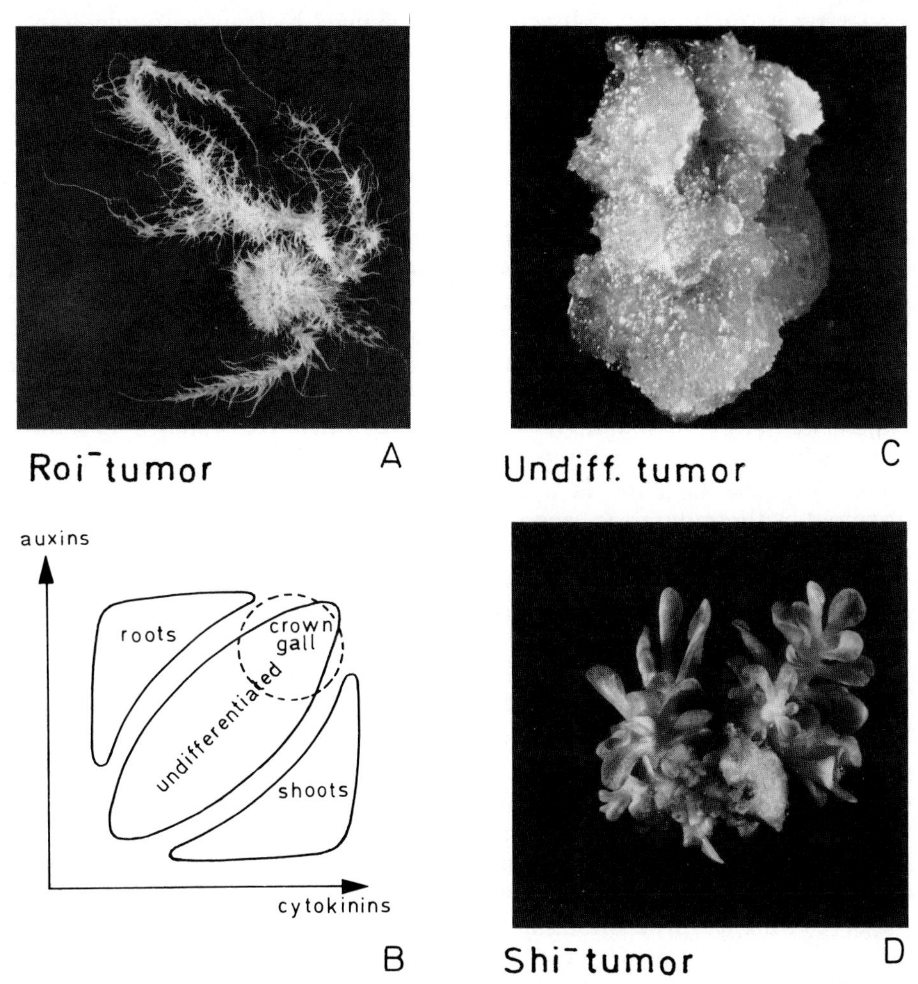

Fig. 3. Tumor morphology of T-DNA mutants and the auxin/cytokinin ratio. This figure is an illustration of the correlation between T-DNA mutations, and changes in the ratio of plant hormones (see diagram B) in producing the morphological characteristics shown in the photographs. Host plants are carrot (A) and tobacco (C and D). Abbreviations: Roi⁻, deficient in root inhibition; Shi⁻, deficient in shoot inhibition; Undiff., undifferentiated tobacco tumor

More recently, Inzé *et al.* (1984) have reported that the Shi⁻ phenotype of mutants in gene *1* can be suppressed by the exogenous addition of intermediates in auxin biosynthesis. They found that α-naphthalene acetamide, which has no auxin activity by itself, can restore the wild-type phenotype of mutants in gene *1*, but not of mutants in gene *2*. This indicates that the role of gene *1* is, directly or indirectly, to produce a natural analogue to α-naphthalene acetamide. Gene *2* in turn, has been shown to encode an enzyme that converts α-naphthalene acetamide, or similar compounds, into an active auxin (Schröder *et al.*, 1984). It was found that *E. coli* and *Agrobacterium* extracts from cells containing gene *2* hydrolyze indole-3-acetamide into the plant hormone indole-3-acetic acid. The same reaction was found in tobacco crown gall cells, but not in nontransformed habituated plant cells, or in *E. coli* and *Agrobacterium* strains lacking the gene. This amidohydrolase apparently also converts α-naphthalene acetamide into α-naphthalene acetic acid in the test system used by Inzé *et al.* (1984).

These results confirm that the T-DNA induces transformation by introducing into plant cells a new enzymatic pathway which cannot be regulated by the host and therefore results in the production of an abnormal concentration of growth regulators.

The functions of the other transcripts in the octopine and nopaline T-DNAs are still unclear. Garfinkel *et al.* (1981) have reported that octopine mutants in gene *6*b induce larger tumors (the Tml⁻ phenotype) on *Kalanchoe daigremontiana;* however, no significant difference has been found when analogous mutants of a nopaline plasmid were tested on 5 different host plants (Joos *et al.*, 1983a). Varying results have also been obtained with mutants in gene *5*. Mutants in the nopaline gene *5* are slightly attenuated (Joos *et al.*, 1983a) while those of an octopine Ti plasmid were shown to enhance the frequency of transformed shoots produced by a Shi⁻ mutation (Leemans *et al.*, 1982).

Mutations in the genes *a, b, c, d, e* in the left of the nopaline T-DNA, gene *7* in TL, genes *3'* and *4'* in TR of the octopine T-DNA (Garfinkel *et al.*, 1981; Leemans *et al.*, 1982; Joos *et al.*, 1983a; Salomon *et al.*, 1984), and gene *6*a in the common DNA (Garfinkel *et al.*, 1981; Joos *et al.*, 1983a) induce apparently normal tumors. These genes may have a minor role in tumor formation or intervene in the synthesis of as yet unidentified opines.

There are some important conclusions to draw from the analysis of all the T-DNA mutants: no single function is absolutely necessary for tumor induction, only deletions removing several genes (e. g. *1, 2* and *4*) are completely avirulent; furthermore, none of the genes located in the T-DNA are necessary for its transfer and integration into the plant genome since T-DNAs with a deletion of all T-DNA functions (except for opine synthases) are still transferred (Leemans *et al.*, 1982; Zambryski *et al.*, 1983b). No tumor is formed in this case but the wound callus can be tested for opines and is usually positive (see also III.F. and III.H.).

C. Agrobacterium rhizogenes, an Analogous System

Agrobacterium rhizogenes causes the plant disease known as hairy-root. This disease is characterized by the fast proliferation of roots at a wound site infected by the bacterium (Elliot, 1951). Although *Agrobacterium rhizogenes* has been considered in a different class of *pathogens* than *A. tumefaciens,* they share many common characteristics:

1. *Agrobacterium rhizogenes* can induce undifferentiated crown gall-like tumors, either when cytokinin is added during the induction phase (Beiderbeck, 1973), or by infection on some plants (De Cleene and De Ley, 1981).
2. Large Ti-like plasmids called Ri (root-inducing) were shown to be responsible for virulence (Moore *et al.*, 1979; White and Nester, 1980a).
3. The incited roots produce agropine, mannopine and agrocinopines and *Agrobacterium rhizogenes* is able to utilize these opines indicating that the opine concept is at the basis of plant cell transformation by these bacteria (Tepfer and Tempé, 1981; Petit *et al.*, 1983).
4. *Agrobacterium rhizogenes* induces cell proliferation and opine synthesis by transfer of a T-DNA to the plant cell (Chilton *et al.*, 1982; Willmitzer *et al.*, 1982b; Byrne *et al.*, 1983).
5. The Vir-region of Ri plasmids is homologous to the Vir-region of Ti plasmids and can complement avirulent octopine Ti plasmids (Costantino *et al.*, 1980; Hooykaas *et al.*, 1982; Hoekema *et al.*, 1984). The Vir-region is thought to be involved in the DNA transfer from *Agrobacterium* to the plant (see III.B.).

In spite of these striking similarities, there is only limited homology between the Ri plasmids and the Ti plasmids (White and Nester, 1980b, Risuleo *et al.*, 1982; Huffman *et al.*, 1984) and the two plasmid classes are compatible (Costantino *et al.*, 1980; Hooykaas *et al.*, 1982). The most surprising observation is that also the T-DNA regions of the two plasmids share very little homology (Willmitzer *et al.*, 1982b; Huffman *et al.*, 1984; Lahners *et al.*, 1984).

We can conclude that the plasmids of the two agrobacteria might share a common, although distant evolutionary origin and that the basic mechanism of infection and gene transfer is probably similar.

D. Some Speculations About the Origin of the T-DNA

The prokaryotic *Agrobacterium* introduces a segment of its Ti plasmid into the plant genome where it is expressed in the eukaryotic host cell. We have already seen how this is accomplished (see section II.A.): simply by providing these genes with all the signals required for expression in the new host, very similar to what scientists are now trying to achieve by recombinant DNA techniques.

But how did this unit of DNA with eukaryotic expression signals become part of the Ti plasmid? There are several ways to think about this problem. One possibility is that the T-DNA originated as a prokaryotic piece of DNA and later developed plant transcription and translation signals. Alternatively *Agrobacterium* captured the T-DNA genes from a plant.

The same regions of the T-DNA expressed in the plant cell are also transcribed (Gelvin *et al.*, 1981; Janssens *et al.*, 1984) and translated (Schröder *et al.*, 1983, 1984) in *Agrobacterium tumefaciens* and *Escherichia coli*. This implies that the T-region also contains prokaryotic signals, and may indicate that the T-region originally was prokaryotic and evolved to function also in eukaryotic plant cells. However, typical eukaryotic sequences like the mouse β-globin gene (Crepin *et al.*, 1981) and SV40 genes (Greenblatt *et al.*, 1976) are also transcribed by *Escherichia coli* RNA polymerase.

More persuasive evidence favoring the prokaryotic origin of genes *1* and *2* of the T-DNA can be found by comparing these genes to similar ones in other bacteria. The T-DNA genes *1* and *2* are involved in the biosynthesis of indole-3-acetic acid (IAA) presumably via an indole-3-acetamide intermediate (Inzé *et al.*, 1984; Schröder *et al.*, 1984). There are several biosynthetic pathways of IAA known in plants but indole-3-acetamide has not been shown to be an intermediate; the identified biosynthetic pathways proceed via indole-acetaldehyde or indole-acetonitrile (Schneider and Wightam, 1978). The pathway with indole-3-acetamide as an intermediate is known to exist in *Pseudomonas savastanoi,* a bacterium that incites tumorous proliferations on olive or oleander (Kosuge *et al.*, 1966). These genes for IAA production are highly expressed in *Pseudomonas* itself (Comai and Kosuge, 1980) and to the best of our knowledge they are not transferred to the plant cell. Since bacteria readily exchange genetic information even between unrelated species and the genes for IAA synthesis are located on a plasmid in *Pseudomonas savastanoi* (Comai and Kosuge, 1980), the IAA genes carried by the Ti plasmid may have a common origin with those of *Pseudomonas savastanoi.*

On the other hand, an interesting observation that might support the eukaryotic origin of the T-DNA genes was made by White *et al.* (1982). They detected hybridization of the *Agrobacterium rhizogenes* T-DNA to DNA of untransformed *Nicotiana glauca* cells (White *et al.*, 1982, 1983). This represents the first evidence that plant genes might have been transferred to *Agrobacterium,* and are now being used for the benefit of the bacteria, comparable with the acquisition of host oncogenes by some RNA tumor viruses. On the other hand, plant species might contain residual sequences from past infections by *Agrobacterium.* This may sound unlikely at first sight because tumors containing T-DNA genes do not readily regenerate into normal plants capable of reproduction. Tumors induced by *Agrobacterium rhizogenes* or *Agrobacterium tumefaciens* Roi$^-$ and Shi$^-$ mutants, however, produce organized root and leaf structures that can be regenerated into normal-looking plants (De Greve *et al.*, 1982b; Barton *et al.*, 1983; David *et al.*, 1984). The rGV1 plant described by De Greve *et al.* (1982b) is rather interesting from this point of view. This tobacco plant arose spontaneously from a shooting octopine tumor and was found to contain little more than the gene for octopine synthase, all the other T-DNA genes being deleted. It is possible, therefore, that some plants, including the *N. glauca* lines studied by White *et al.*, (1983) and the carrot lines studied by Spanò *et al.* (1982) arose from similar sorts of transformed shoots or roots.

The present data do not allow us to discern whether the T-DNA originated from prokaryotic or eukaryotic sequences.

III. Transfer and Integration of the T-DNA in the Plant Cell Nucleus

A. Agrobacterium Holds the Key

A variety of attempts to transform plant cells using techniques applicable to animal cell cultures were only marginally successful (Sarkar *et al.*, 1974; Fernandez *et al.*, 1978; Loesch-Fries and Hall, 1980; Fraley, 1983). Even the Ti plasmid itself, when added to plant protoplasts as naked DNA rarely leads to transformation. Transformed lines that do arise from DNA uptake experiments do not contain the proper T-DNA ends suggesting that the normal integration mechanism is not functioning (Davey *et al.*, 1980; Draper *et al.*, 1982; Krens *et al.*, 1982; Krens, 1983). *Agrobacterium*, by comparison, is quite efficient: it is routinely possible to transform 1—10% of tobacco protoplasts by infection with *Agrobacterium tumefaciens* (Herrera-Estrella *et al.*, 1983a; Horsch *et al.*, 1984).

Very little is known about the mechanism by which *Agrobacterium* transfers the T-DNA to the plant nucleus. The process must include at least two stages. First, the interaction of *Agrobacterium* with plant cells initiates a series of events, activating specific genes in the bacteria (Matthysse 1984; Stachel *et al.*, 1984), ultimately leading to the transfer of the whole or part of the Ti plasmid into the plant cell. Subsequently, the T-DNA recombines with nuclear plant DNA resulting in stable integration.

B. Early Interactions Between Agrobacterium and Plant Cells

Although the T-DNA is the only portion of the Ti plasmid stably maintained in tumors, none of the genes responsible for plant cell recognition and interaction, — and for subsequent steps leading to and including T-DNA integration —, are located within this region (Leemans *et al.*, 1982; Zambryski *et al.*, 1983). Several mutations affecting oncogenicity map on the Ti plasmid outside the T-DNA or on the chromosome of *Agrobacterium tumefaciens*.

Chromosomal genes are involved in the primary contact of *Agrobacterium* with the plant cells. Two loci have been identified that are constitutively expressed and that are important for attachment of the bacterium to the plant cell (Douglas *et al.*, 1982, 1984). A detailed chapter about these interactions was published by Matthysse (1984).

A group of genes that is important for the early events in infection are located on the Ti plasmid in the D-region of homology, also called the Vir-region. Most mutations in the Vir-region decrease or eliminate the oncogenicity of the Ti plasmid (Holsters *et al.*, 1980; De Greve *et al.*, 1981; Iyer *et al.*, 1982; Klee *et al.*, 1982, 1983). Klee *et al.* (1982, 1983) defined 10 cistrons in this region by complementation tests using cosmid clones overlapping these mutations. These genes are distributed over five separate loci, *vir*A through *E*.

Stachel *et al.* (1984; S. Stachel, personal communication) have recently constructed fusions between the *lacZ* gene and the *vir* genes and found additional independently transcribed loci in the Vir-region: *vir*F and G. All of the loci tested, except for *vir*A, are induced to a higher level of expression by cocultivation with *Nicotiana tabacum* cell suspensions.

Coinfection of different Vir mutants on the same wound does not restore oncogenicity. This suggests that most of the products encoded by the Vir-region are not diffusible between bacteria and moreover that, if the whole Ti plasmid enters the plant, then the Vir-region is not expressed there because mutations do not complement each other (Iyer *et al.*, 1982, Klee *et al.*, 1982). One possible exception might be located in the *vir*E locus, as recently reported by Otten *et al.* (1984).

The expression of the Vir-region in the bacterium may be important for the induction of cell proliferation early in infection (see below) and/or DNA transfer and integration. Since no functions for T-DNA transfer could be assigned to the T-DNA genes (Joos *et al.*, 1983 a), the Vir-region is a very reasonable candidate for this role.

The phenotype of the tumor, incited by *Agrobacterium* is often influenced by the host plant and the place of inoculation (Garfinkel *et al.*, 1981; Joos *et al.*, 1983 a), presumably because of differences in hormone balance at the sites of infection. The mutants in gene *4*, for example, are more virulent on *Daucus carota* slices than on tobacco (G. Gheysen, unpublished results), perhaps because carrot root cells normally produce a cytokinin-like function (Sembdner *et al.*, 1980). Mutants in gene *1* or *2*, on the other hand, give a better response when inoculated along the stem of a complete plant than on top of a decapitated tobacco plant (Ooms *et al.*, 1981; Binns *et al.*, 1982). The elimination of the auxin-producing apical meristem in the latter case may be the reason for a weaker response.

Agrobacterium itself produces hormones such as auxins. Several independent pathways contribute to the auxin production in the bacterium. The T-region (genes *1* and *2*) encodes enzymes in the biosynthetic pathway of auxins (Schröder *et al.*, 1984) (see also II.B.). In addition Liu *et al.* (1982) describe a gene necessary for IAA production in *Agrobacterium* that maps on the Ti plasmid but to the left of the T-DNA. Agrobacteria without Ti plasmids continue to synthesize IAA (Liu *et al.*, 1982; Schröder *et al.*, 1983) (for more details, see Kado, 1984). Some or all of these may influence the growth of the plant cells very early in infection before the T-DNA is integrated and expressed in the plant cells.

C. T-Region and T-DNA Border Sequences

When an insertion is made in the T-region (e. g. by transposon mutagenesis), the additional DNA is cotransferred and integrated in the plant genome (Hernalsteens *et al.*, 1980). The T-DNA in this tumor is enlarged by the presence of the transposon suggesting that a specific mechanism recognizes the borders of the T-region and integrates the DNA that is situated in between (Lemmers *et al.*, 1980; Holsters *et al.*, 1982).

Molecular cloning and sequencing of the T-DNA/plant junctions from transformed plant cell lines has also revealed that the integration event is rather precise (Thomashow *et al.*, 1980b; Zambryski *et al.*, 1980, 1982; Holsters *et al.*, 1982, 1983; Simpson *et al.*, 1982; Yadav *et al.*, 1982). This is especially striking at the right border: 3 nopaline borders end at exactly the same bp (Zambryski *et al.*, 1982), one border differs by 1 bp (Zambryski *et al.*, 1982), and 2 octopine borders end in the 10 bp surrounding this point (Holsters *et al.*, 1983) (see Fig. 4). Comparison of this sequence with the sequences surrounding the left border of the T-region in the Ti plasmid reveals at both sides a direct-repeat of 25 bp with some mismatches. This repeat flanks the nopaline T-region and also the TL- and TR-regions of the octopine Ti plasmid (Zambryski *et al.*, 1982; Barker *et al.*, 1983; Gielen *et al.*, 1984).

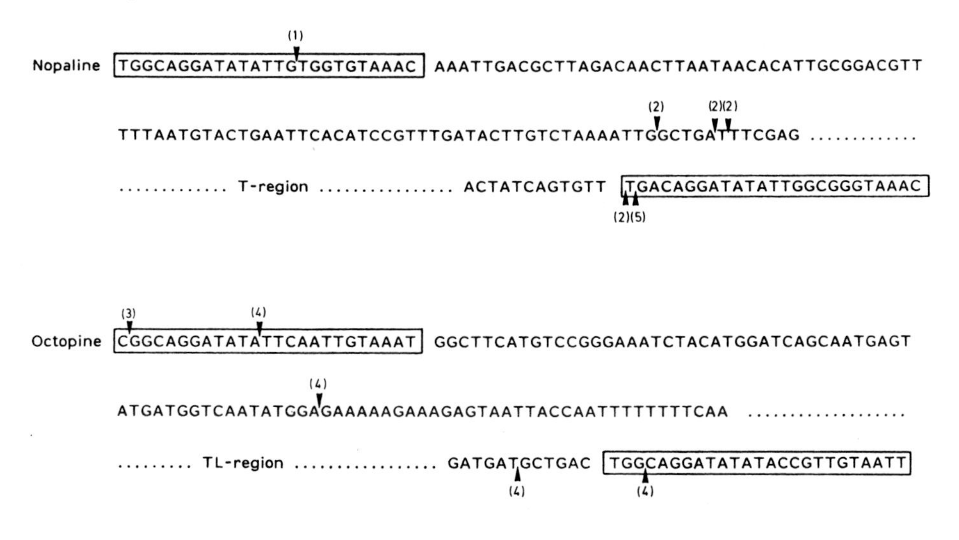

(1) Yadav et al., 1982; (2) Zambryski et al., 1982; (3) Simpson et al., 1982; (4) Holsters et al., 1983;
(5) 3 clones from Zambryski et al. (1982)

Fig. 4. T-region boundary sequences from nopaline and octopine Ti plasmids. The T-region is shown from left to right. The boxed areas represent the 25-bp repeats. The arrows indicate the position of isolated T-DNA/plant junctions

The junction between the plant DNA and the T-DNA at the left is much more variable. Here the divergence points of T-region and T-DNA are located in the 25-bp box in only 3 of the studied cases (2 octopine and 1 nopaline clone) (Simpson *et al.*, 1982; Yadav *et al.*, 1982; Holsters *et al.*, 1983). The other 4 (1 octopine and 3 nopaline clones) that have been sequenced have a shorter T-DNA ending as far as 100 bp to the right, inside of this 25-bp repeat (Zambryski *et al.*, 1982; Holsters *et al.*, 1983).

Genetic analysis of the nopaline Ti plasmid also demonstrates the importance of the right 25-bp sequence. Deletions of the left border have very little, if any, effect on T-DNA transfer (Joos *et al.*, 1983a). Elimination of the right border, by contrast, drastically reduces the Ti plasmid's capacity to transfer or insert its T-DNA (Holsters *et al.*, 1980; Joos *et al.*, 1983a; Wang *et al.*, 1984). It is possible that the right end is not only necessary but also sufficient for transfer and/or integration. This is indicated by the following experiments (Caplan *et al.*, 1985): the nopaline synthase gene closely linked to only the right border of the T-region is efficiently transferred to the plant cell even if inserted on an independent replicon, and complemented *in trans* by the Vir-region (Van Haute, 1984).

Furthermore, the border sequences seem to be recognized wherever they are situated in the bacterial genome, since a T-DNA inserted in the bacterial chromosome is also transferred into the plant cell (Depicker *et al.*, in preparation). However, it is not known whether the border sequences are only recognized in the bacteria or also in the plant cell nor if they are involved in transfer or integration, or both.

D. The 25-bp Repeat Sequence

The importance of the 25-bp border sequence has recently been confirmed in a more direct experiment. Wang *et al.* (1984) substituted the right T-DNA border region of the nopaline pTiC58 plasmid for pBR322 which resulted in a nononcogenic Ti plasmid. This mutant Ti plasmid was then used as an acceptor plasmid to test for sequences which restore oncogenicity. The results show that reintroduction of only the 25-bp sequence in the orientation normally found in the Ti plasmid is sufficient to restore tumor formation. The other orientation only gives tumors with a low efficiency. Thus, the 25-bp sequence seems to be sufficient to promote transfer and, furthermore, this sequence is preferentially active in one orientation.

This „border test system" would allow testing the functionality of sequences similar to the right consensus sequence such as the left 25-bp T-region sequence.

A search through the T-region sequences of the octopine and nopaline Ti plasmids reveals several sequences that show partial homology with the 25-bp repeat (Gielen *et al.*, 1984; M. Van Lijsebettens and R. Villarroel, unpublished results). These sequences could be an explanation for the aberrant T-DNAs that are sometimes observed (Thomashow *et al.*, 1980a; De Beuckeleer *et al.*, 1981; De Greve *et al.*, 1982b; Hepburn *et al.*, 1983a; Van Lijsebettens *et al.*, in preparation). For instance, the sequence present in the 3′-untranslated region of the octopine synthase gene might be recognized as a left-terminus sequence resulting in hormone-dependent cells that only express the octopine synthase gene, like the rGV1 plant discussed earlier (section II.D.) (De Greve *et al.*, 1982b). Alternatively, if this sequence is recognized as a right-terminus sequence, tumor lines containing a shorter T-DNA which do not synthesize octopine would be formed (De Beuckeleer *et al.*, 1981). From this point of view, abnormal T-DNAs could result from faulty recognition before or during integration, rather than from deletions occurring in the plant cell during subsequent cultivation, although deletion formation can occur in tissue cultures (Yang and Simpson, 1981).

These aberrant T-DNAs are also present in normal hormone-independent opine-positive tumors but, since *in vivo*-made tumors are a mixture of several clones, each derived from separate integration events, their phenotype is not visible (Ooms *et al.*, 1982a; Hepburn *et al.*, 1983a). Van Lijsebettens *et al.* (in preparation) studied these aberrant phenotypes in nopaline tumors, infecting *Nicotiana* protoplasts with *Agrobacterium* C58. They found clones that were differentiating into shoots (and therefore missing gene *1* or *2*, or both) and not expressing agrocinopine synthase with a frequency of 10%. Analysis of 3 T-DNAs from this experiment shows that they have a left border in the *Hind*III-22 fragment of pTiC58; this fragment contains 2 sequences that show 57—64% homology with the consensus sequence of the 25-bp repeat. Another T-DNA has a left border in *Hind*III-15; since the sequence of this fragment is still incomplete, we cannot yet make a comparison with the 25-bp repeat. It may be relevant that Hepburn *et al.* (1983a) also have characterized shorter T-DNAs that end in the *Hind*III-22 and *Hind*III-15 fragments of the T-region of pTiT37, a nopaline plasmid which is very homologous to pTiC58.

The mutants that are missing the right border are not always completely nononcogenic. Mutant pGV3101 has a Tn*1* substituting the right-terminus sequence, yet still makes tumors (Holsters *et al.*, 1980), although at a very low frequency (Joos *et al.*, 1983a). In these tumor lines another sequence must substitute for the missing border.

E. T-DNA Integration Compared to Other Mobile Elements

Sequence rearrangements at the junctions between target and inserted DNA can give information to build some theories about the insertion mechanism. The discovery of direct-repeats generated in the target DNA by a transposon insertion is one of the observations, that are the basis of the transposition models by Grindley and Sherratt (1979), Shapiro (1979), Harshey and Bukhari (1981), and Galas and Chandler (1981). The present sequence data on junction fragments between T-DNA and the host chromosome do not yet allow to propose such models for T-DNA integration. However, the following observations may be significant.

First of all, there is evidence that the left borders which do not end in or near the 25-bp sequences are secondary borders and that the corresponding T-DNAs were once part of a longer T-DNA copy. The junction fragment of a nopaline T-DNA tandem array (containing right border — additional DNA sequences — left border) was sequenced; although this left border stops in the T-region approximately 100 bp away from the 25-bp sequence, a sequence of 14 bp was found in this junction fragment which is identical to a sequence about 20 bp away from the 25-bp sequence (Zambryski *et al.*, 1982). Furthermore, the junctions of tandem arrays contain many repeated sequences, derived from Ti plasmid or plant-DNA sequences (Zambryski *et al.*, 1980, 1982; Holsters *et al.*, 1983). The presence of plant DNA in these junctions indicates that the rearrangements probably occur during or after integration into the plant genome (Holsters *et al.*, 1983).

The formation of tandem duplications may be analogous to the generation of repeated integrated copies of SV40 and adenovirus following transformation by these tumor viruses (Sambrook *et al.*, 1980; Doerfler, 1982). The fact that the 25-bp direct-repeats are, unlike the ends of prokaryotic and eukaryotic transposons (Calos and Miller, 1980), not transferred intactly into the plant genome, also resembles the adenovirus system: although these viruses have well-conserved DNA termini that are preserved when the virus replicates as an episome, these sequences are lost upon integration (Deuring *et al.*, 1981).

The T-DNA is similar to transposons in its ability to integrate into different nonhomologous places in the plant genome. This was shown by hybridization of the cloned T-DNA/plant junction fragments to nontransformed plant DNA: not only different classes of repetitive DNA but also unique DNA sequences hybridize to the cloned plant-DNA (Zambryski *et al.*, 1982; Holsters *et al.*, 1983). One feature that the T-DNA may have in common with all movable DNA including bacteriophage λ (Pinkham *et al.*, 1980), prokaryotic (Calos and Miller, 1980) and eukaryotic transposons (Dunsmuir *et al.*, 1980) and retroviruses (Roeder *et al.*, 1980), is the preference to integrate into sites with a high AT content (Zambryski *et al.*, 1982).

F. Crossing the Cellular and Genetic Barriers

There are several models consistent with the data presented here that can account for the introduction of T-DNA into the plant cell. One possibility is that the T-region is excised from the Ti plasmid within the bacteria, and subsequently transferred alone to the plant cells. Since no T-DNA genes are essential for integration (Joos *et al.*, 1983a) either plant genes must be activated to provide the functions for stable integration, or one or more proteins e. g. encoded by the Vir-region must cotransfer with the T-DNA. Alternatively, most or all of the Ti plasmid or even the complete bacterial genome might be transferred to the plant cell but only the T-DNA is integrated. Plant and/or bacterial genes may be involved with stabilization of the T-DNA in this model.

A series of plasmids were constructed to study these hypotheses (Joos *et al.*, 1983b). The plasmids are derived from a mutant C58 Ti plasmid, pGV3850, which contains the two T-region borders and the nopaline synthase gene but none of the *onc* genes (Zambryski *et al.*, 1983b). This plasmid transfers the nopaline synthase gene but is unable to produce tumors.

A T-region segment lacking the border sequences was translocated to different sites of pGV3850 (Joos *et al.*, 1983b). The resulting plasmids conferred oncogenicity although with a lower efficiency than normal. This indicates that the borderless T-region is transferred to the plant cell, regardless of its place on the Ti plasmid. However, when the borderless T-region is separated on an independent replicon *in trans*-complemented by pGV3850, the strain can no longer induce tumors. Since the Vir-region can complement the normal T-DNA *in trans* (Hoekema *et al.*, 1983; Hille *et al.*, 1984b), the *cis*-acting functions in pGV3850 are most likely the border regions.

The fact that the borderless T-DNA can be integrated in the plant genome suggests that more than just the DNA in between the borders can be introduced into the plant cell. It is not known if such transfer is biologically meaningful as part of the transfer process, or if it simply reflects errors in it.

It has been suggested that transfer functions essential for bacterial conjugation would also be involved in the transfer of the T-DNA to plant cells (Tempé *et al.*, 1977). This is unlikely since most of the Tra^- mutations are still oncogenic indicating that the capability to introduce plasmid DNA into plant cells is independent from the bacterial conjugation system (Koekman *et al.*, 1979; Holsters *et al.*, 1980; De Greve *et al.*, 1981). However since bacteria often transfer DNA by conjugation, the introduction of T-DNA into the plant cells may be affected by similar but different functions. In particular, the „conjugative transposition" model proposed by Gawron-Burke and Clewell (1982) may be applicable to T-DNA transfer. Wang *et al.* (1984) recently proposed a model consistent with the available data. The first step would be the introduction of a nick by a specific enzyme that recognizes the right 25-bp terminus sequence. From this point DNA replication would occur to make a T-DNA copy, a second cut would terminate replication at the left 25-bp terminus sequence. Although this model is far from proven, it can be used as a working hypothesis to answer the following questions: (i) is the transfer process replicative; (ii) is it possible to isolate an enzyme which binds to the 25-bp sequence and perhaps cuts it; (iii) which nucleotides are necessary or important in this consensus sequence; (iv) where do these steps occur, in the bacteria or in the plant cell, or in both?

G. T-DNA Stability

The characteristic properties of crown gall cells, tumorous growth and opine synthesis, are stably inherited even after many years of passage in tissue culture. The obvious reason for this is the covalent integration of the T-DNA in the plant DNA. This T-DNA is organized in nucleosomes within the chromatin of the host cell (Schäfer *et al.*, 1984). Does this mean that the T-DNA is now a new locus in the plant genome, completely stable through mitosis and meiosis?

Braun and his collaborators (Braun and Wood, 1976; Turgeon *et al.*, 1976) grafted a nopaline-producing T37 teratoma shoot onto a decapitated tobacco plant and observed that the tumorous phenotype was suppressed: the shoot developed into a normal-looking plant that flowered and set seed. When tissue of this plant, however, is grown *in vitro*, it gives rise to typical crown gall teratoma cultures indicating that the neoplastic properties have not been lost but have been turned off in some way (Wood *et al.*, 1978). Histological and physiological studies of the grafted plants

(Braun and Wood, 1976; Turgeon, 1981) have demonstrated that the cells from stems and leaves are indistinguishable in morphology and function from those of normal tobacco shoots of comparable age and stage of development. This means that the morphogenetic mechanisms directing differentiation during normal development can completely suppress the tumorous state of the transformed cells. When these cells are separated from the plant, the T-DNA is released from the morphogenetic restraints present in an intact organism so that the tissues dedifferentiate again into tumorous cells. The reversible state of the morphogenetic control suggests that either the expression or the effects of the T-DNA functions can under certain conditions be suppressed by the plant host, but that the T-DNA is usually stably maintained through mitosis.

The meiotic transmission of the T-DNA through seeds however could not be achieved initially. F1 progeny of the grafted plants discussed above have been found to be completely normal and free of T-DNA sequences as shown by DNA hybridization (Lemmers *et al.*, 1980; Yang *et al.*, 1980). Completely normal shoots have been obtained after kinetin treatment of a tobacco tumor (Yang and Simpson, 1981). In this case, the loss of the tumorous phenotype and nopaline synthesis is accompanied by the loss of most of the T-DNA. These revertant plants and their progeny do retain however sequences homologous to the ends of the T-DNA (Yang and Simpson, 1981).

The next step was made when De Greve *et al.* (1982b) isolated normal plants that still expressed octopine synthase. The gene for this enzymatic activity was transmitted through the seeds and inherited as a dominant Mendelian marker (Otten *et al.*, 1981). Since these plants only retain the region of the T-DNA that encodes the octopine synthase (De Greve *et al.*, 1982b) it was argued that the T-DNA is able to go through meiosis if devoid of its *onc* genes. Further experiments indicate that the *onc* genes (genes *1, 2* and *4*) can sometimes also be transmitted through meiosis. Plants regenerated from tumors induced by *Agrobacterium rhizogenes* (David *et al.*, 1984) and an *Agrobacterium tumefaciens* root-forming mutant (Barton *et al.*, 1983) transmit full-length copies of their T-DNA to the offspring. Wöstemeyer *et al.* (1984) have described the sexual transmission of an active gene *4,* which is responsible for the suppression of root formation. By crossing different grafted teratoma shoots, they attain two types of seedlings: normal seedlings that are opine-negative and abnormal opine-positive seedlings unable to develop a root system. These observations show that the tumor traits, opine synthesis and suppression of root formation are linked and sexually transmitted in a Mendelian way.

Variability of nopaline expression in the F1 progeny of regenerated plants is reported by Barton *et al.* (1983). The lack of correlation between the number of T-DNA copies and the level of nopaline could be due to regulation by methylation as suggested by Hepburn *et al.* (1983b). The loss of octopine synthesis in some tumors as found by Van Slogteren *et al.* (1983), is also not caused by T-DNA deletions because some shoots originating from these tumors are again positive. The absence of octopine synthase activity in this case was also correlated with methylation (Van Slogteren, 1983). In both cases, the opine production can be restored by adding 5-azacytidine, an *in vivo* demethylating agent.

We can conclude that, although deletions have been reported (Yang and Simpson, 1981; Wöstemeyer *et al.*, 1983) the T-DNA is usually very stable and is transmitted through mitosis and meiosis as an inherent part of the plant genome.

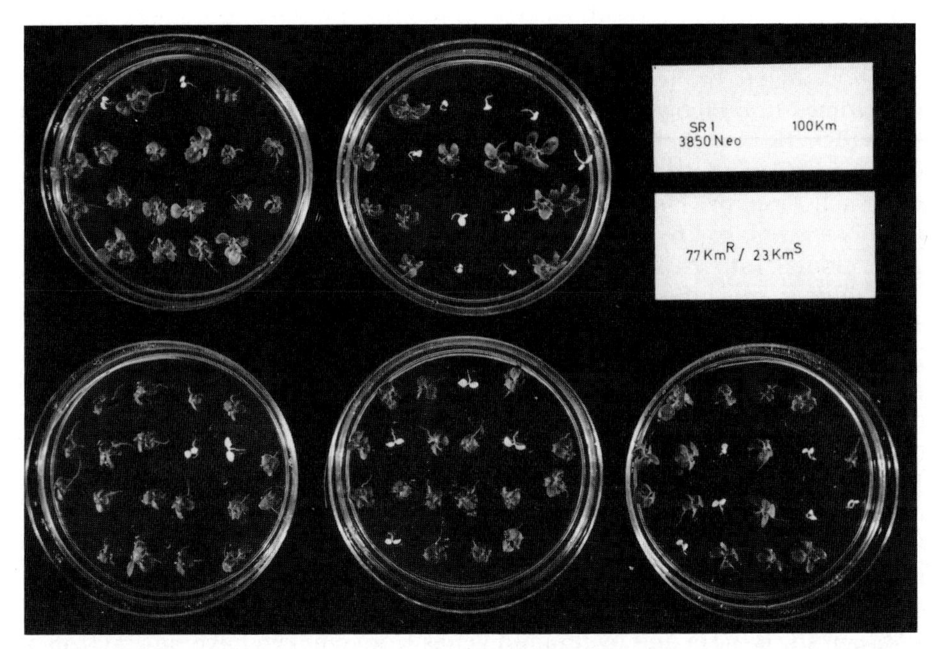

Fig. 5. Germination of 100 F1 seeds of a self-fertilized kanamycin-resistant tobacco plant (SR1 3850 Neo). The seeds were placed on plant growth medium containing 100 μg/ml kanamycin. The resistant seeds germinated and formed plantlets. The sensitive seeds also germinated but the seedlings did not grow further, they etiolated and died. The 3 : 1 ratio of resistance demonstrates that the resistance marker is dominant, that the mother plant is hemizygous for this trait, and that the resistance is transmitted as a single Mendelian factor (De Block *et al.*, 1984), see also III.H.

H. Domestication of the Ti Plasmid

Agrobacterium tumefaciens inserts foreign genes into plants where they are expressed. Not all the properties of the T-DNA system are useful from a practical point of view: there is no need for plants that express opines and we usually do not want tumorous plant tissue. A better understanding of the mechanisms that are responsible for transfer and expression of the T-DNA enabled us to modify the T-DNA into a useful and efficient plant vector system.

An ideal T-DNA vector system should contain:

1. all signals needed to transfer DNA and stably integrate it in the plant nucleus;
2. all products needed to mediate these processes;
3. no functions that might interfere with differentiation into normal plants;
4. a system to express the introduced genes in plant cells;
5. a marker that allows selection of transformed cells; and finally,
6. it should provide a simple way to introduce foreign DNA into this vector.

As has been discussed earlier, only the borders are necessary for transfer and integration if the Vir functions are supplied *in cis* or *in trans*. When we delete the rest of the T-DNA, it is still capable of integrating but it becomes nononcogenic, so that the resulting „transformed" plant cells keep their full regenerative capacity.

Two different ways have been developed to introduce foreign DNA into such a vector. The first one uses a Ti plasmid in which the oncogenes have been replaced by a pBR322 copy (Zambryski *et al.*, 1983b). A foreign gene cloned into pBR322 can be cointegrated with this Ti plasmid via homologous recombination. The second system is a binary vector system which consists of a broad-host range plasmid with the cloned T-DNA borders an a complementing plasmid containing the Ti Vir functions. In order to be suitable for different cloning experiments, several unique and convenient restriction sites should lie in between the borders. Prototypes of this system have already been shown to work (de Framond *et al.*, 1983; Hoekema *et al.*, 1983). Current research in our and other laboratories is directed to design and test similar improved plant vectors.

Attempts to directly express foreign genes in plants, using the antibiotic resistance genes of prokaryotic plasmids or genes from eukaryotic organisms such as the yeast alcohol dehydrogenase gene (Barton *et al.*, 1983), or the α-actin and ovalbumin genes from chicken have failed (Koncz *et al.*, 1984). Presumably specific transcription factors or signals required for their expression are not present in the plant cells.

The nopaline synthase promoter has been shown to function in plant cells and was readily available from Ti plasmid subclones to be fused with the coding sequence of a foreign gene. Another advantage of this promoter was also that it is functional in callus and in most plant organs (Zambryski *et al.*, 1983b). Most of the other plant genes that have been isolated such as the leghemoglobin gene (Wiborg *et al.*, 1982) and members of the zein (Pedersen *et al.*, 1982) and phaseolin (Sun *et al.*, 1981) gene families are only expressed at particular developmental stages. It was therefore decided to construct "chimeric genes" consisting of the nopaline synthase promoter, the coding sequence of a bacterial gene, preferentially a selectable marker, and the termination signals of the nopaline synthase gene. When introduced into plants via the Ti plasmid, the chimeric genes with the coding sequences for the aminoglycoside phosphotransferase (APH[3′]II from Tn *5*), or the chloramphenicol acetyltransferase (CAT from Tn *9*) were shown to be expressed (Herrera-Estrella *et al.*, 1983a, 1983b; Bevan *et al.*, 1983b; Fraley *et al.*, 1983). Furthermore, plant cells containing these chimeric genes became resistant to the toxic drugs, indicating that these new constructs could be used for the selection of cells that acquire foreign genes. Indeed, these chimeric genes have been used as dominant selectable markers to select for transformed plant cells (Herrera-Estrella *et al.*, 1983a; Horsch *et al.*, 1984), from which phenotypically normal and fertile plants have been regenerated (De Block *et al.*, 1984). These plants express the resistance gene in all tissues.

The efficiency of transformation obtained by this modified *Agrobacterium* system outshines so far all attempts to transfer DNA into plant cells by other means. Having developed a practical and efficient way to transfer genes into plants, more sophisticated constructions can now be made to study gene expression and regulation and for the modification of agricultural crops.

One of the first examples is the light-regulated expression of the gene encoding the small subunit (*ss* gene) of the ribulose-1,5-bis phosphate carboxylase (Berry-Lowe *et al.*, 1982; Cashmore, 1983). Broglie *et al.* (1984) have shown that the pea *ss* gene is expressed and light-regulated when introduced into *Petunia* cells. Herrera-Estrella *et al.* (1984) were able to bring the bacterial chloramphenicol acetyl transferase gene under light-inducible control by fusing the bacterial gene to the pea *ss* gene promoter. This chimeric gene is expressed and light-inducible in tobacco cells, indicating that the 5′-flanking sequence of the *ss* gene is sufficient for light-induction. Future experiments, deleting or substituting sequences upstream of the transcription start point may reveal the sequences that are responsible for this type of regulation.

The demonstration that chimeric gene constructions are specifically regulated has many practical applications. It will be useful to construct genes whose expression is restricted to specific types of plant organs, such as seeds, leaves or roots.

A major problem for general applicability of the *Agrobacterium* system in the modification of crop plants is the fact that monocotyledonous plants, which include the important cereal food crops, are thus far not susceptible to *Agrobacterium* infection. [Note added in proof: Recently, tumor induction by *A. tumefaciens* on the monocotyledonous plant *Asparagus officinalis* has been reported (Hernalsteens *et al.*, 1984).] This may be due to the inability of the bacterium to interact efficiently with monocotyledonous plant cells, or to the inability of the infected plant cells to grow as neoplasia. The availability of nononcogenic Ti plasmid vectors in combination with eukaryotic selectable markers may be the first step to study this problem.

IV. Conclusion

The *Agrobacterium* crown gall phenomenon is to date the most thoroughly studied and best documented of the known bacteria/plant interactions. No doubt the successful application of the Ti plasmid as vector for the introduction and expression of foreign genetic material in plant cells has attracted much of the interest. One should, on the other hand, also appreciate the fundamental contribution of crown gall research to the relatively new, but rapidly developing field of plant molecular biology. As such, *Agrobacterium* is and will continue to be an example of how basic and applied aspects of science are tightly connected and mutually stimulating.

V. Acknowledgements

We acknowledge A. Caplan, A. Depicker, G. Engler, J. Simpson, R. Villarroel, and P. Zambryski for advice and critical reading of the manuscript. We thank M. De Cock and R. Maenhaut for typing the manuscript, and A. Verstraete and K. Spruyt for preparing the figures. We are grateful to A. Caplan, A. Depicker, S. Stachel, E. Van Haute, and M. Van Lijsebettens for the permission to include unpublished results. This work is supported by grants from the "A.S.L.K.-Kankerfonds", from the "Instituut tot Aanmoediging van het Wetenschappelijk Onderzoek in Nijverheid en Landbouw", from the Services of the Prime Minister (O.O.A. 12052179), from the "Fonds voor Geneeskundig Wetenschappelijk Onderzoek" (F.G.W.O. 3.001.82), to MVM and JS, and is carried out under Research Contract no. GVI-4-017-B (RS) of the Biomolecular Engineering Programme of the Commission of the European Communities. GG is Research Assistant to the National Fund for Scientific Research (Belgium).

VI. References

Akiyoshi, D. E., Morris, R. O., Hinz, R., Mischke, B. S., Kosuge, T., Garfinkel, D. J., Gordon, M. P., Nester, E. W., 1983: Cytokinin/auxin balance in crown gall tumors is regulated by specific loci in the T-DNA. Proc. Natl. Acad. Sci., U.S.A. **80,** 407—411.

Amasino, R. M., Miller, C. O., 1982: Hormonal control of tobacco crown gall tumor morphology. Plant Physiol. **69,** 389—392.

Barker, R. F., Idler, K. B., Thompson, D. V., Kemp, J. D., 1983: Nucleotide sequence of the T-DNA region from the *Agrobacterium tumefaciens* octopine Ti plasmid pTi15955. Plant Mol. Biol. **2,** 335—350.

Barton, K. A., Binns, A. N., Matzke, A. J. M., Chilton, M.-D., 1983: Regeneration of intact tobacco plants containing full length copies of genetically engineered T-DNA, and transmission of T-DNA to R1 progeny. Cell **32,** 1033—1043.

Beiderbeck, R., 1973: Wurzelinduktion an Blättern von *Kalanchoë daigremontiana* durch *Agrobacterium rhizogenes* und der Einfluß von Kinetin auf diesen Prozess. Z. Pflanzenphysiol. **68,** 460—467.

Berry-Lowe, S. L., McKnight, T. D., Shah, D. M., Meagher, R. B., 1982: The nucleotide sequence, expression, and evolution of one member of a multigene family encoding the small subunit of ribulose-1,5-bisphosphate carboxylase in soybean. J. Mol. Appl. Genet. **1,** 483—498.

Bevan, M. W., Chilton, M.-D., 1982: Multiple transcripts of T-DNA detected in nopaline crown gall tumors. J. Mol. Appl. Genet. **1,** 539—546.

Bevan, M., Barnes, W. M., Chilton, M.-D., 1983a: Structure and transcription of the nopaline synthase gene region of T-DNA. Nucl. Acids Res. **11,** 369—385.

Bevan, M. W., Flavell, R. B., Chilton, M.-D., 1983b: A chimaeric antibiotic resistance gene as a selectable marker for plant cell transformation. Nature (London) **304,** 184—187.

Binns, A. N., Sciaky, D., Wood, H. N., 1982: Variation in hormone autonomy and regenerative potential of cells transformed by strain A66 of *Agrobacterium tumefaciens.* Cell **31,** 605—612.

Bomhoff, G., Klapwijk, P. M., Kester, H. C. M., Schilperoort, R. A., Hernalsteens, J. P., Schell, J., 1976: Octopine and nopaline synthesis and breakdown genetically controlled by a plasmid of *Agrobacterium tumefaciens.* Mol. Gen. Genet. **145,** 177—181.

Braun, A. C., Mandle, R. J., 1948: Studies on the inactivation of the tumor-inducing principle in crowngall. Growth **12,** 255—269.

Braun, A. C., White, P. R., 1943: Bacteriological sterility of tissues derived from secundary crown gall tumors. Phytopathol. **33,** 85—100.

Braun, A. C., Wood, H. N., 1976: Suppression of the neoplastic state with the acquisition of specialized functions in cells, tissues, and organs of crown gall teratomas of tobacco. Proc. Natl. Acad. Sci., U.S.A. **73,** 496—500.

Breathnach, R., Chambon, P., 1981: Organization and expression of eukaryotic split genes coding for proteins. Ann. Rev. Biochem. **50,** 349—383.

Broglie, R., Coruzzi, G., Fraley, R. T., Rogers, S. G., Horsch, R. B., Niedermeyer, J. G., Fink, C. L., Flick, J. S., Chua, N.-H., 1984: Light-regulated expression of a pea ribulose-1,5-bisphosphate carboxylase small subunit gene in transformed plant cells. Science **224,** 838—843.

Byrne, M. C., Koplow, J., David, C., Tempé, J., Chilton, M.-D., 1983: Structure of T-DNA in roots transformed by *Agrobacterium rhizogenes.* J. Mol. Appl. Genet. **2,** 201—209.

Calos, M. P., Miller, J. H., 1980: Transposable elements. Cell **20,** 579—595.

Caplan, A., Herrera-Estrella, L., Inzé, D., Van Haute, E., Van Montagu, M., Schell, J., Zambryski, P., 1983: Introduction of genetic material into plant cells. Science **222,** 815—821.

Caplan, A. B., Van Montagu, M., Schell, J., 1985: Genetic analysis of integration mediated by single T-DNA borders. J. Bacteriol. **161,** 655—664.

Cashmore, A. R., 1983: Nuclear genes encoding the small subunit of ribulose-1,5-bisphosphate carboxylase. In: Kosuge, T., Meredith, C. P., Hollaender, A. (eds.), Genetic Engineering of Plants — an agricultural perspective, pp. 29—38. New York: Plenum Press.

Chilton, M.-D., Drummond, M. H., Merlo, D. J., Sciaky, D., Montoya, A. L., Gordon, M. P., Nester, E. W., 1977: Stable incorporation of plasmid DNA into higher plant cells: the molecular basis of crown gall tumorigenesis. Cell **11,** 263—271.

Chilton, M.-D., Saiki, R. K., Yadav, N., Gordon, M. P., Quetier, F., 1980: T-DNA from *Agrobacterium* Ti plasmid is in the nuclear DNA fraction of crown gall tumor cells. Proc. Natl. Acad. Sci., U.S.A. **77,** 4060—4064.

Chilton, M.-D., Tepfer, D. A., Petit, A., David, C., Casse-Delbart, F., Tempé, J., 1982: *Agrobacterium rhizogenes* inserts T-DNA into the genomes of the host plant root cells. Nature (London) **295,** 432—434.

Chilton, W. S., Tempé, J., Matzke, M., Chilton, M.-D., 1984: Succinamopine: a new crown gall opine. J. Bacteriol. **157,** 357—362.

Comai, L., Kosuge, T., 1980: Involvement of plasmid deoxyribonucleic acid in indoleacetic acid synthesis in *Pseudomonas savastanoi.* J. Bacteriol. **143,** 950—957.

Costantino, P., Hooykaas, P. J. J., den Dulk-Ras, H., Schilperoort, R. A., 1980: Tumor formation and rhizogenicity of *Agrobacterium rhizogenes* carrying Ti-plasmids. Gene **11,** 79—87.

Crepin, M., Triadou, P., Lelong, J. C., Gros, F., 1981: Identification of transcription initiation sites for bacterial RNA polymerase and eukaryotic RNA polymerase B on the 5′ end of the mouse β-globin gene. Eur. J. Biochem. **118,** 371—377.

Davey, M. R., Cocking, E. C., Freeman, J., Draper, J., Pearce, N., Tudor, I., Hernalsteens, J. P., De Beuckeleer, M., Van Montagu, M., Schell, J., 1980: The use of plant protoplasts for transformation by *Agrobacterium* and isolated plasmids. In: Ferenczy, L., Farkas, G. L. (eds.), Advances in protoplast research, pp. 425—430. Oxford: Pergamon Press.

David, C., Chilton, M.-D., Tempé, J., 1984: Conservation of T-DNA in plants regenerated from hairy root cultures. Bio/technology **2,** 73—76.

De Beuckeleer, M., De Block, M., De Greve, H., Depicker, A., De Vos, R., De Vos, G., De Wilde, M., Dhaese, P., Dobbelaere, M. R., Engler, G., Genetello, C., Hernalsteens, J. P., Holsters, M., Jacobs, A., Schell, J., Seurinck, J., Silva, B., Van Haute, E., Van Montagu, M., Van Vliet, F., Villarroel, R., Zaenen, I., 1978: The use of the Ti-plasmid as a vector for the introduction of foreign DNA into plants. In: Ridé, M. (ed.), Proceedings IVth International Conference on Plant Pathogenic Bacteria, pp. 115—126. Angers: I. N. R. A.

De Beuckeleer, M., Lemmers, M., De Vos, G., Willmitzer, L., Van Montagu, M., Schell, J., 1981: Further insight on the transferred-DNA of octopine crown gall. Mol. Gen. Genet. **183,** 283—288.

De Block, M., Herrera-Estrella, L., Van Montagu, M., Schell, J., Zambryski, P., 1984: Expression of foreign genes in regenerated plants and their progeny. EMBO J. **3,** 1681—1689.

De Cleene, M., De Ley, J., 1981: The host range of infectious hairy-root. Bot. Rev. **47,** 147—194.

de Framond, A. J., Barton, K. A., Chilton, M.-D., 1983: Mini-Ti: a new vector strategy for plant genetic engineering. Bio/technology **1,** 262—269.

De Greve, H., Decraemer, H., Seurinck, J., Van Montagu, M., Schell, J., 1981: The functional organization of the octopine *Agrobacterium tumefaciens* plasmid pTiB6S3. Plasmid **6,** 235—248.

De Greve, H., Dhaese, P., Seurinck, J., Lemmers, M., Van Montagu, M., Schell, J., 1982a: Nucleotide sequence and transcript map of the *Agrobacterium tumefaciens* Ti plasmid-encoded octopine synthase gene. J. Mol. Appl. Genet. **1,** 499—512.

De Greve, H., Leemans, J., Hernalsteens, J. P., Thia-Toong, L., De Beuckeleer, M., Willmitzer, L., Otten, L., Van Montagu, M., Schell, J., 1982b: Regeneration of normal and fertile plants that express octopine synthase, from tobacco crown galls after deletion of tumour-controlling functions. Nature (London) **300,** 752—755.

Depicker, A., De Wilde, M., De Vos, G., De Vos, R., Van Montagu, M., Schell, J., 1980: Molecular cloning of overlapping segments of the nopaline Ti-plasmid pTiC58 as a means to restriction endonuclease mapping. Plasmid **3,** 193—211.

Depicker, A., Stachel, S., Dhaese, P., Zambryski, P., Goodman, H. M., 1982: Nopaline synthase: transcript mapping and DNA sequence. J. Mol. Appl. Genet. **1,** 561—574.

Depicker, A., Van Montagu, M., Schell, J., 1983: Plant cell transformation by *Agrobacterium* plasmids. In: Kosuge, T., Meredith, C. P., Hollaender, A. (eds.), Genetic Engineering of Plants — an agricultural perspective, pp. 143—176. New York: Plenum Press.

Deuring, R., Winterhoff, U., Tamanoi, F., Stabel, S., Doerfler, W., 1981: Site of linkage between adenovirus type 12 and cell DNAs in hamster tumour line CLAC3. Nature (London) **293,** 81—84.

De Vos, G., De Beuckeleer, M., Van Montagu, M., Schell, J., 1981: Restriction endonuclease mapping of the octopine tumor inducing pTiAch5 of *Agrobacterium tumefaciens.* Plasmid **6,** 249—253.

Dhaese, P., De Greve, H., Gielen, J., Seurinck, J., Van Montagu, M., Schell, J., 1983a: Identification of sequences involved in the polyadenylation of higher plant nuclear transcripts using *Agrobacterium* T-DNA genes as models. EMBO J. **2,** 419—426.

Dhaese, P., Seurinck, J., Van Montagu, M., Schell, J., 1983b: Control signals for the expression of *Agrobacterium* Ti plasmid genes in plant cells are typically eukaryotic. Arch. Int. Physiol. Biochim. **91,** B13—B14.

Doerfler, W., 1982: Uptake, fixation, and expression of foreign DNA in mammalian cells: the organization of integrated adenovirus DNA sequence. In: Graf, T., Jarnisch, R. (eds.), Tumorviruses, Neoplastic Transformation and Differentiation (Current Topics in Microbiology and Immunology 101), pp. 128—194. Berlin - Heidelberg - New York: Springer-Verlag.

Douglas, C. J., Halperin, W., Nester, E. W., 1982: *Agrobacterium tumefaciens* mutants affected in attachment to plant cells. J. Bacteriol. **152,** 1265—1275.

Douglas, C. J., Halperin, W., Nester, E. W., 1984: *Agrobacterium* chromosomal genes involved in attachment to plant cells. J. Cell. Biochem. Suppl. **8B,** 251.

Draper, J., Davey, M. R., Freeman, J. P., Cocking, E. C., Cox, B. J., 1982: Ti plasmid homologous sequences present in tissues from *Agrobacterium* plasmid-transformed Petunia protoplasts. Plant and Cell Physiol. **23,** 451—458.

Dunsmuir, P., Brorein, W. J. Jr., Simon, M. A., Rubin, G. M., 1980: Insertion of the *Drosophila* transposable element copia generates a 5 base pair duplication. Cell **21,** 575—579.

Elliot, C., 1951: Manual of Bacterial Plant Pathogens, 2nd rev. ed. Waltham, Mass.: Chronica Botanica.

Ellis, J. G., Murphy, P. J., 1981: Four new opines from crown gall tumours — their detection and properties. Mol. Gen. Genet. **181,** 36—43.

Ellis, J. G., Ryder, M. H., Tate, M. E., 1984: *Agrobacterium tumefaciens* T_R-DNA encodes a pathway for agropine biosynthesis. Mol. Gen. Genet. **195,** 466—473.

Engler, G., Depicker, A., Maenhaut, R., Villarroel-Mandiola, R., Van Montagu, M., Schell, J., 1981: Physical mapping of DNA base sequence homologies between an octopine and a nopaline Ti-plasmid of *Agrobacterium tumefaciens.* J. Mol. Biol. **152,** 183—208.

Fernandez, S. M., Lurquin, P. F., Kado, C. I., 1978: Incorporation and maintenance of recombinant-DNA plasmid vehicles pBR313 and pCR1 in plant protoplasts. FEBS Lett. **87,** 277—282.

Firmin, J. L., Fenwick, G. R., 1978: Agropine — a major new plasmid-determined metabolite in crown gall tumours. Nature (London) **276,** 842—844.

Fraley, R. T., 1983: Liposome-mediated delivery of tobacco mosaic virus RNA into petunia protoplasts. Plant Mol. Biol. **2,** 5—14.

Fraley, R. T., Rogers, S. G., Horsch, R. B., Sanders, P. R., Flick, J. S., Adams, S. P., Bittner, M. L., Brand, L. A., Fink, C. L., Fry, J. S., Galluppi, G. R., Goldberg, S. B., Hoffman, N. L., Woo, S. C., 1983: Expression of bacterial genes in plant cells. Proc. Natl. Acad. Sci., U.S.A. **80,** 4803—4807.

Galas, D. J., Chandler, M., 1981: On the molecular mechanism of transposition. Proc. Natl. Acad. Sci., U.S.A. **78,** 4858—4862.

Garfinkel, D. J., Simpson, R. B., Ream, L. W., White, F. F., Gordon, M. P., Nester, E. W., 1981: Genetic analysis of crown gall: fine structure map of the T-DNA by site-directed mutagenesis. Cell **27,** 143—153.

Gawron-Burke, C., Clewell, D. B., 1982: A transposon in *Streptococcus faecalis* with fertility properties. Nature (London) **300,** 281—284.

Gelvin, S. B., Gordon, M. P., Nester, E. W., Aronson, A. I., 1981: Transcription of *Agrobacterium* Ti plasmid in the bacterium and in crown gall tumors. Plasmid **6,** 17—29.

Gelvin, S. B., Thomashow, M. F., McPherson, J. C., Gordon, M. P., Nester, E. W., 1982: Sizes and map positions of several plasmid-DNA-encoded transcripts in octopine-type crown gall tumors. Proc. Natl. Acad. Sci., U.S.A. **79,** 76—80.

Gielen, J., De Beuckeleer, M., Seurinck, J., Deboeck, F., De Greve, H., Lemmers, M., Van Montagu, M., Schell, J., 1984: The complete nucleotide sequence of the TL-DNA of the *Agrobacterium tumefaciens* plasmid pTiAch5. EMBO J. **3,** 835—846.

Goldberg, R. B., Hoschek, G., Vodkin, L. O., 1983: An insertion sequence blocks the expression of a soybean lectin gene. Cell **33,** 465—475.

Goldberg, S. B., Flick, J. S., Rogers, S. G., 1984: Nucleotide sequence of the *tmr* locus of *Agrobacterium tumefaciens* pTiT37 T-DNA. Nucl. Acid Res. **12,** 4665—4677.

Goldmann, A., Tempé, J., Morel, G., 1968: Quelques particularités de diverses souches d' *Agrobacterium tumefaciens.* C. R. Soc. Biol. (Paris) **162,** 623—631.

Greenblatt, J. F., Allet, B., Weil, R., 1976: Synthesis of the tumour antigen and the major capsid protein of Simian virus SV40 in a cell-free system derived from *Escherichia coli.* J. Mol. Biol. **108,** 361—379.

Grindley, N. D. F., Sherratt, D. J., 1979: Sequence analysis at IS *1* insertion sites: models for transposition. Cold Spring Harbor Symp. Quant. Biol. **43,** 1257—1261.

Guyon, P., Chilton, M.-D., Petit, A., Tempé, J., 1980: Agropine in "null type" crown gall tumors: evidence for the generality of the opine concept. Proc. Natl. Acad. Sci., U.S.A. **77,** 2693—2697.

Harshey, R. M., Bukhari, A., 1981: A mechanism of DNA transposition. Proc. Natl. Acad. Sci., U.S.A. **78,** 1090—1094.

Heidekamp, F., Dirkse, W. G., Hille, J., van Ormondt, H., 1983: Nucleotide sequence of the *Agrobacterium tumefaciens* octopine Ti plasmid-encoded *tmr* gene. Nucl. Acid Res. **11,** 6211—6223.

Hepburn, A. G., Clarke, L. E., Blundy, K. S., White, J., 1983a: Nopaline Ti plasmid, pTiT37, T-DNA insertions into a flax genome. J. Mol. Appl. Genet. **2,** 211—224.

Hepburn, A. G., Clarke, L. E., Pearson, L., White, J., 1983b: The role of cytosine methylation in the control of nopaline synthase gene expression in a plant tumor. J. Mol. Appl. Genet. **2,** 315—329.

Hernalsteens, J. P., Van Vliet, F., De Beuckeleer, M., Depicker, A., Engler, G., Lemmers, M., Holsters, M., Van Montagu, M., Schell, J., 1980: The *Agrobacterium tumefaciens* Ti plasmid as a host vector system for introducing foreign DNA in plant cells. Nature (London) **287,** 654—656.

Hernalsteens, J.-P., Thia-Toong, L., Schell, J., Van Montagu, M., 1984: An *Agrobacterium*-transformed cell culture from the monocot *Asparagus officinalis.* EMBO J. **3,** 3039—3041.

Herrera-Estrella, L., De Block, M., Messens, E., Hernalsteens, J.-P., Van Montagu,

M., Schell, J., 1983a: Chimeric genes as dominant selectable markers in plant cells. EMBO J. **2,** 987—995.

Herrera-Estrella, L., Depicker, A., Van Montagu, M., Schell, J., 1983b: Expression of chimaeric genes transferred into plant cells using a Ti-plasmid-derived vector. Nature (London) **303,** 209—213.

Herrera-Estrella, L., Van den Broeck, G., Maenhaut, R., Van Montagu, M., Schell, J., Timko, M., Cashmore, A., 1984: Light-inducible and chloroplast-associated expression of a chimaeric gene introduced into *Nicotiana tabacum* using a Ti plasmid vector. Nature (London) **310,** 115—120.

Hille, J., Hoekema, A., Hooykaas, P., Schilperoort, R. A., 1984a: Gene organisation of the Ti-plasmid. In: Verma, D.P.S., Hohn, Th. (eds.), Genes Involved in Microbe-Plant Interactions (Plant Gene Research, Vol. 1), pp. 287—309, Wien - New York: Springer.

Hille, J., Van Kan, J., Schilperoort, R., 1984b: Trans-acting virulence functions of the octopine Ti-plasmid from *Agrobacterium tumefaciens.* J. Bacteriol. **158,** 754—756.

Hoekema, A., Hirsch, P. R., Hooykaas, P. J. J., Schilperoort, R. A., 1983: A binary plant vector strategy based on separation of *vir-* and T-region of the *Agrobacterium tumefaciens* Ti plasmid. Nature (London) **303,** 179—181.

Hoekema, A., Hooykaas, P., Schilperoort, R., 1984: Transfer of the octopine T-DNA segment to plant cells mediated by different types of *Agrobacterium* tumor- or root-inducing plasmids: Generality of virulence systems. J. Bacteriol. **158,** 383—385.

Holsters, M., Silva, B., Van Vliet, F., Genetello, C., De Block, M., Dhaese, P., Depicker, A., Inzé, D., Engler, G., Villarroel, R., Van Montagu, M., Schell, J., 1980: The functional organization of the nopaline *A. tumefaciens* plasmid pTiC58. Plasmid **3,** 212—230.

Holsters, M., Villarroel, R., Van Montagu, M., Schell, J., 1982: The use of selectable markers for the isolation of plant-DNA/T-DNA junction fragments in a cosmid vector. Mol. Gen. Genet. **185,** 283—289.

Holsters, M., Villarroel, R., Gielen, J., Seurinck, J., De Greve, H., Van Montagu, M., Schell, J., 1983: An analysis of the boundaries of the octopine TL-DNA in tumors induced by *Agrobacterium tumefaciens.* Mol. Gen. Genet. **190,** 35—41.

Hooykaas, P. J., Schilperoort, R. A., 1984: The molecular genetics of crown gall tumorigenesis in molecular genetics of plants. In: Scandalios, J. G., (ed.), Molecular Genetics of Plants (Advances in Genetics, Vol. 22), pp. 209—283. New York: Academic Press.

Hooykaas, P. J. J., Ooms, G., Schilperoort, R. A., 1982: Tumors induced by different strains of *Agrobacterium tumefaciens.* In: Kahl G., Schell, J. (eds.), Molecular Biology of Plant Tumors, pp. 373—390. New York: Academic Press.

Horsch, R. B., Fraley, R. T., Rogers, S. G., Sanders, P. R., Lloyd, A., Hoffman, N., 1984: Inheritance of functional foreign genes in plants. Science **223,** 496—498.

Hu, N.-T., Peifer, M. A., Heidecker, G., Messing, J., Rubenstein, I., 1982: Primary structure of a genomic zein sequence of maize. EMBO J. **1,** 1337—1342.

Huffman, G. A., White, F. F., Gordon, M. P., Nester, E. W., 1984: Hairy-root-inducing plasmid: physical map and homology to tumor-inducing plasmids. J. Bacteriol. **157,** 269—276.

Inzé, D., Follin, A., Van Lijsebettens, M., Simoens, C., Genetello, C., Van Montagu, M., Schell, J., 1984: Genetic analysis of the individual T-DNA genes of *Agrobacterium tumefaciens;* further evidence that two genes are involved in indole-3-acetic acid synthesis. Mol. Gen. Genet. **194,** 265—274.

Iyer, V. N., Klee, H. J., Nester, E. W., 1982: Units of genetic expression in the virulence region of a plant tumor-inducing plasmid of *Agrobacterium tumefaciens*. Mol. Gen. Genet. **188,** 418—424.

Janssens, A., Engler, G., Zambryski, P., Van Montagu, M., 1984: The nopaline C58 T-DNA region is transcribed in *Agrobacterium tumefaciens*. Mol. Gen. Genet. **195,** 341—350.

Joos, H., Inzé, D., Caplan, A., Sormann, M., Van Montagu, M., Schell, J., 1983a: Genetic analysis of T-DNA transcripts in nopaline crown galls. Cell **32,** 1057—1067.

Joos, H., Timmerman, B., Van Montagu, M., Schell, J., 1983b: Genetic analysis of transfer and stabilization of *Agrobacterium* DNA in plant cells. EMBO J. **2,** 2151—2160.

Kado, C. I., 1984: Phytohormone-mediated tumorigenesis by plant pathogenic bacteria. In: Verma, D. P. S., Hohn, Th. (eds.), Genes Involved in Microbe-Plant Interactions (Plant Gene Research, Vol. 1), pp. 311—336. Wien - New York: Springer.

Kahl, G., Schell, J., 1982: Molecular Biology of Plant Tumors, pp. 615. New York: Academic Press.

Karcher, S. J., DiRita, V. J., Gelvin, S. B., 1984: Transcript analysis of T_R DNA in octopine-type crown gall tumors. Mol. Gen. Genet. **194,** 159—165.

Kemp, J. D., 1982: Enzymes in octopine and nopaline metabolism. In: Kahl, G., Schell, J. (eds.), Molecular Biology of Plant Tumors, pp. 461—474. New York: Academic Press.

Kerr, A., 1969: Transfer of virulence between isolates of *Agrobacterium*. Nature (London) **223,** 1175—1176.

Kerr, A., 1971: Acquisition of virulence by non-pathogenic isolation of *Agrobacterium radiobacter*. Physiol. Plant Pathol. **1,** 241—246.

Klee, H. J., Gordon, M. P., Nester, E. W., 1982: Complementation analysis of *Agrobacterium tumefaciens* Ti plasmid mutations affecting oncogenicity. J. Bacteriol. **150,** 327—331.

Klee, H. J., White, F. F., Iyer, V. N., Gordon, M. P., Nester, E. W., 1983: Mutational analysis of the virulence region of an *Agrobacterium tumefaciens* Ti plasmid. J. Bacteriol. **153,** 878—883.

Klee, H., Montoya, A., Horodyski, F., Lichtenstein, C., Garfinkel, D., Fuller, S., Flores, C., Peschon, J., Nester, E., Gordon, M., 1984: Nucleotide sequence of the *tms* genes of the pTiA6NC octopine Ti plasmid: two gene products involved in plant tumorigenesis. Proc. Natl. Acad. Sci., U.S.A. **81,** 1728—1732.

Koekman, B. P., Ooms, G., Klapwijk, P. M., Schilperoort, R. A., 1979: Genetic map of an octopine Ti-plasmid. Plasmid **2,** 347—357.

Koncz, C., Kreuzaler, F., Kalman, Zs., Schell, J., 1984: A simple method to transfer, integrate and study expression of foreign genes, such as chicken ovalbumin and α-actin in plant tumors. EMBO J. **3,** 1029—1037.

Kosuge, T., Heskett, M. G., Wilson, E. E., 1966: Microbial synthesis and degradation of indole-3-acetic acid. I. The conversion of L-tryptophan. J. Biol. Chem. **241,** 3738—3744.

Krens, F. A., 1983: Studies on transformation of tobacco leaf protoplasts: Ti-plasmid DNA transformation and cocultivation with *Agrobacterium tumefaciens*. PhD. thesis, Leiden.

Krens, F. A., Molendijk, L., Wullems, G. J., Schilperoort, R. A., 1982: *In vitro* transformation of plant protoplasts with Ti-plasmid DNA. Nature (London) **296,** 72—74.

Lahners, K., Byrne, M. C., Chilton, M.-D., 1984: T-DNA fragments of hairy root plasmid pRi8196 are distantly related to octopine and nopaline Ti-plasmid T-DNA. Plasmid **11,** 130—140.

Leemans, J., Deblaere, R., Willmitzer, L., De Greve, H., Hernalsteens, J. P., Van Montagu, M., Schell, J., 1982: Genetic identification of functions of TL-DNA transcripts in octopine crown galls. EMBO J. **1,** 147—152.

Lemmers, M., De Beuckeleer, M., Holsters, M., Zambryski, P., Depicker, A., Hernalsteens, J. P., Van Montagu, M., Schell, J., 1980: Internal organization, boundaries and integration of Ti-plasmid DNA in nopaline crown gall tumours. J. Mol. Biol. **144,** 353—376.

Lichtenstein, C., Klee, H., Montoya, A., Garfinkel, D., Fuller, S., Flores, C., Nester, E., Gordon, M., 1984: Nucleotide sequence and transcript mapping of the *tmr* gene of the pTiA6NC octopine Ti-plasmid: a bacterial gene involved in plant tumorigenesis. J. Mol. Appl. Genet. **2,** 354—362.

Lioret, C., 1956: Sur la mise en évidence d'un acide aminé non identifié particulier aux tissus de crown gall. Bull. Soc. Fr. Physiol. vég. **2,** 76.

Liu, S.-T., Perry, K. L., Schardl, C. L., Kado, C. I., 1982: *Agrobacterium* Ti-plasmid indoleacetic acid gene is required for crown gall oncogenesis. Proc. Natl. Acad. Sci., U.S.A. **79,** 2812—2816.

Loesch-Fries, L. S., Hall, T. C., 1980: Synthesis, accumulation and encapsidation of individual brome mosaic virus RNA components in barley protoplasts. J. Gen. Virol. **47,** 323—332.

Matthysse, A., 1984: *Agrobacterium*-plant surface interactions. In: Verma, D. P. S., Hohn, Th. (eds.), Genes Involved in Microbe-Plant Interactions (Plant Gene Research, Vol. 1), pp. 33—54. Wien - New York: Springer.

Moore, L., Warren, G., Strobel, G., 1979: Involvement of a plasmid in the hairy root disease of plants caused by *Agrobacterium rhizogenes.* Plasmid **2,** 617—626.

Morel, G., 1956: Métabolisme de l'arginine par les tissus de crown gall de topinambour. Bull. Soc. Fr. Physiol. vég. **2,** 75.

Murai, N., Kemp, J. D., 1982: Octopine synthase mRNA isolated from sunflower crown gall callus is homologous to the Ti plasmid of *Agrobacterium tumefaciens.* Proc. Natl. Acad. Sci., U.S.A. **79,** 86—90.

Ooms, G., Hooykaas, P. J., Moleman, G., Schilperoort, R. A., 1981: Crown gall plant tumors of abnormal morphology, induced by *Agrobacterium tumefaciens* carrying mutated octopine Ti plasmids; analysis of T-DNA functions. Gene **14,** 33—50.

Ooms, G., Bakker, A., Molendijk, L., Wullems, G. J., Gordon, M. P., Nester, E. W., Schilperoort, R. A., 1982a: T-DNA organization in homogenous and heterogenous octopine-type crown gall tissues of *Nicotiana tabacum.* Cell **30,** 589—597.

Ooms, G., Hooykaas, P. J. J., Van Veen, R. J. M., Van Beelen, P., Regensburg-Tuink, T. J. G., Schilperoort, R. A., 1982b: Octopine Ti plasmid deletion mutants of *Agrobacterium tumefaciens* with emphasis on the right side of the T-region. Plasmid **7,** 15—29.

Otten, L., De Greve, H., Hernalsteens, J. P., Van Montagu, M., Schieder, O., Straub, J., Schell, J., 1981: Mendelian transmission of genes introduced into plants by the Ti plasmids of *Agrobacterium tumefaciens.* Mol. Gen. Genet. **183,** 209—213.

Otten, L., De Greve, H., Leemans, J., Hain, R., Hooykaas, P., Schell, J., 1984: Restoration of virulence of Vir region mutants of *Agrobacterium tumefaciens* strain B6S3 by coinfection with normal and mutant *Agrobacterium* strains. Mol. Gen. Genet. **195,** 159—163.

Pedersen, K., Devereux, J., Wilson, D. R., Sheldon, E., Larkins, B. A., 1982: Cloning and sequence analysis reveal structural variation among related zein genes in maize. Cell **29,** 1015—1026.

Petit, A., Tempé, J., 1985: The function of T-DNA in nature. In: Van Vloten-Doting, L., Groot, G. S. P., Hall, T. C., (eds.), Molecular form and function of the plant genome, in press. New York: Plenum Press.

Petit, A., Delhaye, S., Tempé, J., Morel, G., 1970: Recherches sur les guanidines des tissus de crown gall. Mise en évidence d'une relation biochimique spécifique entre les souches d'*Agrobacterium* et les tumeurs qu'elles induisent. Physiol. vég. **8,** 205—213.

Petit, A., David, C., Dahl, G. A., Ellis, J. G., Guyon, P., Casse-Delbart, F., Tempé, J., 1983: Further extension of the opine concept: plasmids in *Agrobacterium rhizogenes* cooperate for opine degradation. Mol. Gen. Genet. **190,** 204—214.

Pinkham, J. L., Platt, T., Enquist, L. W., and Weisberg, R. A., 1980: The secundary attachment site for bacteriophage λ in the proA/B gene of *Escherichia coli.* J. Mol. Biol. **144,** 587—592.

Rafalski, J. A., Scheets, K., Metzler, M., Peterson, D. M., Hedgcoth, C., Söll, D. G., 1984: Developmentally regulated plant genes: the nucleotide sequence of a wheat gliadin genomic clone. EMBO J. **3,** 1409—1415.

Risuleo, G., Baltistoni, P., Costantino, P., 1982: Regions of homology between tumorigenic plasmids of *Agrobacterium rhizogenes* and *Agrobacterium tumefaciens.* Plasmid **7,** 45—51.

Roeder, G. S., Farabaugh, P. J., Chaleff, D. T., Fink, G. R., 1980: The origins of gene instability in yeast. Science **209,** 1375—1380.

Salomon, F., Deblaere, R., Leemans, J., Hernalsteens, J.-P., Van Montagu, M., Schell, J., 1984: Genetic identification of functions of TR-DNA transcripts in octopine crown galls. EMBO J. **3,** 141—146.

Sambrook, J., Botchan, M., Hu, S. L., Mitchison, T., Stringer, J., 1980: Integration of viral DNA sequences in cells transformed by adenovirus 2 or SV40. Proc. R. Soc. Lond. B **210,** 423—435.

Sarkar, S., Upadhya, M. D., Melchers, G., 1974: A highly efficient method of inoculation of tobacco mesophyll protoplasts with ribonucleic acid of tobacco mosaic virus. Mol. Gen. Genet. **135,** 1—9.

Schäfer, W., Weising, K., Kahl, G., 1984: T-DNA of a crown gall tumor is organized in nucleosomes. EMBO J. **3,** 373—376.

Schell, J., Van Montagu, M., De Beuckeleer, M., De Block, M., Depicker, A., De Wilde, M., Engler, G., Genetello, C., Hernalsteens, J. P., Holsters, M., Seurinck, J., Silva, B., Van Vliet, F., Villarroel, R., 1979: Interactions and DNA transfer between *Agrobacterium tumefaciens,* the Ti-plasmid and the plant host. Proc. R. Soc. Lond. B **204,** 251—266.

Schneider, E. A., Wightam, F., 1978: Auxines. In: Letham, D. S., Goodwin, P. B., Higgens, T. J. V. (eds.), Phytohormones and related compounds (A comprehensive treatise, Vol. I.), pp. 29—92. Amsterdam, Elsevier/North-Holland.

Schröder, G., Schröder, J., 1982: Hybridization selection and translation of T-DNA encoded mRNAs from octopine tumors. Mol. Gen. Genet. **185,** 51—55.

Schröder, J., Schröder, G., Huisman, H., Schilperoort, R. A., Schell, J., 1981: The mRNA for lysopine dehydrogenase in plant tumor cells is complementary to a Ti-plasmid fragment. FEBS Lett. **129,** 166—168.

Schröder, G., Klipp, W., Hillebrand, A., Ehring, R., Koncz, C., Schröder, J., 1983: The conserved part of the T-region expresses four proteins in bacteria. EMBO J. **2,** 403—409.

Schröder, G., Waffenschmidt, S., Weiler, E. W., Schröder, J., 1984: The T-region of Ti plasmids codes for an enzyme synthesizing indole-3-acetic acid. Eur. J. Biochem. **138,** 387—391.

Sembdner, G., Gross, D., Liebisch, H.-W., Schneider, G., 1980: Biosynthesis and metabolism of plant hormones. In: MacMillan, J. (ed.), Hormonal regulation of development I (Encyclopedia of Plant Physiology, New Series, Vol. 9), pp. 281—444. Berlin - Heidelberg - New York: Springer-Verlag.

Shapiro, J. A., 1979: Molecular model for the transposition and replication of bacteriophage Mu and other transposable elements. Proc. Natl. Acad. Sci., U.S.A. **76,** 1933—1937.

Simpson, R. B., O'Hara, P. J., Kwok, W., Montoya, A. L., Lichtenstein, C., Gordon, M. P., Nester, E. W., 1982: DNA from the A6S/2 crown gall tumors contains scrambled Ti-plasmid sequences near its junctions with the plant DNA. Cell **29,** 1005—1014.

Skoog, F., Miller, C. O., 1957: Chemical regulation of growth and organ formation in plant tissues cultured *in vitro.* Symp. Soc. Exp. Biol. **11,** 118—131.

Spanò, L., Pomponi, M., Costantino, P., Van Slogteren, G. M. S., Tempé, J., 1982: Identification of T-DNA in the root-inducing plasmid of the agropine type *Agrobacterium rhizogenes* 1855. Plant Mol. Biol. **1,** 291—300.

Stachel, S., An, G., Nester, E., 1984: Inducible expression of the virulence genes of the A6 *Agrobacterium tumefaciens* Ti-plasmid. J. Cell. Biochem. Suppl. **8 B,** 64.

Sun, S. M., Slightom, J. L., Hall, T. C., 1981: Intervening sequences in a plant gene — Comparison of the partial sequence of cDNA and genomic DNA of French bean phaseolin. Nature (London) **289,** 37—41.

Tempé, J., Petit, A., 1982: Opine utilization by *Agrobacterium.* In: Kahl, G., Schell, J. (eds.), Molecular Biology of Plant Tumors, pp. 451—459. New York: Academic Press.

Tempé, J., Schell, J., 1977: Is crown gall a natural instance of gene transfer? In: Legocki, A. B. (ed.), Translation of natural and synthetic polynucleotides, pp. 416—420. Poznan: University of Agriculture.

Tempé, J., Petit, A., Holsters, M., Van Montagu, M., Schell, J., 1977: Thermosensitive step associated with transfer of the Ti-plasmid during conjugation: possible relation to transformation in crown gall. Proc. Natl. Acad. Sci., U.S.A. **74,** 2848—2849.

Tempé, J., Estrade, C., Petit, A., 1978: The biological significance of opines. II. The conjugative activity of the Ti-plasmids of *Agrobacterium tumefaciens.* In: Ridé, M. (ed.), Proceedings IVth International Conference on Plant Pathogenic Bacteria, pp. 153—160. Angers: I.N.R.A.

Tempé, J., Petit, A., Farrand, S. K., 1984: Induction of cell proliferation by *Agrobacterium tumefaciens* and *Agrobacterium rhizogenes:* a parasite's point of view. In: Verma, D. P. S., Hohn, Th. (eds.), Genes Involved in Microbe-Plant Interactions (Plant Gene Research, Vol. 1), pp. 271—286. Wien - New York: Springer.

Tepfer, D. A., Tempé, J., 1981: Production d'agropine par des racines formées sous l'action *d'Agrobacterium rhizogenes,* souche A4. C. R. Acad. Sc. Paris **292,** Série III, 153—156.

Thomashow, M. F., Nutter, R., Montoya, A. L., Gordon, M. P., Nester, E. W., 1980a: Integration and organisation of Ti-plasmid sequences in crown gall tumors. Cell **19,** 729—739.

Thomashow, M. F., Nutter, R., Postle, K., Chilton, M.-D., Blattner, F. R., Powell, A., Gordon, M. P., Nester, E. W., 1980b: Recombination between higher plant

DNA and the Ti plasmid of *Agrobacterium tumefaciens*. Proc. Natl. Acad. Sci., U.S.A. **77,** 6448—6452.

Turgeon, R., 1981: Structure of grafted crown gall teratoma shoots of tobacco: regulation of transformed cells. Planta **153,** 42—48.

Turgeon, R., Wood, M. N., Braun A. C., 1976: Studies on the recovery of crown gall tumor cells. Proc. Natl. Acad. Sci., U.S.A. **73,** 3562—3564.

Van Haute, E., 1984: Genetic study of the functions for nopaline catabolism, replication and tumor induction of the *Agrobacterium tumefaciens* plasmid pTiC 58. Ph. D. thesis, State University Ghent.

Van Larebeke, N., Engler, G., Holsters, M., Van den Elsacker, S., Zaenen, I., Schilperoort, R. A., Schell, J., 1974: Large plasmid in *Agrobacterium tumefaciens* essential for crown gall-inducing ability. Nature (London) **252,** 169—170.

Van Larebeke, N., Genetello, C., Schell, J., Schilperoort, R. A., Hermans, A. K., Hernalsteens, J. P., Van Montagu, M., 1975: Acquisition of tumour-inducing ability by non-oncogenic agrobacteria as a result of plasmid transfer. Nature (London) **255,** 742—743.

Van Slogteren, G. M. S., 1983: Characterization of octopine crown gall tumorcells. — A study on expression of *Agrobacterium tumefaciens* derived T-DNA genes. Ph. D. thesis, University Leiden.

Van Slogteren, G. M. S., Hoge, J. H. C., Hooykaas, P. J. J., Schilperoort, R. A., 1983: Clonal analysis of heterogeneous crown gall tumor tissue induced by wild-type and shooter mutant strains of *A. tumefaciens*. Plant Mol. Biol. **2,** 321—333.

Velten, J., Willmitzer, L., Leemans, J., Ellis, J., Deblaere, R., Van Montagu, M., Schell, J., 1983: TR genes involved in agropine production. In: Pühler, A. (ed.), Molecular genetics of the bacteria plant interaction, pp. 303—312. Berlin - Heidelberg - New York: Springer-Verlag.

Wang, K., Herrera-Estrella, L., Van Montagu, M., Zambryski, P., 1984: Right 25-bp terminus sequences of the nopaline T-DNA is essential for and determines direction of DNA transfer from *Agrobacterium* to the plant genome. Cell **38,** 455—462.

Watson, B., Currier, T. C., Gordon, M. P., Chilton, M.-D., Nester, E. W., 1975: Plasmid required for virulence of *Agrobacterium tumefaciens*. J. Bacteriol. **123,** 255—264.

White, P. R., Braun, A. C., 1942: A cancerous neoplasm of plants. Autonomous, bacteria-free crown-gall tissue. Cancer Res. **2,** 597—617.

White, F. F., Nester, E. W., 1980a: Hairy root: plasmid encodes virulence traits in *Agrobacterium rhizogenes*. J. Bacteriol. **141,** 1134—1141.

White, F. F., Nester, E. W., 1980b: Relationship of plasmids responsible for hairy root and crown gall tumorigenicity. J. Bacteriol. **144,** 710—720.

White, F. F., Ghidossi, G., Gordon, M. P., Nester, E. W., 1982: Tumor induction by *Agrobacterium rhizogenes* involves the transfer of plasmid DNA to the plant genome. Proc. Natl. Acad. Sci., U.S.A. **79,** 3193—3197.

White, F. F., Garfinkel, D. J., Huffman, G. A., Gordon, M. P., Nester, E. W., 1983: Sequences homologous to *Agrobacterium rhizogenes* T-DNA in the genomes of uninfected plants. Nature (London) **301,** 348—350.

Wiborg, O., Hyldig-Nielsen, J., Jensen, E., Paludan, K., Marcker, K., 1982: The nucleotide sequences of two leghemoglobin genes from soybean. Nucl. Acids Res. **10,** 3487—3494.

Willmitzer, L., De Beuckeleer, M., Lemmers, M., Van Montagu, M., Schell, J., 1980: DNA from Ti-plasmid is present in the nucleus and absent from plastids of plant crown-gall cells. Nature (London) **287,** 359—361.

Willmitzer, L., Schmalenbach, W., Schell, J., 1981: Transcription of T-DNA in octopine and nopaline crown gall tumours is inhibited by low concentrations of α-amanitin. Nucl. Acids Res. **9,** 4801—4812.

Willmitzer, L., Simons, G., Schell, J., 1982a: The TL-DNA in octopine crown gall tumours codes for seven well-defined polyadenylated transcripts. EMBO J. **1,** 139—146.

Willmitzer, L., Sanchez-Serrano, J., Buschfeld, E., Schell, J., 1982b: DNA from *Agrobacterium rhizogenes* is transferred and expressed in axenic hairy root plant tissues. Mol. Gen. Genet. **186,** 16—22.

Willmitzer, L., Dhaese, P., Schreier, P. H., Schmalenbach, W., Van Montagu, M., Schell, J., 1983: Size, location, and polarity of T-DNA-encoded transcripts in nopaline crown gall tumors; evidence for common transcripts present in both octopine and nopaline tumors. Cell **32,** 1045—1056.

Winter, J. A., Wright, R. L., Gurley, W. B., 1984: Map locations of five transcripts homologous to T_R-DNA in tobacco and sunflower crown gall tumors. Nucl. Acids Res. **12,** 2391—2406.

Wood, H. N., Binns, A. N., Braun, A. C., 1978: Differential expression of oncogenicity and nopaline synthesis in intact leaves derived from crown gall teratomas of tobacco. Differentiation **11,** 175—180.

Wöstemeyer, A., Otten, L., De Greve, H., Hernalsteens, J. P., Leemans, J., Van Montagu, M., Schell, J., 1983: Regeneration of plants from crown gall cells. In: Lurquin, P., Kleinhofs, A. (eds.), Genetic Engineering in Eukaryotes (NATO ASI Series A, Vol. 61), pp. 137—151. New York: Plenum Press.

Wöstemeyer, A., Otten, L. A. B. M., Schell, J., 1984: Sexual transmission of T-DNA in abnormal tobacco regenerants transformed by octopine and nopaline strains of *Agrobacterium tumefaciens.* Mol. Gen. Genet. **194,** 500—507.

Yadav, N. S., Vanderleyden, J., Bennet, D. R., Barnes, W. M., Chilton, M.-D., 1982: Short direct repeats flank the T-DNA on a nopaline Ti plasmid. Proc. Natl. Acad. Sci., U.S.A. **79,** 6322—6326.

Yang, F., Simpson, R. B., 1981: Revertant seedlings from crown gall tumors retain a portion of the bacterial Ti plasmid DNA sequences. Proc. Natl. Acad. Sci., U.S.A. **78,** 4151—4155.

Yang, F., Montoya, A. L., Merlo, D. J., Drummond, M. H., Chilton, M.-D., Nester, E. W., Gordon, M. P., 1980: Foreign DNA sequences in crown gall teratomas and their fate during the loss of the tumorous traits. Mol. Gen. Genet. **177,** 707—714.

Zaenen, I., Van Larebeke, N., Teuchy, H., Van Montagu, M., Schell, J., 1974: Supercoiled circular DNA in crown gall inducing *Agrobacterium* strains. J. Mol. Biol. **86,** 109—127.

Zambryski, P., Holsters, M., Kruger, K., Depicker, A., Schell, J., Van Montagu, M., Goodman, H. M., 1980: Tumor DNA structure in plant cells transformed by *A. tumefaciens.* Science **209,** 1385—1391.

Zambryski, P., Depicker, A., Kruger, K., Goodman, H., 1982: Tumor induction by *Agrobacterium tumefaciens:* analysis of the boundaries of T-DNA. J. Mol. Appl. Genet. **1,** 361—370.

Zambryski, P., Goodman, H. M., Van Montagu, M., Schell, J., 1983a: *Agrobacterium* tumor induction. In: Shapiro, J. A. (ed.), Mobile Genetic Elements, pp. 505—535. New York: Academic Press.

Zambryski, P., Joos, H., Genetello, C., Leemans, J., Van Montagu, M., Schell, J., 1983b: Ti plasmid vector for the introduction of DNA into plant cells without alteration of their normal regeneration capacity. EMBO J. **2,** 2143—2150.

Section II.
Movement of Genetic Information Between the Plant Organelles

Chapter 3

Movement of Genetic Material Between the Chloroplast and Mitochondrion in Higher Plants

David M. Lonsdale

Cytogenetics Department, Plant Breeding Institute, Maris Lane, Trumpington, Cambridge CB2 2LQ, U.K.

With 2 Figures

Contents

The organelles within the eukaryotic cell have originated in one of two ways: I, the mitochondrion and chloroplast originally descended from free living bacteria-like organisms which entered into an endosymbiotic relationship within a host cell having a nuclear genome, or II, that the nuclear and organelle genomes became physically compartmentalized and functionally specialized within a single cell. The arguments relating to these two evolutionary hypotheses have been discussed in detail elsewhere (Gray and Doolittle, 1982) and will not be subject to discussion here, though it does appear that the endosymbiont hypothesis is more acceptable for the origin of both the mitochondrion and chloroplast.

I. Inter-Organelle DNA Transposition

The present day size of the organelle genomes, the appreciation of their genetic capacity and knowledge that the majority of the structural and functional polypeptides of the chloroplast and mitochondrion are now nuclear gene products, indicates that if the endosymbiont hypothesis is to be believed then a massive transfer of genetic material must have taken place from the chloroplasts and mitochondria into the nucleus since the endosymbiotic relationship was first established.

Evidence that this process of inter-organelle DNA transposition can occur has only recently been obtained. Mitochondrial and chloroplast DNA sequences from a whole variety of species have been detected in the nuclear genome (van den Boogaart, 1982; Farrelly and Butow, 1983; Wright and Cummings, 1983; Gellisen *et al.,* 1983; Kemble *et al.,* 1983; Timmis and Scott, 1983; Jacobs *et al.,* 1983; Hadler *et al.,* 1983; Tsuzuki *et al.,* 1983; Scott and Timmis, 1984). The acquisition of mitochondrial and chloroplast gene sequences by the nucleus allows the mitochondrion and chloroplast to dispense with these functions provided that the transposed nuclear sequence remains functional or can be made to function in its new environment. In the Mex-1 mutant of Podospora such a transfer has apparently taken place. The mitochondrial α-sen DNA sequence, which carries the cytochrome c oxidase subunit I gene, has been transposed to the nucleus without any apparent loss in mitochondrial function, and the mitochondrial DNA now lacks this sequence (Wright and Cummings, 1983).

As well as the mitochondrial and chloroplast DNA sequences which have been detected in the nuclear genome, chloroplast DNA sequences are commonly found in the mitochondrial genomes of higher plants (Stern and Lonsdale, 1982; Stern *et al.,* 1983; Stern and Palmer, 1984) and in at least one instance a chloroplast DNA sequence has been detected in both the mitochondrial and nuclear DNAs (Timmis and Scott, 1983). This proven promiscuity of DNA within the eukaryotic cell has not yet been established for the transfer of nuclear DNA sequences to the chloroplast or the mitochondrion nor the transfer of mitochondrial sequences to the chloroplast. However, it has recently been suggested that the plasmid DNA species of plant mitochondria, which have no sequence homology to the mitochondrial chromosome may be nuclear in origin but have been sequestered by the mitochondrion in forms which have the ability to replicate (Lonsdale, 1984).

The apparent lack of observed foreign DNA sequences in the chloroplast genomes of higher plants probably reflects the lack of available sites into which these sequences can insert without causing a lethal mutation. This is supported to some extent by the fact that chloroplast genomes of higher plants display a remarkable constancy in size, structure and organisation, in many ways analogous to mammalian mitochondrial genomes. This is in contrast to the nuclear and mitochondrial genomes which contain large and variable amounts of non-coding sequences into which sequences can readily insert without causing genetic disruption.

II. Sequences Homologous to Chloroplast DNA in Higher Plant Mitochondrial Genomes

A. The Genome of Zea mays

Probing of restriction digests of mitochondrial DNA with nick-translated chloroplast DNA, identified a discrete set of restriction fragments in the mitochondrial DNA restriction profile which hybridised strongly to the chloroplast DNA probe. Many of these fragments apparently co-migrate with chloroplast DNA restriction fragments (Stern and Londsdale, 1982). These restriction fragments were also identified in cloned sequences of mitochondrial DNA. Detailed restriction mapping of these clones revealed three major chloroplast DNA homologous sequences present in the mitochondrial genome:

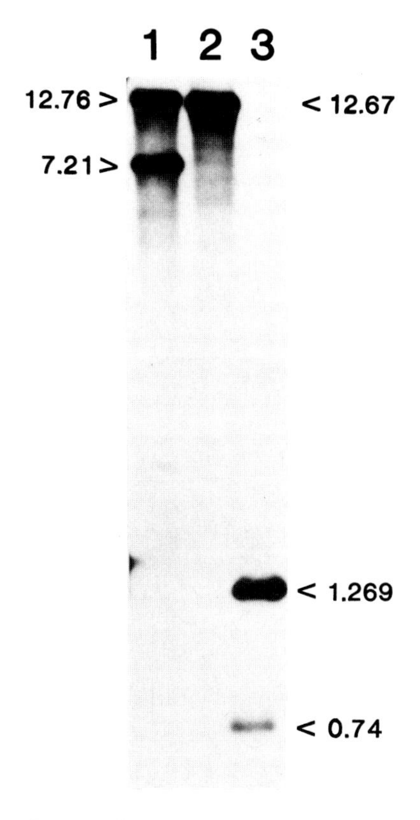

Fig. 1. Autoradiograph of restriction digests chloroplast DNA of *Zea mays* probed with the mitochondrial cosmid 2 cl. Chloroplast DNA was digested with the restriction endonucleases Bam H 1 (1), Eco RI (2) and Hind III (3). Restriction fragments were separated in a 1 % agarose gel and transferred to nitrocellulose. The nitrocellulose filter was subsequently probed using nick-translated 2 cl, a mitochondrial DNA sequence cloned in a cosmid vector (Lonsdale *et al.*, 1981). Sizes of fragments are given in kilobase pairs (kb)

1. A 12 kb sequence from the chloroplast DNA inverted repeat (Stern and Lonsdale, 1982). This sequence contains the chloroplast coding sequences for the 3′-exon of $tRNA^{ala}$, $tRNA^{ile}$, $tRNA^{val}$ and $tRNA^{leu}$ as well as the coding sequence for the 16 S ribosomal RNA (Larrinua *et al.*, 1983; Selden *et al.*, 1983).
2. A sequence between 1.9 kb and 2.7 kb in length having the entire coding sequence for the large subunit of ribulose-1,5-bisphosphate carboxylase (Lonsdale *et al.*, 1983).
3. A sequence of not more than 2 kb in length which in the chloroplast genome hybridises to Hind III fragments of 1269 bp and 740 bp, Bam H 1 fragments 1 and 4 and Eco RI-A (Figure 1; Edwards and Kossel, 1981; Larrinua *et al.*, 1983). In the chloroplast genome, this region codes for the 3′-end of the 23 S ribosomal RNA, the 4.5 S and 5 S RNAs and the tRNAs for arginine, asparagine-2 and Glycine-1 (Larrinua *et al.*, 1983; Selden *et al.*, 1983).

These three major sequences homologous to chloroplast DNA have been mapped onto the 570 kb mitochondrial master chromosome (Figure 2). There is approximately 17 kb between the 12 kb chloroplast inverted repeat homology and the large subunit pseudogene sequence, which is itself approximately 117 kb from the sequence homologous to the 3′-end of the 23 S ribosomal RNA.

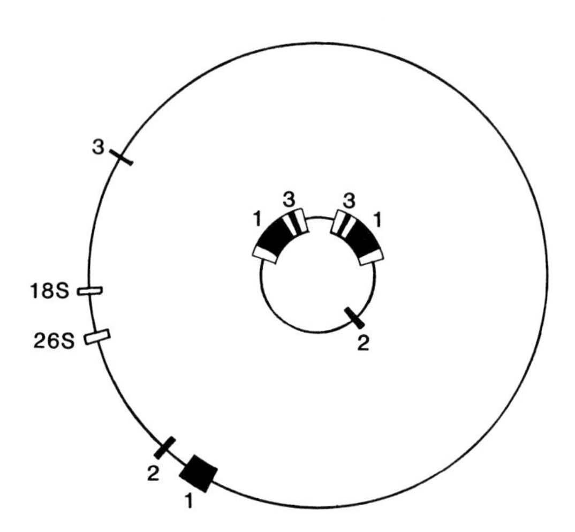

Fig. 2. The circular maps of the mitochondrial 'master chromosome' (outer circle, 570 kb) and the chloroplast genome (inner circle, 139 kb) of *Zea mays*. The positions of the mitochondrial sequences homologous to 12 kb of the chloroplast inverted repeat (1), the large subunit gene of ribulose-1,5-bisphosphate carboxylase (2) and 1269 bp and 740 bp Hind III fragments, see Fig. 1 and text (3) are shown relative to the position of the 26 S and 18 S mitochondrial ribosomal RNA genes

B. Other Higher Plant Species

The transfer of chloroplast DNA sequences to the mitochondrion seems to be a general phenomenon in higher plants. Analysis of mungbean, spinach and pea mitochondrial DNAs (Stern *et al.*, 1983; Stern and Palmer, 1984) as well as the mitochondrial DNAs of watermelon, zucchini squash, cucumber and muskmelon (Stern *et al.*, 1983) revealed a more or less random distribution of chloroplast DNA sequences in these mitochondrial genomes.

Mitochondrial genome size in higher plants is extremely variable ranging from 200 kb to 2400 kb (Palmer and Shields, 1984; Ward *et al.*, 1981; Bendich, 1982). In a single family, for example the cucurbits, the genome size can range from 300 kb (watermelon) to 2400 kb (muskmelon). That chloroplast DNA sequence are found in mitochondrial DNA led to the suggestion that the ability of the mitochondrion to take up foreign DNA could in part account for the large and variable sizes of plant mitochondrial genomes (Lonsdale *et al.*, 1983).

However, there would appear to be no correlation between mitochondrial genome size and the amount of chloroplast DNA homologous sequences within the cucurbits (Stern *et al.*, 1983).

III. Functionality of the Chloroplast Pseudogene Sequences in the Mitochondrial Genome

It is unclear what biological function, if any, these chloroplast sequences play in the mitochondrion. Alterations in the genetic code between the two organelles (Fox and Leaver, 1981), differences in transcriptional initiation and termination and translational (Chao *et al.*, 1983) signals would most likely prevent them from functioning as chloroplast genes would in a chloroplast environment. In the *Zea mays* mitochondrial genome, extensive alterations including deletions in the 12 kb chloroplast inverted repeat sequence which characterises the male sterile lines led to the speculation that cytoplasmic male sterility may be correlated with the partial loss of these specific chloroplast DNA sequences (Stern and Lonsdale, 1982). If one makes the assumption that all plant mitochondrial genomes encode the same mitochondrial polypeptides then only those sequences common to all mitochondrial genomes are likely to be a functional necessity. Therefore as the chloroplast sequence insertions are apparently random (Stern *et al.*, 1983; Stern and Palmer, 1984) they can at best be considered to be superfluous to mitochondrial function. However, there may be some local effect on DNA stability, replication or transcription because of the local increase in the average AT content of these sequences (chloroplast DNA, 38 % G+C; mitochondrial DNA, 47 % G+C).

IV. Mechanism of Sequence Transfer

There are several possible mechanisms which would enable chloroplast DNA to be transferred and incorporated into the mitochondrial genome. The first mechanism might be the direct interaction of the two organelles, leading to fusion and intramolecular recombination between the chloroplast and mitochondrial genomes. Phasecontrast cinematography and microscopy studies provide evidence that the chloroplast and mitochondrion could fuse (Wildman *et al.,* 1962; Wildman *et al.,* 1974). However, more detailed studies using transmission electron microscopy have generally shown that mitochondria can form very close associations with chloroplasts, lying in deep invaginations of the chloroplast membrane under low light intensities (Montes and Bradbeer, 1976; Ballantine and Ford, 1970): It is only in the meristematic tissue of the Albostrians barley mutant that transmission electron microscopy has provided direct evidence for fused organelles (Wellburn and Wellburn, 1979) and some support for Wildman's (1962) original claims.

Outer membrane continuities between chloroplasts and mitochondria have been occasionally observed in several species (Crotty and Ledbetter, 1973; Dalvayrac *et al.,* 1981). If the inner membranes are permeable to large molecules, such as DNA, then a passive transfer of nucleic acids can be envisaged. Alternatively the presence of DNA in the cytoplasm from lysed, broken or degenerating chloroplasts may be taken up by the mitochondrion in a process analogous to bacterial transformation. This is probably the simplest and most plausible explanation and may be facilitated if the mitochondria are within the chloroplast lumen (Brown *et al.,* 1983).

Another method of transfer might be through specific vector mediated systems. In certain strains of yeast the var 1 – cob/box petite sequence present in nuclear DNA is flanked by a tandem pair of Ty elements (Farrelly and Butow, 1983). Whether these Ty elements have participated in the conveyance of this petite sequence into the nuclear genome or whether their proximity is entirely co-incidental cannot be determined. Little evidence is currently available as to whether transposable elements in general have sequence homology to, or can transpose to organelle genomes other than the nucleus, although the Ac element of maize has no sequence homology to either the chloroplast or mitochondrial genomes (Harris, Lonsdale and Flavell, unpublished data).

V. Rate of Sequence Transposition and Selection of Novel Genotypes

The large number of inter-organelle DNA transpositions detected to date (van den Boogaart, 1982; Stern and Lonsdale, 1982; Farrelly and Butow, 1983; Wright and Cummings, 1983; Gellisen *et al.,* 1983; Timmis and Scott, 1983; Kemble *et al.,* 1983; Stern *et al.,* 1983; Jacobs *et al.,* 1983; Hadler *et al.,* 1983; Tsuzuki *et al.,* 1983; Stern and Palmer, 1984; Scott and Timmis,

1984), and the high degree of sequence conservation suggest that inter-organelle DNA transposition are an ongoing evolutionary process. Inter-organelle DNA transposition at the level of a single cell cannot be detected so that it is not possible to estimate its frequency. The chance of any transposition event becoming fixed is significantly greater for transposition into the nuclear genome than it is for the mitochondrial genome, where the ploidy level is some hundred or perhaps a thousand times greater than the nucleus. In the cytoplasm, a new mitochondrial genotype which initially is represented by a single molecule in one mitochondrion of the cell, must be differentially amplified in some way so that it eventually constitutes the majority of the mitochondrial DNA population in the next maternal germplasm. Mechanisms, accounting for the apparently rapid shifts in mitochondrial genotypes based on stochastic segregation have been proposed (Birky, 1978; Birky *et al.*, 1982; Birky, 1983; Olivo *et al.*, 1983). Other mechanisms which may complement and enhance stochastic segregation can be envisaged. One of these is the observation that mitochondria and mtDNA are amplified during mammalian oocyte development (Piko and Matsumoto, 1976; Michaels *et al.*, 1982). Assuming cytoplasmic sectoring and mitochondrial amplification a genetic 'founder effect' may be operative. This differential amplification would lead to a rapid shift in the mitochondrial genotype. Such shifts have been described in Holstein cow maternal lineages (Hauswirth and Laipis, 1982a, 1982b; Olivo *et al.*, 1983) and have been observed in fertile revertants of male sterile plants (Levings *et al.*, 1980; Gengenbach *et al.*, 1981; Kemble *et al.*, 1982). Yet another type of mtDNA segregation has been described in mouse-rat hybrid cell line (Hayashi *et al.*, 1983). In these lines the pattern of mtDNA segregation was determined by the segregation of the chromosomes within the chimeric nucleus. Once chromosomal segregation had occurred, the segregated chromosomes and mtDNA were of the same species. It is therefore possible that in a normal eukaryotic cell a chromosomal mutation could well dictate the segregation and selective amplification of a novel mitochondrial genotype.

In the fertility reversion phenomena of the S and T male sterile cytoplasms of maize such a mechanism may be operating. Cytoplasmic male sterility, an inability to produce fertile pollen, can be regarded as the interaction of a foreign cytoplasm with a normal nucleus or a normal cytoplasm with a foreign nucleus (Hanson and Conde, 1984). This incompatability can therefore be due to either the cytoplasm or to the nucleus. In maize, the S and T cytoplasms contain mitochondrial alterations which are responsible for the incompatibility. In fertility reversion, a change in the nuclear genotype could well select compatible forms of the mitochondrial genome from amongst the incompatible genomes; the new nuclear/cytoplasmic combination results, in this instance, in normal pollen production. Such altered mitochondrial genotypes may well pre-exist albeit at low levels within the mitochondrial genome pool. In the event of such a nuclear mutation, selection would be rapid and many of the selected genotypes would be similar if not identical. Analysis of the mtDNAs of the S (Schardl *et al.*, 1984) and T (Gengenbach *et al.*, 1981; Kemble *et al.*, 1982) fertile revertants reveals that many display similar if not identical alterations in their mitochondrial genomes, suggesting that such a mechanism may in fact be operating.

If in fact such a powerful selection mechanism can operate in plant cytoplasm systems in selecting novel mitochondrial genotypes, then the ability to control and manipulate this mechanism is imperative if effective control of genetic engineering of the cytoplasm (i. e. the mitochondrion) is ever to be achieved.

VI. References

Ballantine, J. E. M., Forde, B. J., 1970: The effect of light intensity and temperature on plant growth and chloroplast ultrastructure in soyabean. Amer. J. Bot. **57,** 1150—1159.

Bendich, A. J., 1982: Plant mitochondrial DNA: The last frontier. In: Slonimski, P., Borst, P., Attardi, G. (eds.), Mitochondrial Genes, pp. 477—481. Cold Spring Harbor, N. Y.

Birky Jr., C. W., 1978: Transmission genetics of mitochondria and chloroplasts. Ann. Rev. Genet. **12,** 471—512.

Birky Jr., C. W., 1983: Relaxed cellular controls and organelle heredity. Science **222,** 468—475.

Birky Jr., C. W., Acton, A. R., Dietrich, R., Carver, M., 1982: Mitochondrial transmission genetics: replication, recombination, and segregation of mitochondrial DNA and its inheritance in crosses. In: Slonimski, P., Borst, P., Attardi, G. (eds.), Mitochondrial Genes, pp. 333—348. Cold Spring Harbor, N. Y.

Brown, R. H., Rigsby, L. L., Akin, D. E., 1983: Enclosure of mitochondria by chloroplasts. Plant Physiol. **71,** 437—439.

Calvayrac, R., Laval-Martin, D., Briand, J., Farineau, J., 1981: Paramylon synthesis by *Euglene gracilis* photoheterotrophically grown under low O_2 pressure. Planta **153,** 6—13.

Chao, S., Sederoff, R. R., Levings III, C. S., 1983: Partial sequence analysis of the 5S to 18S rRNA gene region of the maize mitochondrial genome. Plant Phys. **71,** 190—193.

Crotty, W. J., Ledbetter, M. C., 1973: Membrane continuities involving chloroplasts and other organelles in plant cells. Science **182,** 839—841.

Edwards, K., Kossel, H., 1981: The rRNA operon from *Zea mays* chloroplasts: nucleotide sequence of 23S rDNA and its homology with *E. coli* 23S rRNA. Nucleic Acids Res. **9,** 2853—2869.

Farrelly, F., Butow, R. A., 1983: Rearranged mitochondrial genes in the yeast nuclear genome. Nature **301,** 296—301.

Fox, T. D., Leaver, C. J., 1981: The *Zea mays* mitochondrial gene coding cytochrome oxidase subunit II has an intervening sequence and does not contain TGA codons. Cell **26,** 315—323.

Gellissen, G., Bradfield, J. Y., White, B. N., Wyatt, G. R., 1983: Mitochondrial DNA sequences in the nuclear genome of a locust. Nature **301,** 631—634.

Gengenbach, B. G., Connelly, J. A., Pring, D. R., Conde, M. F., 1981: Mitochondrial DNA variation in maize plants regenerated during tissue culture selection. Theor. Appl. Genet. **59,** 161—167.

Gray, M. W., Doolittle, W. F., 1982: Has the endosymbiont hypothesis been proven? Microbiol. Rev. **46,** 1—42.

Hadler, H. I., Dimitrijevic, B., Mahalingam, R., 1983: Mitochondrial DNA and nuclear DNA from normal rat liver have a common sequence. Proc. Natl. Acad. Sci., U.S.A. **80,** 6495—6499.

Hanson, M. R., Conde, M. F., 1985: Functioning and variation of cytoplasmic genomes: lessons from cytoplasmic-nuclear interactions affecting male fertility in plants. Intern. Rev. Cytology (in press).

Hauswirth, W. W., Laipis, P. J., 1982a: Mitochondrial DNA polymorphism in a maternal lineage of Holstein cows. Proc. Natl. Acad. Sci., U.S.A. **79,** 4686—4690.

Hauswirth, W. W., Laipis, P. J., 1982b: Rapid variation in mammalian mitochondrial genotypes: implications for the mechanism of maternal inheritance. In: Slonimski, P., Borst, P., Attardi, G. (eds.), Mitochondrial Genes, pp. 137—141. Cold Spring Harbor, N. Y.

Hayashi, J.-I., Tagashira, Y., Yoshida, M. C., Ajiro, K., Sekiguchi, T., 1983: Two distinct types of mitochondrial DNA segregation in mouse-rat hybrid cells. Stochastic segregation and chromosome-dependent segregation. Exp. Cell Res. **147,** 51—61.

Jacobs, H. T., Posakony, J. W., Grula, J. W., Roberts, J. W., Xin, J.-H., Britten, R. J., Davidson, E. H., 1983: Mitochondrial DNA sequences in the nuclear genome of *Strongylocentrotus purpuratus.* J. Mol. Biol. **165,** 609—632.

Kemble, R. J., Flavell, R. B., Brettell, R. I. S., 1982: Mitochondrial DNA analyses of fertile and sterile maize plants derived from tissue culture with the Texas male sterile cytoplasm. Theor. Appl. Genet. **62,** 213—217.

Kemble, R. J., Mans, R. J., Gabay-Laughnan, S., Laughnan, J. R., 1983: Sequences homologous to episomal mitochondrial DNAs in the maize nuclear genome. Nature **304,** 744—747.

Larrinua, I. M., Muskavitch, K. M. T., Gubbins, E. J., Bogorad, L., 1983: A detailed restriction endonuclease site map of the *Zea mays* plastid genome. Plant Mol. Biol. **2,** 129—140.

Levings III, C. S., Kim, B. D., Pring, D. R., Conde, M. F., Mans, R. J., Laughnan, J. R., Gabay-Laughnan, S. J., 1980: Cytoplasmic reversion of cms-S in maize: association with a transpositional event. Science **209,** 1021—1023.

Lonsdale, D. M., 1984: A review of the structure and organisation of the mitochondrial genome of higher plants. Plant Mol. Biol. **3,** 201—206.

Lonsdale, D. M., Thompson, R. D., Hodge, T. P., 1981: The integrated forms of the S1 and S2 DNA elements of maize male-sterile mitochondrial DNA are flanked by a large repeated sequence. Nucl. Acid Res. **9,** 3657—3669.

Lonsdale, D. M., Hodge, T. P., Howe, C. J., Stern, D. B., 1983: Maize mitochondrial DNA contains a sequence homologous to the ribulose-1,5-bisphosphate carboxylase large subunit gene of chloroplast DNA. Cell **34,** 1007—1014.

Michaels, G. S., Hauswirth, W. W., Laipis, P. J., 1982: Mitochondrial DNA copy number in bovine oocytes and somatic cells. Dev. Biol. **94,** 246—251.

Montes, G., Bradbeer, J. W., 1976: An association of chloroplasts and mitochondria in *Zea mays* and *Hyptis suaveolens.* Plant Sci. Lett. **6,** 35—41.

Olivo, P. D., van de Walle, M. J., Laipis, P. J., Hauswirth, W. W., 1983: Nucleotide sequence evidence for rapid genotypic shifts in the bovine mitochondrial DNA D-loop. Nature **306,** 400—402.

Palmer, J. D., Shields, C. R. 1984: Tripartite structure of the *Brassica campestris* mitochondrial genome. Nature **307,** 437—440.

Piko, L., Matsumoto, L., 1976: Number of mitochondria and some properties of mitochondrial DNA in the mouse egg. Dev. Biol. **49,** 1—10.

Schardl, C. L., Pring, D. R., Fauron, C. M.-R., Lonsdale, D. M., 1984: Mitochondrial DNA rearrangements resulting in fertile revertants of S-type male sterile maize submitted.

Scott, N. S., Timmis, J. N., 1984: Homologies between nuclear and plastid DNA in spinach. Theor. Appl. Genet. **67,** 279—288.

Selden, R. F., Steinmetz, A., McIntosh, L., Bogorad, L., Burkard, G., Mubumbila, M., Kuntz, M., Crouse, E. J., Weil, J. H., 1983: Transfer RNA genes of *Zea mays* chloroplast DNA. Plant Mol. Biol. **2,** 141—153.

Stern, D. B., Lonsdale, D. M., 1982: Mitochondrial and chloroplast genomes of maize have a 12-kilobase DNA sequence in common. Nature **229,** 698—702.

Stern, D. B., Palmer, J. D., 1983: Extensive and widespread homologies between mitochondrial DNA and chloroplast DNA in plants. Proc. Natl. Acad. Sci., U.S.A. **81,** 1946—1950.

Stern, D. B., Palmer, J. D., Thompson, W. F., Lonsdale, D. M., 1983: Mitochondrial DNA sequence evolution and homology to chloroplast DNA in angiosperms. In: Goldberg, R. B. (ed.), UCLA Symposia on Molecular and Cellular Biology (New Series, Vol. 12), pp. 467—477. New York: Alan R. Liss Inc.

Timmis, J. N., Scott, N. S., 1983: Sequence homology between spinach nuclear and chloroplast genomes. Nature **305,** 65—67.

Tsuzuki, T., Nomiyama, H., Setoyama, C., Maeda, S., Shimada, K., 1983: Presence of mitochondrial-DNA-like sequences in the human nuclear DNA. Gene **25,** 223—229.

van den Boogaart, P., Samallo, J., Asgsteribb, E., 1982: *Neurospora crassa* possesses in addition to the nuclear gene for a dicyclohexyl-carbodiimide-binding protein of the mitochondrial ATPase a similar gene on mitochondrial DNA. Nature **298,** 187—189.

Ward, B. L., Anderson, R. S., Bendich, A. J., 1981: The mitochondrial genome is large and variable in a family of plants (Cucurbitaceae). Cell **25,** 793—803.

Wellburn, F. A. M., Wellburn, A. R., 1979: Conjoined mitochondria and plastids in the barley mutant 'Albostrians'. Planta **147,** 178—179.

Wildman, S. G., Hongladarom, T., Honda, S. I., 1962: Chloroplasts and mitochondria in living plant cells: cinephotomicrographic studies. Science **138,** 434—436.

Wildman, S. G., Jope, C., Atchison, B. A., 1974: Role of mitochondria in the origin of chloroplast starch grains. Description of the phenomenon. Plant Physiol. **54,** 231—237.

Wright, R. M., Cummings, D. J., 1983: Integration of mitochondrial gene sequences within the nuclear genome during senescence in a fungus. Nature **302,** 86—88.

Chapter 4

Movement of Genetic Information Between the Chloroplast and Nucleus

J. N. Timmis[1] and N. Steele Scott[2]

[1] Department of Genetics, University of Adelaide, GPO Box 498, Adelaide, South Australia

[2] CSIRO Division of Horticultural Research, GPO Box 350, Adelaide, South Australia

With 6 Figures

The discovery of cytoplasmic inheritance in plants at the turn of the century (Correns and Baur as described in Kirk and Tilney-Basset, 1978), culminated in the demonstration of plastid (pt) DNA in the late 1960's. At the same time there was mounting biochemical and genetic evidence which showed that most of plastid biogenesis and function was controlled by nuclear genes and involved proteins synthesized on cytoplasmic ribosomes (see reviews by Kirck and Tilney-Bassett, 1978; Ellis, 1983). The expression of plastid DNA and the use of plastid ribosomes to synthesize large amounts of particular plastid proteins has only been described in the special case of the photosynthetically competent plastid, the chloroplast (Scott and Possingham, 1980). We wish to distinguish between the general term *plastid* which describes a family of related plant cell organelles of which the most commonly studied and perhaps the most numerous and important are *chloroplasts*. In general we will use the name plastid and only use the term chloroplast in specific instances. It appears however that chloroplasts and all other plastids carry an identical subgenome which has been called the plastome. In this article we will briefly describe the interaction of the genetic information from nucleus and plastid that is involved in the formation of chloroplasts and other plastid forms and go on to discuss in more detail the recent observations which indicate that plastids and nuclei share extensive DNA sequence homology (Timmis and Scott, 1983).

The properties of the plastid subgenome have been extensively reviewed (e. g. Bohnert *et al.*, 1982) and will only be summarized here. In higher plants the plastome is a circular DNA molecule of about 150 kbp, containing in most cases an inverted repeat of 20—25 kbp which includes the plastid rRNA genes. The plastome contains enough DNA to code for only about 100 proteins, of the many hundreds required for plastid synthesis. Several of these plastid located genes have been identified and sequenced. They include the enzymatically active large subunit of the CO_2 fixing enzyme Ribulose-1,5-bis phosphate carboxylase/oxygenase (Rubisco), proteins of the ptATPase complex and a complete set of genes for tRNA similar to those found in prokaryotic cells (Bohnert *et al.*, 1982; Bottomley and Bohnert, 1982). Different plastid types contain various numbers of identical copies of the plastome, from as few as 10—20 copies in root plastids and mature chloroplasts, up to hundreds of copies in young chloroplasts (Scott and Possingham, 1980) and potato amyloplasts (Scott *et al.*, 1984). In the higher plants examined, most of which are cultivated species, a striking feature of ptDNA is strong intraspecific homogeneity though there is minor considerable sequence variation between species and families and also interspecies variation in plastome size and the arrangement of the DNA repeats, particularly in lower plants (Bohnert *et al.*, 1982).

The biosynthesis of a functional chloroplast requires the cooperative expression of both genome and plastome (Scott and Possingham, 1980; Ellis, 1982). Most chloroplast proteins are nuclear encoded, synthesized in the cytoplasm and imported into chloroplasts. Certain key chloroplast proteins are complexes of more than one subunit, with some constituent polypeptides being synthesized on cytoplasmic ribosomes and transported into the chloroplast to combine with other polypeptides encoded and synthesized inside the chloroplast. The best known example of such collaboration is the CO_2 fixing enzyme Rubisco. A precursor to the small subunit of this protein, larger than the mature form, is encoded by nuclear genes and synthesized on cytoplasmic ribosomes. The precursor is directed to the chloroplast by an additional 55 amino acid transit sequence which is precisely removed during passage into the organelle. Within the chloroplast the small subunit associates with equal numbers of the chloroplast synthesized large subunit polypeptides to form the active hexadecameric protein.

There are several genomic copies of the gene for the small subunit of Rubisco in petunia and although DNA sequencing reveals 8—9% divergence between four separate genes, the amino acid sequences directed by each gene are probably identical (Dunsmuir *et al.*, 1983). There may be as many as 12 genes in all, of which the four referred to above are known to be transcribed. Photosynthetic leaf cells on average contain 3C nuclei (Lamppa *et al.*, 1980; Scott and Possingham, 1980) which therefore contain 30—40 genes for the small subunit in each nucleus. There is only one gene for the large subunit of Rubisco per plastome, but in leaf cells there are between 5,000 and 10,000 copies of the plastome per cell and the large subunit polypeptide is the major product of translation on chloroplast ribosomes (Scott and Possingham, 1980; Bottomley and Bohnert, 1982) and is the most abundant protein of green plants. This complex cooperative pattern of protein synthesis (summarized in Figure 1a), where the products of 30—40 nuclear copies of the small subunit gene and at least 5,000 plastid copies of the large subunit gene are combined in equimolar amounts to produce a single mature protein, is a remarkable feat of gene regulation. The utilization of the nuclear gene would appear therefore to be notably more efficient than that of the plastid gene. It seems likely that the synthesis and turnover of most other proteins and protein complexes, such as the ptATPase complex and plastid ribosomes, which involve expression of both plastome and genome will be tightly regulated in the same manner.

The majority of proteins concerned with the formation of plastids are totally directed by the nuclear/cytoplasmic system (illustrated in Fig. 1 b) with the suggestion that nuclear DNA directs ptDNA synthesis and con-

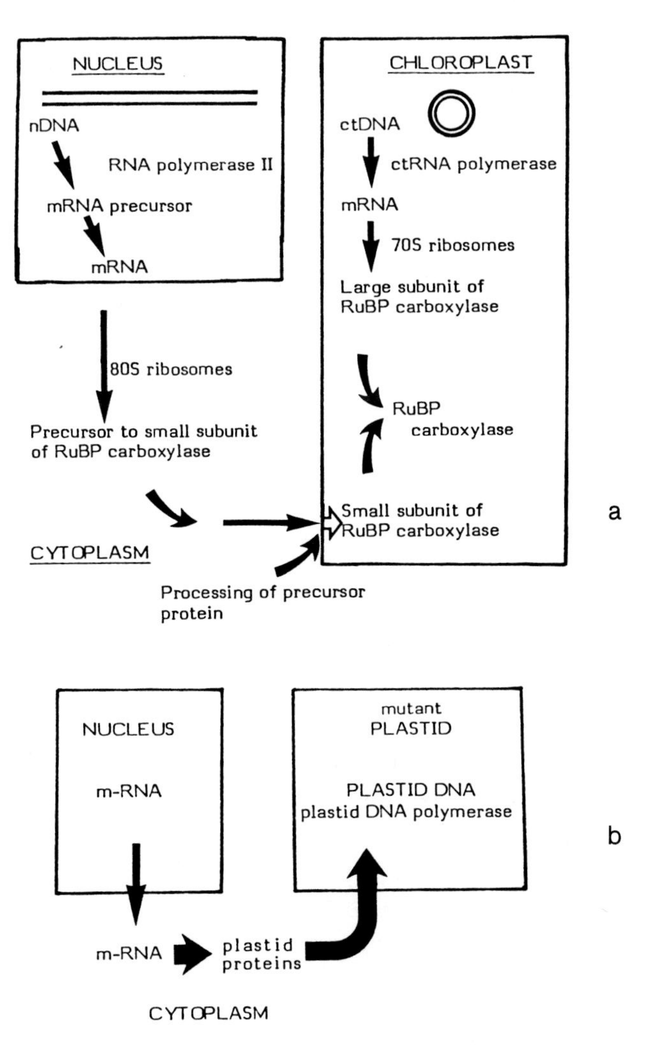

Fig. 1. Schematic representation of the interaction between nuclear/cytoplasmic and plastid genetic systems

trols the production of a basic plastid skeleton. The suggestion is based on the observation that no ribosomes could be detected in plastids found in albino tissue of mutant *Pelargonium* (Knoth *et al.*, 1974), barley (Borner *et al.*, 1976) and maize (Walbot and Coe, 1979) or in heat bleached leaves of rye (Feirabend and Schrader-Reichardt, 1976). The presence of ptDNA in the albino tissue was shown by autoradiography in *Pelargonium* (Knoth *et al.*, 1952) and by restriction enzyme analysis in maize (Walbot and Coe, 1979) and heat treated rye (Herrmann and Feirabend, 1980). In the albino tissue of the albostrians mutant of barley (Scott *et al.*, 1982), the replication of ptDNA and the biogenesis of the undifferentiated plastid structure must be controlled only by the nuclear-cytoplasmic system since the absence of ribosomes from the albino plastids precludes expression of the ptDNA.

This complex cooperation between the separate genetic systems of the nucleus and chloroplast may have arisen by gene transfer in either or both directions after a fusion of independent organisms (Margulis, 1970). The reduction in size of the plastome would decrease the energy required for the maintenance of a fully autonomous system. The greater sophistication and efficiency of the nuclear/cytoplasmic system is itself a good reason for nuclear control of plastid biogenesis. Perhaps an important consideration is that these former prokaryotic genes are now able to participate in a diploid genetic system with the advantages of allelic variation, meiosis and recombination. A convenient hypothesis is the simple one, that the many genes responsible for chloroplast biogenesis and function that are now found in the nucleus were transferred during evolution from a once free-living photosynthetically competent plastid progenitor such as a blue-green alga. Clearly, considerable reorganization of any prokaryotic sequences acquired by the nucleus in this way would be necessary to enable them to function in the eukaryotic transcription and translation system and for the cytoplasmically synthesized proteins to be directed towards, and integrated into plastids. It seems unlikely that such reorganizations would have been accomplished by specific transfer of individual genes and are more likely to have involved many variations, perhaps splicing together quite random segments of the pre-chloroplast and nuclear genome.

Our assumptions on undertaking the following experiments were that, if this sort of gene transfer and reorganization had occurred as a mechanism of origin for the nuclear involvement in chloroplast biogenesis, the remnants, or in botanical terms "the leaf scars", of the transfer may still be found in the nucleus as sequences related to the present day chloroplast genome. It seemed unlikely that the latter would have been specifically withheld from any such sequence transpositions, and more likely that complete, or representative fragments of the entire pre-chloroplast genome may have been transposed but that only a proportion, albeit large, of its functions taken over by the nucleus. This scenario also implies that unused sections of the pre-chloroplast genome were rapidly lost from the plastid. There was another interesting possibility, namely that DNA transfer is not a rare event requiring geological time periods, but a continuing dynamic property of genetic material related more generally to evolution of eukaryotic genomes.

With these ideas in mind we (Timmis and Scott, 1983; Scott and Timmis, 1984) utilized the spinach chloroplast (ct) DNA clone bank of Palmer and Thompson (1981) as probes to identify homologous sequences in nuclear DNA. Of the 145 kbp ctDNA genome (Bohnert *et al.,* 1982) more than 99 % was available as 13 fragments cloned into either the PstI site of pBR322 or the XhoI site of pACYC177 (Palmer and Thompson, 1981). Each fragment was individually hybridized to Southern blots of restricted homologous spinach ptDNA and high loadings of nuclear DNA. In initial experiments a range of restriction enzymes were used but relatively few were found to cut spinach nDNA preparations efficiently. Each PstI clone, for example, while yielding its "parent" restriction band on hybridization to ptDNA digested with PstI also hybridized to nDNA which was largely undigested by PstI and consequently the homologies were not resolved into identifiable bands by electrophoresis. Other enzymes ranged from more efficient (BamHI) to very efficient (EcoRI) cutters of nDNA samples.

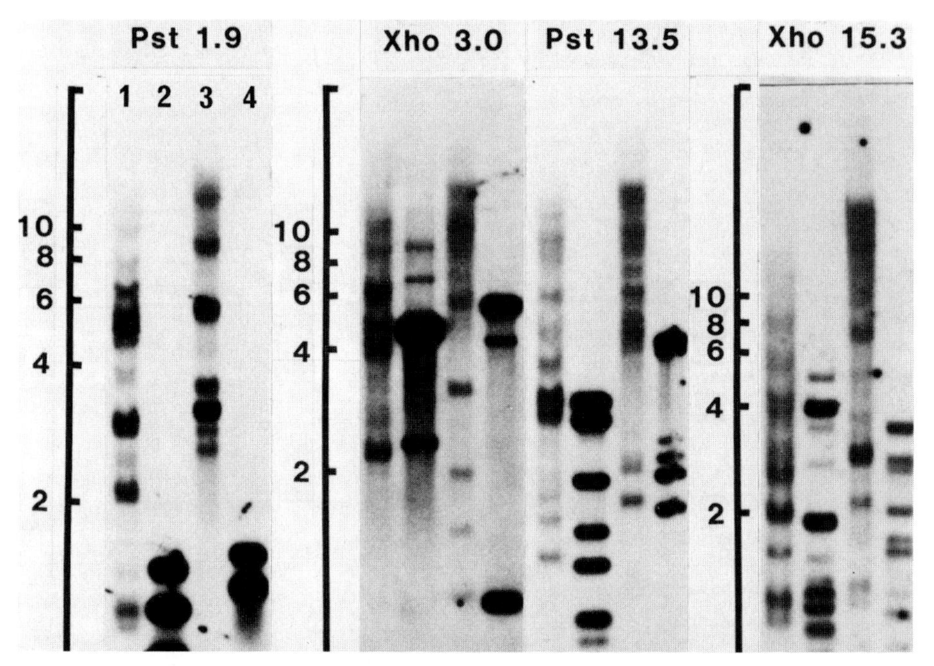

Fig. 2. Homologies between ptDNA and nDNA. The hybridization of individual clones from a spinach ptDNA clone bank (Palmer and Thomson, 1981) to nDNA and ptDNA digested with EcoRI and BamHI is shown. Each set of four gel tracks is labelled with the ptDNA clone used as probe and comprises from *left to right:* 1. Nuclear DNA digested with EcoRI; 2. Plastid DNA digested with EcoRI; 3. Nuclear DNA digested with BamHI and 4. Plastid DNA digested with BamHI. The molecular weight scale on the left of the tracks is in kbp. Nuclear DNA tracks were loaded with 5 μg nDNA and ctDNA tracks contain between 1.9 and 8.2 ng ctDNA dependent upon the number of fragments hybridized. Reproduced in modified form from Scott and Timmis (1984) with permission

Fig. 2 shows four examples of Southern hybridizations (Southern, 1975) of ptDNA clones to nDNA and ctDNA restricted with either BamHI or EcoRI. These examples are typical of results obtained for all the ptDNA clones, with each segment of the plastome hybridizing to a specific set of nDNA fragments in a pattern markedly different and usually more complex than that produced on hybridization to homologous ptDNA re-

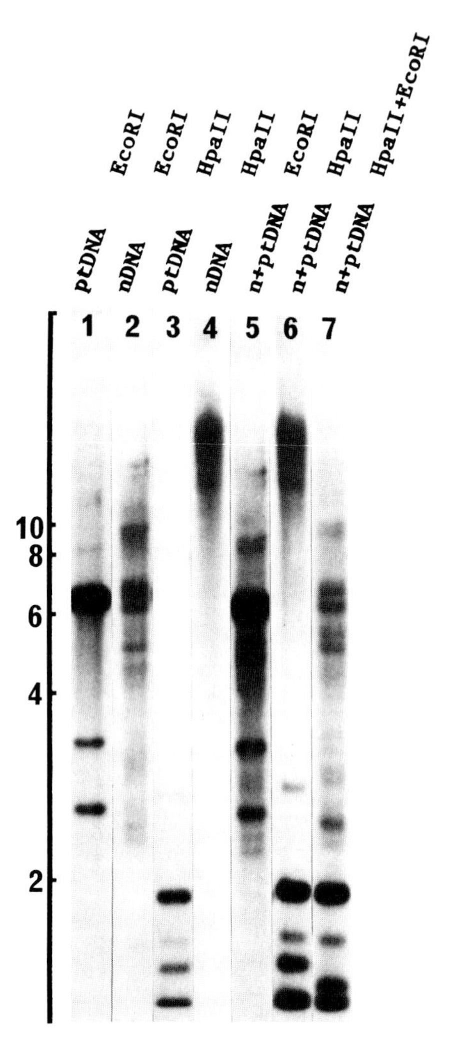

Fig. 3. Resolution of ptDNA and nDNA by methylation sensitive restriction. Detection of sequence homologies between nuclear and ptDNA in spinach with a cloned 7.7 kbp PstI fragment of spinach DNA. The source of DNA and the enzyme(s) used for digestion prior to Southern transfer and hybridization appear at the head of each gel lane. The ctDNA loading (4.9 ng per lane) compared with the nDNA loading (5.0 μg per lane) is equivalent to 6 copies of ctDNA per haploid spinach genome. The scale on the left is the molecular weight in kbp. Reproduced in modified form from Timmis and Scott (1983) with permission

stricted with the same enzyme. The lack of correlation between paired tracks of nDNA and ptDNA cut with the same enzyme and hybridized with the same probe, and the generally higher molecular weight of the nDNA bands, suggests that nDNA contains sequences homologous to ptDNA in tracts which are not very long. After allowing for the slight mobility differences caused by unequal loadings, in a few cases there appears to be some similarity in the two patterns. For example, the EcoRI bands at 2.6 and 2.9 kbp in the hybridization of the 13.5 PstI fragment (Fig. 2), correspond with ctDNA bands suggesting the possibility of integration of some longer ctDNA related tracts in nDNA.

Since both EcoRI and BamHI cut ptDNA into a large number of comparatively small fragments they are useful in defining the nature of the integration of ptDNA into nDNA. Individual chloroplast sequences transposed to and integrated into nDNA will be bordered by sequences bearing a new pattern of EcoRI or BamHI sites, thus generating new restriction enzyme fragments with homology to ptDNA. The multiple bands of homology in nDNA shown for the probes in Fig. 2, which are generally larger than those in ptDNA, suggest that the majority of hybridization involves a variety of small (1—2 kbp) related fragments integrated into the nucleus at several different specific sites. The alternative possibility of loss of EcoRI and BamHI restriction sites in the nuclear homologies seems less likely since there are so few exact molecular weight homologies, and there must also be an equal chance of generation of new restriction sites.

The difference in efficiency of cleavage of nDNA between EcoRI and BamHI can be seen in the ethidium bromide stained gels and is also apparent for the sequences which hybridize to chloroplast probes (Fig. 2). The nDNA BamHI tracks invariably show bands which are superimposed on a more pronounced "smear" of hybridization at high average molecular weight. The sensitivity of BamHI to base modification may well explain both these results, and the more extreme lack of digestion of nDNA by PstI referred to earlier. Consistent with these observations, methylation of C residues in spinach nuclear DNA can be as high as 30% (Scott and Possingham, 1980), whereas there is no detectable methylation of ptDNA in higher plants (Bohnert *et al.*, 1982). The 7.7 kbp PstI fragment of ptDNA was utilized for studies using this differential base modification to resolve nuclear from plastid sequences. In Fig. 3 EcoRI digests of ptDNA and nDNA (tracks 1 and 2) show significant differences in restriction pattern when hybridized with the 7.7 kbp probe. When similar DNA samples are digested with HpaII there is a more striking difference in hybridization pattern (compare tracks 3 and 4) as ptDNA is cleaved into small fragments whereas nDNA is essentially uncut by this enzyme. When a mixture of ptDNA and nDNA is prepared and digested with these two enzymes separately and together (tracks 5—7) further strong evidence for the nuclear origin of the EcoRI hybridization bands is obtained. Track 5 shows the homologous ptDNA pattern dominating the lane while track 6 shows the high molecular weight smear derived from nDNA plus the expected low molecular weight fragments of ptDNA resulting from HpaII digestion of the mixture of DNAs. In track 7, which is digested with both enzymes, the EcoRI fragments are indistinguishable from those in nDNA, whereas the combined enzymes cleave ptDNA into small fragments. These experiments clearly resolve the methylated nuclear fragments hybridizing to the 7.7 kpb probe from *bona fide* unmodified ptDNA and rule out contamination of nDNA with ptDNA and differential partial restriction as explanations of the results in Fig. 2.

The resolution of nDNA from ptDNA by restriction enzyme digestion allows parallel experiments to be carried out utilizing total tissue DNA,

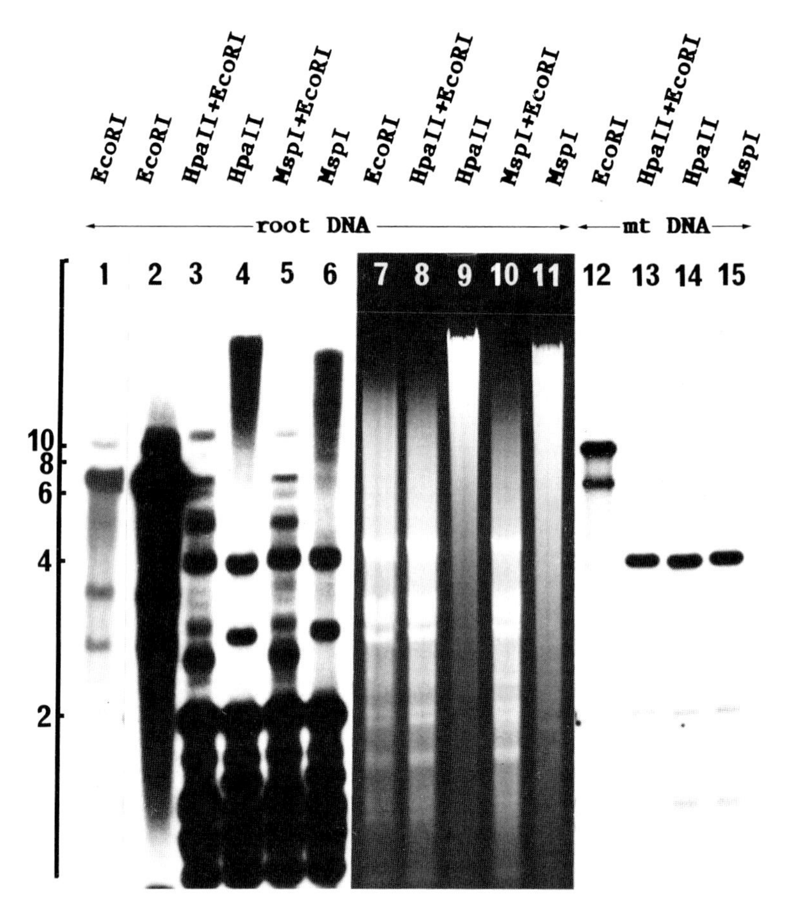

Fig. 4. Detection of sequence homologies in total root DNA and mtDNA with the cloned 7.7 kbp PstI fragment of spinach ctDNA. Tracks 1—6 and 12—15 are autoradiographs after Southern transfer and hybridization. Tracks 7—11 are the ethidium bromide stained gel lanes which gave rise to tracks 1—6. The source of DNA and the enzyme(s) used for digestion appear at the head of each lane. Lane 1 is a short exposure of lane 2. Root DNA was loaded at 5 µg per lane while mtDNA was loaded at 0.2 µg per lane. The scale on the left is molecular weight in kbp. Reproduced in part from Timmis and Scott (1983) with permission

obviating the necessity for physical purification of organelles. Fig. 4 indicates the value of using total DNA extracts which are higher in mean molecular weight with consequent improvement in quality of the Southern hybridizations. EcoRI digestion of total root DNA is dominated by homologous hybridization to root plastid DNA (track 2) which, as a shorter exposure of the autoradiograph (track 1) shows is essentially the same as the digestion pattern of ptDNA. There is, however, the significant addition of a prominent band of 8.8 kbp, which is present in large amounts compared with nDNA bands (cf. Fig. 3). These results (Fig. 4) duplicate those of Fig. 3 revealing a HpaII insensitive nuclear portion of total DNA and the predicted plastid bands hybridizing to the 7.7 kbp probe (track 4). The HpaII insensitive component is digested by EcoRI to yield bands similar to those found in purified nDNA (track 3). Both HpaII and HpaII + EcoRI digestion yield a prominent band of 3.9 kbp not seen in either nDNA or ptDNA. Similar results are obtained when MspI is substituted for HpaII (Fig. 4, tracks 5 and 6). The latter enzyme cleaves the sequence C/CGG but not the methylated sequence C/^{m}CGG, whereas MspI cleaves both these tetranucleotides but not ^{m}C/CGG (Doerfler, 1983). As the two enzymes generate indistinguishable fragments with sequences homologous to the 7.7 kbp probe we conlude that both cytosine residues are methylated at any CCGG sites in these regions of the genome.

From ethidium bromide stained gels (Fig. 4, tracks 7—11) it appears that the methylation status of the nDNA sequences homologous to the 7.7 kbp probe is consistent with that most of the nDNA. Neither HpaII nor MspI cleave nDNA extensively, although MspI is significantly more efficient in cleaving total root DNA than HpaII (compare tracks 4, 6, 9 and 11). Again, EcoRI (tracks 7, 8, and 10) is confirmed as extremely efficient in digestion. The strong correspondence between tracks 3 and 5, on the other hand, suggests that the homologous sequences to the 7.7 probe contain disproportionately fewer MspI sites compared with the average of nDNA. It should be remembered that nDNA comprises >98 % of total root DNA.

The bands of 8.8 kbp in EcoRI digests and 3.9 kbp in HpaII and MspI digests of root DNA are interesting as they represent hybridization of the 7.7 chloroplast probe to mitochondrial (mt)DNA (Fig. 4, tracks 12—15). Partially purified mtDNA contains this additional 8.8 kbp fragment as a prominent band as well as three minor contaminating ptDNA fragments when digested with EcoRI (lane 12). The former is reduced to 3.9 kbp when digested additionally with HpaII (lane 13), the same as found in single HpaII or MspI digests (lanes 14 and 15). The chloroplast contaminant component of mtDNA preparations is reduced to the expected range of small fragments by digestion with either HpaII or MspI. The 8.8 kbp mtDNA EcoRI fragment has been cloned and a 2.7 kbp BamHI subfragment defined which is homologous to a 1.7 kbp KpnI subfragment of the 7.7 kbp PstI clone (Whisson and Scott, unpublished results). This chloroplast fragment contains the gene for the P700 chlorophyl-a apo-protein (Westhoff *et al.*, 1983) and both mtDNA and ptDNA subfragments hybridize to a reduced set of the nDNA fragments also homologous to the 7.7 kbp PstI probe (Whisson and Scott, unpublished results). The 7.7 kbp probe therefore contains a region which has homologous counterparts in all three genetic compartments of the cell.

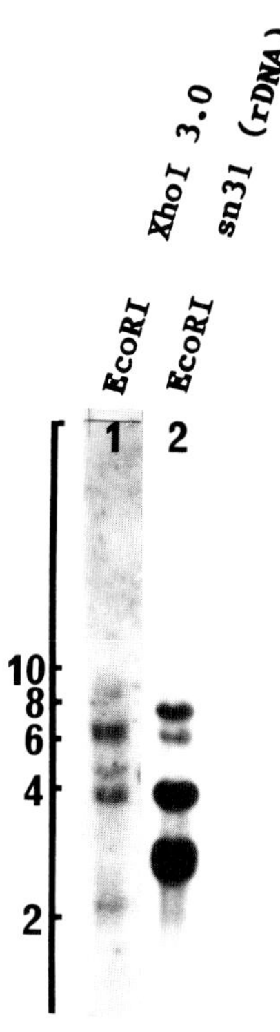

Fig. 5. Cross-homology of the spinach chloroplast and nuclear ribosomal RNA genes. A spinach nDNA sample (5 μg) was digested with EcoRI, fragments separated electrophoretically, Southern blotted and hybridized first with a cloned ctDNA XhoI fragment containing part of the chloroplast 23 S rRNA gene (lane 1), and after removal of the first probe, rehybridized with the complete cloned nuclear rDNA repeat unit (lane 2). The scale on the left is molecular weight in kbp. Reproduced in modified form from Timmis and Scott (1983) with permission

Experiments using other ctDNA probes, based on restriction enzyme resolution of differentially methylated nDNA and ptDNA have given similar results to those using the 7.7 kbp PstI fragment (Timmis and Scott, 1983; Scott and Timmis, 1984). Probing with the 3.0 kbp XhoI fragment yielded four major nucleus specific EcoRI fragments one of which appears to be sensitive to HpaII indicating that it is disproportionately undermethylated compared with the majority of nDNA. The 3.0 kbp XhoI fragment is also interesting as it contains part of the chloroplast 23 S rRNA gene and allows comparison of functionally similar genes from the chloroplast and nucleus. If the observed ptDNA/nDNA homologies were due to cross-hybridization between functionally similar genes rather than transposition of sequences, this probe would be expected to hybridize to the same bands as the 25S or 18S nuclear rRNA genes. Figure 5 shows nDNA probed first with the 3.0 kbp XhoI fragment (track 1) and secondly with the complete spinach nuclear rDNA repeat unit (Timmis and Scott, 1983). The two probes generally hybridize to a different set of fragments and, where there are similarities in the mobility of hybridizing bands, the lack of equimolarity shows that cross-hybridization is unimportant as an explanation of the homologies, even in this extreme case of a highly reiterated nuclear gene. The repetition of rRNA genes (Ingle *et al.,* 1975) would also be

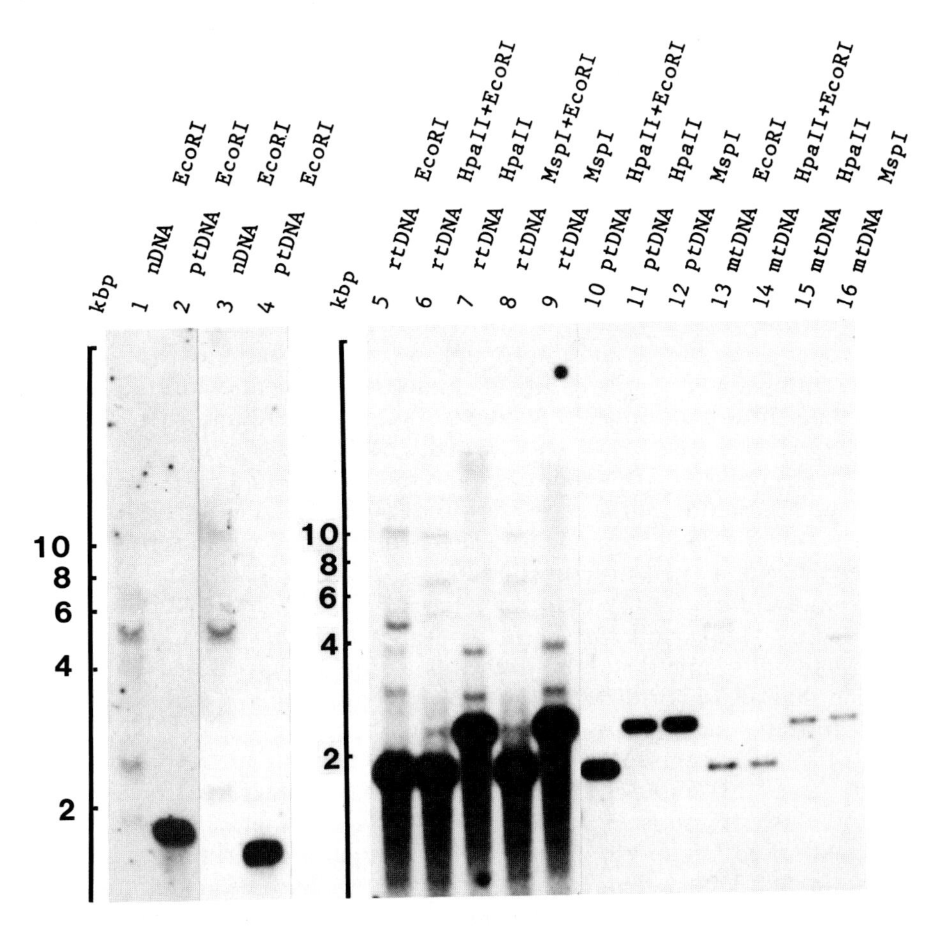

Fig. 6. Hybridization of nDNA, mtDNA, ctDNA and root (rt)DNA with specific cloned ctDNA genes. Nuclear DNA, rtDNA (5 μg), ctDNA (4 ng) and mtDNA (0.2 μg) were digested with restriction enzymes, fragments separated electrophoretically, Southern blotted and hybridized with the cloned gene for the β subunit of ATPase (lanes 1 and 2) or the gene for the large subunit of Rubisco (lanes 3—16). The source of DNA and the enzyme(s) used for digestion appear at the head of each gel lane. The scale on the left is molecular weight in kbp. Reproduced from Scott and Timmis (1984) with permission

expected to increase the signal of any cross-hybridization between functionally similar genes, but the 3.0 kbp XhoI fragment shows a similar degree of homology to nDNA as the rest of the ptDNA circle (Scott and Timmis, 1984). Similarly there is no evidence that this, or the other homologies are due to low level non-specific hybridization to any non rDNA repetitious sequences containing EcoRI sites which are evident as intense bands superimposed on the heterogeneous ethidium bromide stained fragments in digested root DNA (Fig. 4).

Two essentially complete and contiguous genes from ctDNA cloned as separate EcoRI fragments in pBR325 (Zurawski *et al.*, 1981) were also used as probes to search for specific homologies in nDNA. These were the genes for the β subunit of ATPase (Fig. 6, lanes 1 and 2) and the gene for the large subunit of Rubisco (tracks 3 and 4). The two probes each hybridize to their respective 'parent' fragments in EcoRI digested ctDNA but to a major common nDNA fragment of about 5 kbp as well as to differential nDNA fragments in the region of 1—3 kbp. This result suggest the homologies in this fragment are close together in nDNA but are not separated by the EcoRI site found in ptDNA.

Application of the differential restriction enzymes EcoRI, HpaII and MspI to nDNA show a complex set of homologies to the large subunit of Rubisco gene (Fig. 6, tracks 5—16). The use of total DNA preparations from roots as a source of nDNA improved the Southern hybridization quality and showed larger bands of homology of extra-chloroplast origin other than those seen in nDNA preparations (track 3), and there are four major bands between 3 and 10 kbp in EcoRI digested root DNA (Fig. 6, track 5). In contrast to the homologies to the 7.7 PstI and 3.0 XhoI ctDNA fragments, these homologies are very undermethylated compared with the majority of spinach genomic DNA. The EcoRI bands are all sensitive (and similarly sensitive) to both HpaII and MspI (Fig. 6, tracks 6 and 8). On the other hand there are only two fragments of about 3 and 4 kbp in nDNA bounded by HpaII and MspI sites with evidence that some other HpaII and MspI sites are distantly spaced leaving a high molecular weight fraction that also hybridizes to the probe (Fig. 6, tracks 7 and 9).

It appears that there is a low level of homology to mtDNA (tracks 13—16) for the Rubisco large subunit gene. Bands of 4.5 kbp in EcoRI digests (track 13) and 3.9 kbp in MspI digests of mtDNA (track 16) are present which are not seen when HpaII is used. Presumably HpaII digests the 4.0 kbp EcoRI band to low molecular weight or superimposes it on the contaminant ctDNA fragment suggesting the presence of a site in mtDNA which is sensitive to HpaII but insensitive to MspI, a very unusual form of methylation (Doerfler, 1983). While the more prominent bands in these gel lanes at 2.4 and 3.0 kbp are probably due to contamination of ctDNA, the possibility cannot be excluded that this is the same sequence bounded by the same restriction sites in both mtDNA and ctDNA and that these stronger bands are of both chloroplast and mitochondrial origin. We know (from Fig. 4) that there is significant ctDNA contamination of this mtDNA sample but on the other hand maize mtDNA contains a promiscuous (Ellis,

1982) Rubisco large subunit gene which has been cloned and used in an *E. coli* transcription/translation system to direct the formation of a product which is recognised by the antibody to wheat chloroplast Rubisco (Lonsdale *et al.*, 1983).

From all these experiments we conclude that the majority of ctDNA has homologous counterparts in the nucleus where it is present as relatively short sequences which may be integrated at several different chromosomal locations. The sequences in nDNA are not likely to differ markedly from those in ptDNA as hybridizations were performed at high stringency (mean ptDNA Tm – 20° C) and hybrids were lost from both nDNA and homologous ctDNA to the same extent if this stringency was increased further.

Our evidence suggests that virtually the entire plastome has homology with nDNA (Timmis and Scott, 1983; Scott and Timmis, 1984). All of the clones in the ctDNA clone bank showed significant homology to nuclear DNA and comparison of the autoradiograms of nDNA by densitometry with those of known loadings of ptDNA we calculated that on average there was the equivalent of 5 copies of plastome homology per haploid nuclear genome. This estimate agrees with that of 4 copies obtained from renaturation analyses of total ptDNA with nDNA. The complete renaturation of ptDNA with nDNA was another indication that all the ptDNA was represented by homologies in nDNA.

There is now strong evidence that chloroplasts have derived from a free living prokaryotic precursor organism. Some of the best evidence comes from sequence analysis comparisons between ptDNA genes and functionally similar genes in prokaryotes. In the region between the large and small ribosomal RNA genes in ptDNA and in the DNA of a blue green alga there are two tRNA genes (Williamson and Doolittle, 1983). This arrangement is very similar to that found in the ptDNA of maize (Koch *et al.*, 1981) suggesting that it has been conserved during the evolution of chloroplasts from a blue-green algal ancestor. Sequence similarities between ribosomal RNA genes of prokaryotes and chloroplasts have also been shown (Schwarz and Kossel, 1980).

The existence of homologies between nuclear DNA and these ptDNA sequences shows that the evolutionary process has probably involved incorporation of once prokaryotic genes into the nucleus. Sequence analysis of these homologies and their divergence may allow the direction and evolutionary timing of the DNA movement to be determined. It is not impossible that ptDNA arose following incorporation of a complete prokaryotic genome into a nucleus and subsequent migration of DNA to the organelle. However, it seems more likely, as discussed below, that prokaryotic DNA was incorporated into the cytoplasm within a precursor organelle which later lost DNA during or following acquisition of the relevant prokaryotic function by the nucleus.

Many nuclear encoded proteins concerned in biogenesis of cytoplasmic organelles are synthesised in the cytoplasm as a precursor containing a transit peptide sequence which directs the protein to its required cellular environment (Harmey and Neupert, 1984). If the many genes concerned with chloroplast and mitochondrial biogenesis now present in the nucleus were derived from prokaryotic genes they would have been subject to complex rearrangements and additions of sequences coding for specific transit peptides and controlling nucleus specific expression. Such feats of genetic engineering within the nucleus would depend on the presence of a large reference library of sequences and specific mechanisms of manipulation. In addition, the process would surely involve trial and error, presumably culminating in a selection system ensuring that the final protein products are functional and efficiently controlled and translocated. A similar system would be equally effective for promoting the evolution of all other proteins and would go a long way towards explaining the paradoxically large size of eukaryotic nuclei. Primitive, pre-nuclear development of such abilities has been observed in some archaebacteria (Sapienza and Doolittle, 1982) and the ability of plant mitochondria to accumulate and maintain foreign sequences (Stern and Lonsdale, 1982; Stern and Palmer, 1984; Londsdale *et al.*, 1983) may be a similar phenomenon. Fitting into the same evolutionary pattern is the observation that in the protozoan *Cyanophora paradoxica* the genome of the photosynthetic organelle, the cyanelle, codes for both subunits of Rubisco (Heinhorst and Shively, 1983). The cyanelle is related to the chloroplast by virtue of its function within the protozoan cell yet is almost indistinguishable morphologically from free-living members of the Cyanophyta (Hall and Claus, 1963); though its genome is only slightly larger in size than ptDNA (Herdman and Stanier, 1977).

There is other evidence to support this notion of a prokaryotic origin of nuclear encoded cytoplasmic organellar genes. The first amino acid in the small subunit of Rubisco was found to be exclusively methionine in 15 out of 16 angiosperm families examined (Martin and Jennings, 1983). One out of the 16 families, the Onagraceae, had some species with methionine and some with phenyl alanine at the N-terminus (Martin and Jennings, 1983) suggesting some relaxation of conservation when this gene is located in the nucleus. This finding suggests the general conservation of the original prokaryotic initiation codon for translation even though there are 165 bases added to the gene to code for the transit peptide in the precursor protein at this precise position. While there are some technical difficulties associated with identifying the N terminal amino acid of this protein, sequence analysis of cDNA clones to the mRNA has confirmed that methionine is the first amino acid of the mature protein in several species (Berry-Lowe *et al.*, 1982; Broglie *et al.*, 1983; Smith *et al.*, 1983).

It appears that in many systems there are homologies between nDNA and mtDNA which are comparable with the nDNA/ptDNA homologies (for review see Timmis and Scott, 1984), suggesting an origin for the relationship between nuclei and mitochondria which parallels that of the nucleus and chloroplast. The results from all these systems are consistent with past or continuing movement and rearrangement of sequences with a flow in the direction of the nucleus and a loss of the consequently redundant DNA from the organelle. The loss of organellar DNA may follow the independent development of organelle control in the nucleus. It seems more likely that once one organellar gene has been incorporated into, and has become functional in, the nucleus, then the ancestral duplicate gene in mtDNA or ptDNA may be deleted. This mechanism accounts for the small size of ptDNA and many mtDNAs compared with any of the putative free living precursors of these organelles. The large number (>95%) of mitochondrial and chloroplast functional and structural proteins coded in nDNA and the paucity of examples of evolutionary intermediates indicates that the majority of genes were readily transferred, reorganized and expressed by the nucleus.

The transfer of the oxi-3 gene between the mitochondria and nucleus of *Podospora* and its mex-1 mutant (Wright and Comings, 1983) may be an example of this rare evolutionary event in process. In *Podospora* there are specific mtDNA sequences which are amplified as a multimeric head-to-tail plasmid-like structure within the mitochondria during senescence. These are called α and β senDNA and they may also be transferred to the nucleus. The α senDNA contains the gene for subunit 1 of cytochrome-c oxidase (oxi-3; Wright *et al.*, 1982), and the mex-1 *Podospora* mutant has mtDNA deficient in the region of α senDNA including the oxi-3 gene, but apparently possesses a functional copy of the gene in its nucleus (Wright and Comings, 1983; Vierny *et al.*, 1982). Transposition of genes could also account for the observation that there may be an active gene for the dicyclohexylcarbodiimide binding protein of mtATPase in the mtDNA as well as in nDNA of *Neurospora* (van den Boogaart *et al.*, 1982). In contrast this protein is exclusive to mtDNA in yeast (Orian *et al.*, 1981). The existence of the ptDNA/nDNA homologies in multiple sites indicates that the transfer of the few remaining genes in the plastid and mitochondria to the nucleus has probably occurred more than once, without successful adaptation by the nucleus. It could be that some genes present insurmountable problems in terms of nuclear control necessitating the maintenance of separated genetic systems.

We have tended to favour a specific evolutionary role for these promiscuous sequences but their functions may be more varied. They could be manifestations of more broadly important evolutionary machinery and/or they may have much shorter term functions within the cell such as the induction of male-sterility in plants (Stern and Lonsdale, 1982) or the maintenance of the strict intraspecies homogeneity of organellar genomes mentioned earlier.

The understanding and possible artificial manipulation of these intriguing sequences remains a challenge for the future.

Acknowledgements

We thank David Hayman for valuable discussion.

References

Berry-Lowe, S., McKnight, T. D., Shah, D. M., Meagher, R. B., 1982: The nucleotide sequence, expression, and evolution of one member of a multigene family encoding the small subunit of ribulose-1,5-bisphosphate carboxylase in soybean. J. Mol. Appl. Gen. **1,** 483—498.

Bohnert, H. J., Crouse, E. J., Schmitt, J. M., 1982: Organization and expression of plastid genomes. In: Encyclopaedia of plant physiology, vol. 14B (Parthier, B., Boulter, D., eds.). Berlin - Heidelberg - New York: Springer, pp. 475—530.

Borner, T., Schumann, B., Hagemann, R., 1976: Biochemical studies on a plastid ribosome-deficient mutant of *Hordeum vulgare.* In: Genetics and biogenesis of chloroplasts and mitochondria (Bücher, T. *et al.,* eds.). Amsterdam: Elsevier/North-Holland Biomedical Press, pp. 41—48.

Bottomley, W., Bohnert, H. J., 1982: The biosynthesis of chloroplast proteins. In: Encyclopaedia of plant physiology, vol. 14B (Parthier, B., Boulter, D., eds.). Berlin - Heidelberg - New York: Springer, pp. 531—596.

Broglie, R., Coruzzi, G., Lamppa, G., Keith, B., Chua, N.-H., 1983: Structural analysis of nuclear genes coding for the precursor to the small subunit of wheat ribulose-1,5-bisphosphate carboxylase. Biotechnology **1,** 55—61.

Doerfler, W., 1983: DNA methylation and gene activity. Ann. Rev. Biochem. **52,** 93—124.

Dunsmuir, P., Smith, S., Bedbrook, J., 1983: A number of different nuclear genes for the small subunit of RuBPCase are transcribed in petunia. Nucleic Acids Res. **11,** 4177—4183.

Edleman, M., Swinton, D., Schiff, J. A., Epstein, H. T., Zeldin, D., 1967: DNA of the blue-green algae (Cyanophyta). Bact. Rev. **31,** 315—335.

Ellis, R. J., 1982: Promiscuous DNA — chloroplast genes inside plant mitochondria. Nature **299,** 678—679.

Ellis, R. J., 1982: Chloroplast protein synthesis: principles and problems. Subcell. Biochem. **9,** 237—261.

Feierabend, J., Schrader-Reichardt, U., 1976: Biochemical differentiation of plastids and other organelles in rye leaves with a high temperature-induced deficiency of plastid ribosomes. Planta **129,** 133—145.

Hall, W. T., Claus, T., 1963: Ultrastructural studies on the blue-green algal symbiont in *Cyanophora paradoxa* Korschikoff. J. Cell Biol. **19,** 551—563.

Harmey, M. A., Neuport, W., 1984: The biosynthesis of mitochondria. In: Synthesis and intercellular transport of mitochondrial proteins (Martonosi, A. N., ed.). New York: Plenum (in press).

Heinhorst, S., Shrively, J. M., 1983: Encoding of both subunits of ribulose-1,5-bisphosphate carboxylase by organelle genome of *Cyanophora paradoxa.* Nature **304,** 373—374.

Herdman, M., Stanier, R. Y., 1977: The cyanelle: chloroplast or endosymbiotic prokaryote? FEMS Microbiol. Lett. **1,** 7—11.

Hermann, R. G., Feierabend, J., 1980: The presence of DNA in ribosome-deficient plastids of heat-bleached rye leaves. Eur. J. Biochem. **104,** 603—609.

Ingle, J., Timmis, J. N., Sinclair, J., 1975: The relationship between satellite DNA, ribosomal RNA gene redundancy and genome size in plants. Plant Physiol. **55,** 496—501.

Kirk, J. T. O., Tilney-Bassett, R. A. E., 1978: The plastids. Their chemistry, structure, growth and inheritance. Amsterdam: Elsevier.

Knoth, R., Herrmann, F. H., Bottger, M., Börner, T., 1974: Structur und Funktion der genetischen Information in den Plastiden. Biochem. Physiol. Pflanzen. **166,** 129—148.

Koch, W., Edwards, K., Kössel, H. 1981: Sequencing of the 16S-23S spacer in a ribosomal RNA operon of *Zea mays* chloroplast DNA reveals two split tRNA genes. Cell **25,** 203—213.

Lamppa, G. K., Elliot, L. V., Bendich, A. J., 1980: Changes in chloroplast number during pea leaf development. Planta **148,** 437—443.

Lonsdale, D. M., Hodge, T. P., Howe, C. J., Stern, D. B., 1983: Maize mitochondrial DNA contains a sequence homologous to the ribulose-1,5-bisphosphate carboxylase large subunit gene of chloroplast DNA. Cell **34,** 1007—1014.

Margulis, L., 1970: Origin of eukaryotic cells. New Haven: Yale University Press.

Martin, P. G., Jennings, A. C., 1983: The study of plant phylogeny using amino acid sequences of ribulose-1,5-bisphosphate carboxylase. I. Biochemical methods and patterns of variability. Aust. J. Biol. **31,** 395—409.

Orian, J. M., Murphy, M., Marzuki, S., 1981: Mitochondrially synthesized protein subunits of the yeast mitochondrial adenosine triphosphatase. A reassessment. Biochim. Biophys. Acta **652,** 234—239.

Palmer, J. D. Thompson, W. F., 1981: Clone banks of the mung bean, pea and spinach chloroplast genomes. Gene **15,** 21—26.

Sapienza, C., Rose, M. R., Doolittle, W. F., 1982: High-frequency genomic rearrangements involving archebacterial repeat sequence elements. Nature **299,** 182—185.

Schwarz, Z., Kössel, H., 1980: The primary structure of 16S rDNA from *Zea mays* chloroplast is homologous to *E. coli* 16S rRNA. Nature **283,** 739—742.

Scott, N. S., Cain, P., Possingham, J. V., 1982: Plastid DNA levels in albino and green leaves of the 'albostrians' mutant of *Hordeum vulgare.* Z. Pflanzenphysiol. **108,** 187—191.

Scott, N. S., Possingham, J. V., 1980: Chloroplast DNA in expanding spinach leaves. J. Exp. Bot. **123,** 1081—1092.

Scott, N. S., Timmis, J. N., 1974: Homologies between nuclear and plastid DNA in spinach. Theor. Appl. Genet. **67,** 279—288.

Scott, N. S., Tymms, M. J., Possingham, J. V., 1984: Plastid DNA levels in the different tissues of potato. Planta **161,** 12—19.

Smith, S. M., Bedbrook, J., Speirs, J., 1983: Characterization of three cDNA clones encoding different mRNAs for the precursor to the small subunit of wheat ribulose bisphosphate carboxylase. Nucleic Acids Res. **11,** 8719—8734.

Southern, E. M., 1975: Detection of specific sequences among DNA fragments separated by gel electrophoresis. J. Mol. Biol. **98,** 503—517.

Stern, D. B., Lonsdale, D. M., 1982: Mitochondrial and chloroplast genomes of maize have a 12 kilobase DNA sequence in common. Nature **299,** 698—702.

Stern, D. B., Palmer, J. D., 1984: Extensive and widespread homologies between mitochondrial and chloroplast DNA in plants. Proc. Natl. Acad. Sci., U.S.A. **81,** 1946—1950.

Timmis, J. N., Scott, N. S., 1983: Spinach nuclear and chloroplast DNAs have homologous sequences. Nature **305,** 65—67.

Timmis, J. N., Scott, N. S., 1984: Promiscuous DNA: sequences homologous between DNA of separate organelles. Trends in Biochem. Sci. **9,** 271—273.

van den Boogaart, P., Samallo, J., Agsteribbe, E., 1982: Similar genes for a mitochondrial ATPase subunit in the nuclear and mitochondrial genomes of *Neurospora crassa.* Nature **298,** 187—189.

Vierny, C., Keller, A.-M., Begel, O., Belcour, L., 1982: A sequence of mitochondrial

DNA is associated with the onset of senescence in a fungus. Nature **279,** 157—159.

Walbot, V., Coe, E. H., 1979: Nuclear gene *iojap* conditions a programmed change to ribosome-less plastids in *Zea mays.* Proc. Natl. Acad. Sci., U.S.A. **76,** 2760—2764.

Westhoff, P., Alt, J., Nelson, N., Bottomley, W., Bünemann, H., Herrmann, R. G., 1983: Genes and transcripts for the P_{700} chlorphyll *a* apoprotein and subunit 2 of photosystem 1 reaction centre complex from spinach thylakoid membranes. Pl. Molec. Biol. **2,** 95—107.

Williamson, S. E., Doolittle, W. F., 1983: Genes for $tRNA^{Ile}$ and $tRNA^{Ala}$ in the spacer between the 16S and 23S rRNA genes of a blue-green alga: strong homology to chloroplast tRNA genes and tRNA genes of the *E. coli* rrnD gene cluster. Nucleic Acids Res. **11,** 225—235.

Wright, R. M., Commings, D. J., 1983: Integration of mitochondrial gene sequences within the nuclear genome during senescence in a fungus. Nature **302,** 86—88.

Wright, R. M., Horrum, M. A., Commings, D. J., 1982: Are mitochondrial structural genes selectively amplified during senescence in *Podospora anserina?* Cell **29,** 505—515.

Zurawski, G., Bottomley, W., Whitfield, P. R., 1982: Structures of the genes for the β and ε subunits of spinach chloroplast ATPase indicate a dicistronic mRNA and an overlapping translation stop/start signal. Proc. Natl. Acad. Sci., U.S.A. **79,** 6260—6264.

Zurawski, G., Perrot, B., Bottomley, W., Whitfield, P. R., 1981: The structure of the gene for the large subunit of ribulose 1,5-bisphosphate carboxylase from spinach chloroplast DNA. Nucleic Acids Res. **9,** 3251—3270.

Chapter 5

Movement of Genetic Information Between Plant Organelles: Mitochondria-Nuclei

R. J. Kemble[1], S. Gabay-Laughnan[2] and J. R. Laughnan[2]

[1] Department of Plant Biology, Allelix Inc., 6850 Goreway Drive, Mississauga, Ontario, L4V 1P1, Canada

[2] Present address: Department of Plant Biology, University of Illinois, Urbana, IL 61801, U.S.A.

With 1 Figure

Contents

I. Introduction

The endosymbiont hypothesis for the origin of eukaryotic cells suggests that mitochondria and chloroplasts arose from free-living prokaryotes which entered into, and established a stable symbiotic relationship with, a progenitor eukaryotic cell. Because the majority of mitochondrial and chloroplast proteins are genetically encoded in nuclear DNA (nDNA), the hypothesis predicts that transfer of genes from the two cytoplasmic genomes to the nucleus has occurred. Although this argument has, for the most part, been accepted for many years it is only since 1982 that experiments have provided definitive proof at the molecular level that such a transfer of genes has taken place. DNA sequences that have undergone this movement have been termed "promiscuous" (Ellis, 1982; Farrelly and Butow, 1983).

The aim of this chapter is to discuss the transfer of genetic information from plant mitochondria to nuclei. Because there has been only one published report of common mitochondrial DNA (mtDNA) and nDNA sequences in higher plants — that by Kemble *et al.* (1983) involving *Zea mays* — we will, by necessity, give emphasis to that system in our discussion. It is important to note, however, that common mtDNA and nDNA sequences have been detected in several animal and lower plant species.

II. Organisms Exhibiting Common Mitochondrial and Nuclear DNA Sequences

Table I summarizes all examples of the common mtDNA and nDNA sequences reported to date. Evidence from these studies suggests that the mitochondrial genes in question were duplicated and a copy was transposed to the nucleus where it became stably integrated. With the exception of *Podospora* (Wright and Cummings, 1983), this event probably occurred only once during the evolution of each organism. Since that time, amplification and rearrangement of these DNA sequences has ensued in sea urchin (Jacobs *et al.*, 1983), locust (Gellissen *et al.*, 1983) and maize (Kemble *et al.*, 1983). Divergence of these nuclear mtDNA sequences has been documented in sea urchin (Jacobs *et al.*, 1983), *N. crassa* (van den Boogaart *et al.*, 1982), locust (Gellissen *et al.*, 1983) and yeast (Farrelly and Butow, 1983).

Table 1. Common Mitochondrial and Nuclear DNA Sequences Reported To Date

Organism	*MtDNA Sequence Involved*	*Reference*
Rattus norvegicus rat liver	Less than 3 kb containing part of rRNA genes and D-loop region	Hadler *et al.* (1983)
Zea mays (maize, corn)	S 1 and 1.94 kb plasmid 1.4 kb plasmid	Kemble *et al.* (1983), A. G. Smith (personal communication)
Locusta migratoria (locust fat body)	rRNA genes	Gellissen *et al.* (1983)
Strongylocentrotus purpuratus (sea urchin embryo)	cytochrome oxidase subunit I gene and 3′ end of 16 S rRNA gene	Jacobs *et al.* (1983)
Saccharomyces cerevisiae (yeast)	Part of *var 1* gene, 3′ end of cytochrome b gene and part of *ori/rep* sequence	Farrelly and Butow (1983)
Neurospora crassa (ascomycete fungus)	ATPase DCCD-binding protein gene	van den Boogaart *et al.* (1982)
Podospora anserina (senescing ascomycete fungus)	α-event senDNA (2.6 kb) and β-event senDNA (9.8 kb) plasmids containing cytochrome c oxidase subunit I and III genes respectively	Wright and Cummings (1983)

The *Podospora* system differs from all the other examples in that the movement of DNA from mitochondria to nuclei occurs regularly, at a specific time, during each life cycle. Both α- and β-event senescent DNA sequences (senDNA) are a normal constituent of the mitochondrial chromosome in young cultures. However, as the culture ages and begins to senesce these sequences are excised, they assume a plasmid conformation and become highly amplified in relation to the juvenile mitochondrial chromosome. At this time both types of senDNA sequences become integrated into nDNA (Wright and Cummings, 1983). The *Podospora* system does have some similarity to that of maize in that the mtDNA sequences integrated into nDNA also exist as plasmid or virus-like autonomously replicating DNAs which can be experimentally isolated as separate entities. This is to say that the mitochondrial vectors which are thought to have transferred to the nucleus (probably once in evolutionary time in the case of maize, but once per cellular life cycle in *Podospora*) are still available in present day strains. This suggests that *Podospora* and maize may well be the organisms of choice for further searches for common mtDNA and nDNA sequences. An in-depth molecular examination of petite yeasts would also be beneficial because Farrelly and Butow (1983) have proposed that the mtDNA sequences integrated into yeast nDNA were transported via "petite mtDNAs".

With the exception of *N. crassa*, no evidence is available to suggest that the mitochondrial genes integrated into nDNA are functionally active. In this species the nuclear copy of the mitochondrial ATPase DCCD-binding protein gene is transcribed and the proteins transported into the mitochondria, whereas it appears that the mitochondrial copy of the gene is silent (van den Boogaart *et al.*, 1982).

III. Common Mitochondrial and Nuclear DNA Sequences in Maize

As Table I indicates, one of the homologous mtDNA and nDNA sequences found in maize is termed S 1. To better understand the relevance of this finding it is necessary to briefly discuss the role that S 1 may play in the agronomically important trait of cytoplasmic male sterility *(cms)*.

Plants that do not produce functional pollen are termed "male sterile". *Cms* occurs in plants when the trait is inherited not by Mendelian rules but instead is cytoplasmically inherited via the maternal parent. In maize there are three groups of *cms* cytoplasms: *cms-T*, *cms-C* and *cms-S*. (Male-fertile cytoplasm is designated "normal" and abbreviated as N). *Cms* can be overridden by the presence of certain nuclear male-fertility restorer *(Rf)* genes. In such cases, even though the plants possess *cms* cytoplasms they produce functional pollen and are male-fertile. Although the entire maize *cms* system is an intriguing area of biology in its own right, the remainder of this section will focus only on the molecular aspects of the *cms-S* trait. For a recent review of *cms* in maize the reader is referred to Laughnan and Gabay-Laughnan (1983).

Some strains carrying *cms-S* spontaneously revert to male fertility (Laughnan and Gabay, 1973; Singh and Laughnan, 1972). Such reversions in plantings of *cms* strains are recognized as either entirely male-fertile tassels or as fertile sectors on otherwise sterile tassels. Genetic analyses have indicated that plants revert to male fertility due either to nuclear mutations (such plants are termed nuclear revertants) or, alternatively, due to cytoplasmic mutations (termed cytoplasmic revertants). Nuclear revertants inherit the male-fertility trait in a Mendelian manner and behave as if they have acquired a new nuclear restorer gene. Cytoplasmic revertants inherit the trait uniparentally. To explain these reversions, Laughnan and Gabay (1975; 1978) postulated the existence of an episomal "fertility element" that could be "fixed" either in the cytoplasm (in the case of cytoplasmic revertants) or in the nucleus (in the case of nuclear revertants).

Among the array of DNAs present in maize mitochondria (Kemble and Bedbrook, 1980; Kemble *et al.*, 1980), S1 and S2 seemed to be the most appropriate molecules to fulfill the postulated "fertility element" role. S1 and S2 are linear DNAs of approximately 6.4kb and 5.4kb respectively and they are present in the mitochondria of *cms-S* group lines only (Pring *et al.*, 1977). They bear some resemblance to prokaryotic transposable elements because they possess terminal inverted repeats of 208bp (Levings and Sederoff, 1983) and further regions of sequence homology (Kim *et al.*, 1982). Their structure is similar to that of many animal viruses, *e. g.*, adenovirus, because they have a protein covalently attached to their 5′ termini (Kemble and Thompson, 1982). In adenovirus, the terminal protein is associated with replication of the virus (Challberg *et al.*, 1980; Rekosh *et al.*, 1977; Tamanoi and Stillman, 1982) and it is inviting to speculate that the S1 and S2 terminal proteins play similar roles. S1 and S2 probably evolved by excision from the mitochondrial chromosome of an N, male-fertile, cytoplasm line (Koncz *et al.*, 1981; Lonsdale *et al.*, 1981; Thompson *et al.*, 1980) to give rise to *cms-S* lines in which they are highly amplified as virus-like DNAs.

More direct evidence to suggest that S1 and S2 may play a role in *cms* was provided by mtDNA analysis of *cms-S* lines and male-fertile cytoplasmic revertants derived from them. S1 and S2 are no longer found as virus-like DNAs in mitochondria of the cytoplasmic revertants. It is suggested that they have been transposed into the mitochondrial chromosome (Levings *et al.*, 1980) or, alternatively, that their loss is concomitant with preferential amplification and rearrangement of homologous sequences already present in the mitochondrial chromosome (Kemble and Mans, 1983). For further discussion of the mtDNA rearrangements associated with *cms* the reader is referred to the recent review by Laughnan and Gabay-Laughnan (1983) and to the chapter in this volume by Levings.

The behaviour of S1 and S2 in the cytoplasmic revertants suggested that they may indeed be the physical manifestation of the postulated episomal "fertility elements". If so, Laughnan and Gabay's (1975; 1978) hypothesis predicts that S1 and/or S2 sequences would be detected in the nDNA of male-fertile nuclear revertants but not in the *cms* parents. Furthermore, the authors suggested that because the nuclear revertants act as if they have acquired a new nuclear gene that restores male fertility, the "fertility element" might be identical in structure to *Rf3,* the standard nuclear restorer gene for *cms-S*. Consequently, S 1 and/or S 2 should be present in the nDNA of Rf 3 restorer lines.

Utilizing cloned probes of S 1 and S 2 sequences, Kemble *et al.* (1983) have detected sequences homologous to S 1 integrated into the nDNA of nuclear revertants. However, the same S 1 sequences in the same restriction fragments and in the same stoichiometry were also present in the *cms* and N cytoplasm parental lines and in the cytoplasmic revertants. No unique S 2 sequences were observed in nDNA and no conformations similar to S 1 and S 2 "free" virus-like DNAs were detected in the nucleus. This indicates that S 1 and S 2 had not integrated into nDNA concomitant with the spontaneous nuclear reversion to fertility and, therefore, the most simplistic model of the episomal "fertility element" hypothesis is not correct.

Although the study (Kemble *et al.*, 1983) did not prove a direct association between S 1 and S 2 and the "fertility elements", it did document, for the first time, the presence of homologous mtDNA, nDNA and virus-like DNA sequences in higher eukaryotes. In *Bam* HI digests of nDNA isolated from revertant lines the sequences homologous to S 1 are present in four major fragments of 10.2 kb, 8.9 kb, 7.6 kb and 4.7 kb. The largest fragment contains a substantial region of S 1 and, since there is no *Bam* HI site in virus-like S 1 (Kim *et al.,* 1982), perhaps the entire length. An extensive portion of S 1 but not including the region homologous to S 2 is present in the 8.9 kb and 7.6 kb fragments. The smallest fragment contains sequences homologous to the central region of S 1. It has recently been shown that this central region of S 1 contains extensive sequence homology to the 3′ end of the chloroplast DNA (cpDNA) gene "photogene 32" (P. Ronald *et al.*, personal communication). Since there are approximately five copies of the entire chloroplast genome integrated into each haploid nuclear genome in spinach (Scott and Timmis, 1984) it is possible that the 4.7 kb fragment contains solely *bona fide* "photogene 32" sequences and not sequences which flank this region in S 1. Current data also cannot categorically discount the possibility that the 4.7 kb fragment arose from cpDNA contamination of the nDNA preparations even though the preparations were obtained from etiolated coleoptiles. The 4.7 kb fragment is currently being cloned, mapped and sequenced to clarify both these points.

If the 10.2 kb fragment represents the initial nuclear integration event of a complete copy of S 1, then genomic rearrangements and mutations must have occurred to give rise to the smaller fragments. A hypothetical summary of the postulated (i) excision of S 1 and S 2 from the mitochondrial chromosome, (ii) amplification of S 1 and S 2 as virus-like DNAs, (iii) transfer of S 1 to the nucleus, and (iv) subsequent genomic rearrangements of S 1 sequences is shown in Figure 1. We envisage that these changes occurred long ago in the evolution of maize. An alternative hypothesis is that S 1 sequences present in the mitochondrial chromosomes transferred directly to the nucleus without involving the virus-like conformations.

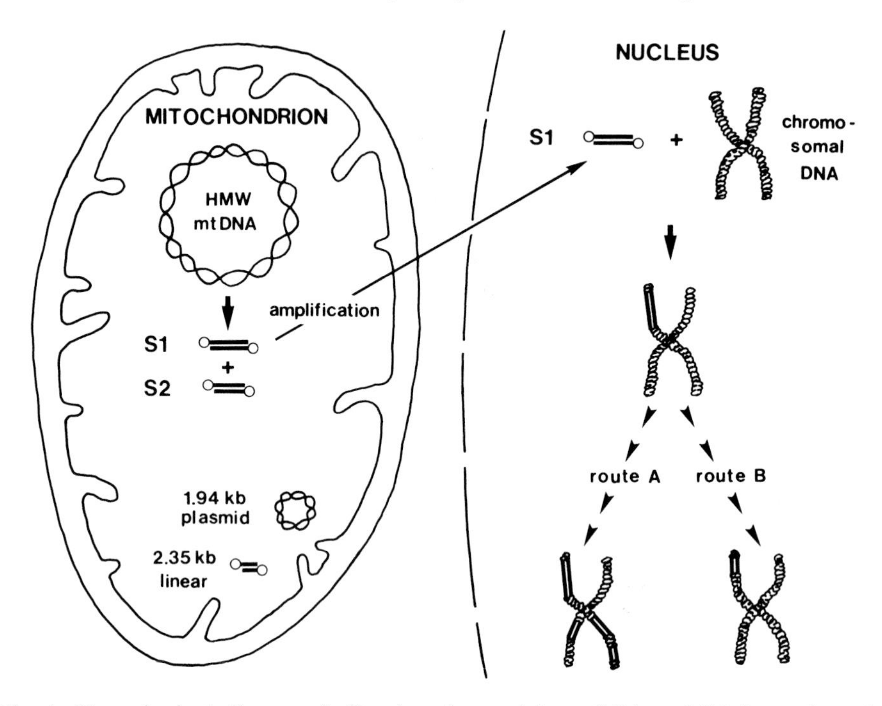

Fig. 1. Hypothetical diagram indicating the excision of S1 and S2 from the mitochondrial chromosome, the transfer of S1 to the nucleus and its integration into nuclear DNA. Although represented on one chromosome, the genomic rearrangements may be dispersed over more than one. For an explantation of routes A and B, see text. HMW refers to the high molecular weight mtDNA or mitochondrial chromosome. The circles at the ends of S1, S2 and the 2.35 kb linear molecule represent 5′ covalently attached proteins. (Not to scale)

If the 4.7 kb fragment contains S1 sequences in addition to the "photogene 32" region then the genomic rearrangements appear to be of two types. As described above, the M825/0h07 nuclear revertants, cytoplasmic revertants and their *cms* and N cytoplasm parental lines exhibited four major nDNA *Bam* HI restriction fragments containing sequences homologous to S1 (Fig. 1 route A). The two inbred lines, W64A and B37, which did not contain *Rf3* or undergo spontaneous reversion to fertility, and the one *Rf3* restorer line analysed, Tr, all exhibited S1 sequence homology to the 4.7 kb *Bam* HI fragment but not to the three larger fragments (Kemble *et al.*, 1983). If, in these three lines, the initial molecular event was also the integration of a complete copy of S1 into nDNA, then different genomic rearrangements must have occurred (Fig. 1, route B). A revision in the model proposed in Figure 1 is necessary if the 4.7 kb fragment does not contain S1 sequences other than those homologous to "photogene 32". In this case, the M825/0h07 lines would continue to follow route A. However, route B would not be relevant because no integration of S1 into nDNA would have occurred in W64A, B37 and Tr inbreds.

An early hypothesis (Laughnan and Gabay, 1973) suggesting that S1 and/or S2 may be similar in structure to the *Rf3* gene is still viable but an amendment is required. Although S1 sequences do not integrate into nDNA concomitant with nuclear reversion to fertility, it is possible that reversion is correlated with rearrangement of S1 sequences already present in the nucleus. A comparison of S1 sequences integrated into nDNA of new restorer lines and their nonrestoring parental lines should clarify this hypothesis.

IV. Concluding Remarks

The ability to genetically engineer plants by inserting individual genes, in a controlled manner, necessitates that suitable DNA vectors, or vehicles, be identified. Although cauliflower mosaic virus and the Ti-plasmid of *Agrobacterium tumefaciens* have undergone intensive study in this light, other candidates for plant transformation vectors should nevertheless be sought. Our laboratories, as well as others (e. g. Howell, 1982), have suggested that because S1 and S2 are naturally-occurring components of maize cells they have a unique potential for use as plant transformation vectors. This potential has obviously greatly increased with the finding of nDNA sequences homologous to S1 (Kemble *et al.,* 1983). It may be possible to transfer foreign DNA from an engineered S1 vector to chromosomal DNA by both transposition and recombination across homologous S1 sequences. Two other maize mtDNAs, the 1.94kb plasmid (Kemble and Bedbrook, 1980; Kemble *et al.,* 1980) and a 1.4kb plasmid, may have similar vector potential because homologous nDNA sequences have also been detected in some nuclear backgrounds (Kemble *et al.,* 1983; A. G. Smith, personal communication).

Although the transfer of mitochondrial S1 sequences to the nucleus is thought to have occurred only once in the evolution of maize it does not rule out the possibility that such interorganellar genetic exchanges may be more frequent. Experiments utilizing additional cloned maize mtDNA fragments could well indicate that sequences other than S1 and the 1.94kb and 1.4kb plasmids are found in both cellular components. Hopefully, the results generated from our maize work will encourage researchers to conduct similar experiments with other plant species. It will be intriguing to know if the transfer of genetic information between mitochondria and nuclei is a general phenomenon in the plant kingdom.

V. References

Challberg, M. D., Desiderio, S. V., Kelly Jr., T. J., 1980: Adenovirus DNA replication *in vitro:* characterization of a protein covalently linked to nascent DNA strands. Proc. Natl. Acad. Sci., U.S.A. **77,** 5105—5109.

Ellis, J., 1982: Promiscuous DNA-chloroplast genes inside plant mitochondria. Nature **299,** 678—679.

Farrelly, F., Butow, R. A., 1983: Rearranged mitochondrial genes in the yeast nuclear genome. Nature **301,** 296—301.

Gellissen, G., Bradfield, J. Y., White, B. N., Wyatt, G. R., 1983: Mitochondrial DNA sequences in the nuclear genome of a locust. Nature **301,** 631—634.

Hadler, H. I., Dimitrijevic, B., Mahalingam, R., 1983: Mitochondrial DNA and nuclear DNA from normal rat liver have a common sequence. Proc. Natl. Acad. Sci., U.S.A. **80,** 6495—6499.

Howell, S. H., 1982: Plant molecular vehicles: potential vectors for introducing foreign DNA into plants. Ann. Rev. Plant Physiol. **33,** 609—650.

Jacobs, H. T., Posakony, J. W., Grula, J. W., Roberts, J. W., Xin, J.-H., Britten, R. J., Davidson, E. H., 1983: Mitochondrial DNA sequences in the nuclear genome of *Strongylocentrotus purpuratus.* J. Mol. Biol. **165,** 609—632.

Kemble, R. J., Bedbrook, J. R., 1980: Low molecular weight circular and linear DNA in mitochondria from normal and male-sterile *Zea mays* cytoplasm. Nature **284,** 565—566.

Kemble, R. J., Gunn, R. E., Flavell, R. B., 1980: Classification of normal and male-sterile cytoplasms in maize. II. Electrophoretic analysis of DNA species in mitochondria. Genetics **95,** 451—458.

Kemble, R. J., Mans, R. J., 1983: Examination of the mitochondrial genome of revertant progeny from S cms maize with cloned S-1 and S-2 hybridization probes. J. Mol. Appl. Genet. **2,** 161—171.

Kemble, R. J., Mans, R. J., Gabay-Laughnan, S., Laughnan, J. R., 1983: Sequences homologous to episomal mitochondrial DNAs in the maize nuclear genome. Nature **304,** 744—747.

Kemble, R. J., Thompson, R. D., 1982: S1 and S2, the linear mitochondrial DNAs present in a male sterile line of maize, possess terminally attached proteins. Nucleic Acids Res. **10,** 8181—8190.

Kim, B. D., Mans, R. J., Conde, M. F., Pring, D. R., Levings III, C. S., 1982: Physical mapping of homologous segments of mitochondrial episomes from S male-sterile maize. Plasmid **7,** 1—14.

Koncz, C., Sumegi, J., Udvardy, A., Racsmany, M., Dudits, D., 1981: Cloning of mtDNA fragments homologous to mitochondrial S2 plasmid-like DNA in maize. Mol. Gen. Genet. **183,** 449—458.

Laughnan, J. R., Gabay, S. J., 1973: Mutations leading to nuclear restoration of fertility in S male-sterile cytoplasm in maize. Theor. Appl. Genet. **43,** 109—116.

Laughnan, J. R., Gabay, S. J., 1975: An episomal basis for instability of S male sterility in maize and some implications for plant breeding. In: Birky Jr., C. W., Perlman, P. S., Byers, T. J. (eds.), Genetics and the Biogenesis of Cell Organelles, pp. 330—349. Columbus: Ohio State Univ. Press.

Laughnan, J. R., Gabay, S. J., 1978: Nuclear and cytoplasmic mutations to fertility in S male-sterile maize. In: Walden, D. B. (ed.), Maize Breeding and Genetics, pp. 427—446. New York: John Wiley.

Laughnan, J. R., Gabay-Laughnan, S., 1983: Cytoplasmic male sterility in maize. Ann. Rev. Genet. **17,** 27—48.

Levings III, C. S., Kim, B. D., Pring, D. R., Conde, M. F., Mans, R. J., Laughnan, J. R., Gabay-Laughnan, S. J., 1980: Cytoplasmic reversion of cms-S in maize: association with a transpositional event. Science **209,** 1021—1023.

Levings, III, C. S., Sederoff, R. R., 1983: Nucleotide sequence of the S-2 mitochondrial DNA from the S cytoplasm of maize. Proc. Natl. Acad. Sci., U.S.A. **80,** 4055—4059.

Lonsdale, D. M., Thompson, R. D., Hodge, T. P., 1981: The integrated forms of the

S1 and S2 DNA elements of maize male sterile mitochondrial DNA are flanked by a large repeated sequence. Nucleic Acids Res. **9,** 3657—3669.
Pring, D. R., Levings III, C. S., Hu, W. W. L., Timothy, D. H., 1977: Unique DNA associated with mitochondria in the "S"-type cytoplasm of male-sterile maize. Proc. Natl. Acad. Sci., U.S.A. **74,** 2904—2908.
Rekosh, D. M. K., Russell, W. C., Bellett, A. J. D., Robinson, A. J., 1977: Identification of a protein linked to the ends of adenovirus DNA. Cell **11,** 283—295.
Scott, N. S., Timmis, J. N., 1984: Homologies between nuclear and plastid DNA in spinach. Theor. Appl. Genet. **67,** 279—288.
Singh, A., Laughnan, J. R., 1972: Instability of S male-sterile cytoplasm in maize. Genetics **71,** 607—620.
Tamanoi, F., Stillman, B. W., 1982: Function of adenovirus terminal protein in the initiation of DNA replication. Proc. Natl. Acad. Sci., U.S.A. **79,** 2221—2225.
Thompson. R. D., Kemble, R. J., Flavell, R. B., 1980: Variations in mitochondrial DNA organization between normal and male-sterile cytoplasms of maize. Nucleic Acids Res. **8,** 1999—2008.
van den Boogaart, P., Samallo, J., Agsteribbe, E., 1982: Similar genes for a mitochondrial ATPase subunit in the nuclear and mitochondrial genomes of *Neurospora crassa.* Nature **298,** 187—189.
Wright, R. M., Cummings, D. J., 1983: Integration of mitochondrial gene sequences within the nuclear genome during senescence in a fungus. Nature **302,** 86—88.

Section III.
Movement of Genetic Information Within Plant Organelles

Chapter 6

Supernumerary DNAs in Plant Mitochondria

R. R. Sederoff and C. S. Levings III

Department of Genetics, North Carolina State University, Raleigh, NC 27695—7614, U.S.A.

Contents

I. Diversity of Genetic Organization in Plant Mitochondrial DNA

Mitochondria DNA (mtDNA) of higher plants is characterized by exceptional diversity. Measurements of mitochondrial genome sizes have shown a bewildering array of variation both in genome size and in the presence of supernumerary DNAs that are smaller than the complete genome. Mitochondrial genomes in higher plants are the largest known and vary over an order of magnitude from about 200 kb to 2400 kb (Table 1). This result is in striking contrast to the conservation of mitochondrial genome size and structure characteristic of animal mitochondrial genomes. Virtually all animal mitochondrial genomes are between 15 kb and 19 kb (Wallace, 1982). Fungi also show great divergence in mitochondrial genome size, ranging from about 17 kb to 108 kb (Clark-Walker and Sriprakash, 1982; Wallace, 1982; Sederoff, 1984). In several protozoa, genome sizes are similar to those of fungi. However, some protozoa show extreme structural variation, particularly in kinetoplast DNA (Borst *et al.*, 1981).

Detailed mapping studies have been done on the mtDNA of maize *(Zea mays* L.*)* (Lonsdale *et al.*, 1983 a, b) and *Brassica campestris* (Palmer and Shields, 1983). The structures of the genomes are complex with multiple chromosomal forms and alternative kinds of organization. In *Brassica,* it has been proposed that there is a tripartite mitochondrial genome consisting of three circular forms. One form consists of a master chromosome of 218 kilobase pairs (kbp). This molecule contains the entire sequence of

Table 1. Genome Sizes in Plant Mitochondria

Species	Common Name	Estimated Size (kb)	Reference
Zea mays	maize	600 MAP	Lonsdale *et al.*, 1983 a, b
Sorghum bicolor	sorghum	300 RES	Pring *et al.*, 1982
Glycine max	soybean	225 RES	Levings *et al.*, 1979 Synenki *et al.*, 1978
Pisum sativum	pea	360 REA	Ward *et al.*, 1981
Vicia faba	broad bean	285 RES	Bendich, 1982
Vicia villosa		375 RES	Bendich, 1982
Triticum aestivum	wheat	210 RES	Quetier and Vedel, 1977
Citrullus vulgaris	watermelon	330 REA, RES	Ward *et al.*, 1981
Cucurbita pepo	zucchini	840 REA, RES	Ward *et al.*, 1981
Cucumis sativus	cucumber	1500 REA	Ward *et al.*, 1981
Cucumis melo	muskmelon	2400 REA	Ward *et al.*, 1981
Nicotiana tabacum	tobacco	390—435 REA	Belliard *et al.*, 1979
Brassica napus		218 MAP	Palmer and Shields, 1984
Solanum tuberosum	potato	135 RES	Quetier and Vedel, 1977
Oenothera berteriana	evening primrose	270—285 RES	Brennicke, 1980
Lactuca sativa	lettuce	210 REA	Wallace, 1982
Parthenocissus tricuspidata	virginia creeper	248 RES	Quetier and Vedel, 1977
Lupinus luteus	lupine	180 RES	Augustyniak *et al.*, 1983
Linum usitatissimum	flax	160 RES	Lockhardt and Levings, unpublished results

Size estimates were determined by different methods. RES refers to estimates made by restriction analysis. MAP refers to estimate obtained by mapping of the genome by overlapping fragment analysis. REA refers to size estimates made by reassociation analysis. Many of the restriction estimates must be considered to be minimum size estimates because many visible bands contain more than one fragment and small fragments may not be detected.

the mitochondrial genome. The master chromosome is composed of unique sequences except for two copies of a 2 kb element present as direct repeats separated asymetrically by 83 kb and 135 kb. Reciprocal recombination between the direct repeats generate two smaller circles of 83 kb and 135 kb, each with a single copy of the repeat. Recombination between the smaller circles reforms the master chromosome. Consequently, the total mtDNA consists of a mixture of molecular forms of several different sizes.

A similar proposal has been made for the larger maize mitochondrial genome (about 600 kb) (Lonsdale *et al.*, 1983 a, b) which also explains the existence of a number of circular molecular forms. A direct repeat of about 3 kb is thought to recombine and give rise to smaller circles of 250 kb and 350 kb. Recombination between other repeated elements would give rise to a variety of large and small circular molecules.

Many studies have shown that plant mtDNAs contain a surprising diversity of supernumerary DNA molecules that are not complete genomes. These mtDNA molecules are remarkably diverse in size and structure. Detailed studies of these molecules have been made in 10 different higher plant species (Table 2). Within a given species many diverse molecular species can be found. The most detailed studies have been made in maize.

Table 2. Plasmid-like DNAs and Supernumerary DNA Molecules in Higher Plant Mitochondria

Species	Length (kb)	Structure	Reference
Zea mays			
N, *cms*-C, T, S	1.9	minicircle	Kemble and Bedbrook, 1980
cms-C	1.5	minicircle	Kemble and Bedbrook, 1980
cms-C	1.4	minicircle	Kemble and Bedbrook, 1980
Zea mays (Cell culture)	1.5	oligomeric series	Dale, 1982
	1.8	oligomeric series	Dale, 1982
Zea mays cms-S	6.397	linear (S-1)	Paillard *et al.,*
	5.453	linear (S-2)	Levings *et al.,* 1983
	2.3	linear (S-3)	Koncz *et al.,* 1981
Zea mays type RU	7.5	linear (R-1)	Weissinger *et al.,* 1982, 1983
	5.4	linear (R-2)	Weissinger *et al.,* 1982, 1983
Zea mays N	67	circle (48 %)	Levings *et al.,* 1979
	49	circle (20 %)	Levings *et al.,* 1979
	99	circle (14 %)	Levings *et al.,* 1979
	13	circle (7.6 %)	Levings *et al.,* 1979
	23	circle (5.8 %)	Levings *et al.,* 1979
	136	circle (3.8 %)	Levings *et al.,* 1979
	30	circle (1 %)	Levings *et al.,* 1979
Zea mays cms-T	82	circle (47 %)	Levings *et al.,* 1979
	56	circle (37 %)	Levings *et al.,* 1979
Zea mays cms-S	54	circle (40 %)	Levings *et al.,* 1979
	117	circle (32 %)	Levings *et al.,* 1979
Zea diploperennis	7.5	linear (D-1)	Timothy *et al.,* 1983
	5.4	linear (D-2)	Timothy *et al.,* 1983
Sorghum bicolor (cms)	5.7	linear (N-1)	Dixon and Leaver, 1982 Pring *et al.,* 1982
	5.3	linear (N-2)	Dixon and Leaver, 1982 Pring *et al.,* 1982
Brassica campestris	11.3	linear	Palmer *et al.,* 1983
Glycine max	20	circle	Synenki *et al.,* 1978
	33	circle	Synenki *et al.,* 1978
	42	circle	Synenki *et al.,* 1978
	54	circle (most abundant)	Synenki *et al.,* 1978
	68	circle	Synenki *et al.,* 1978
	81	circle	Synenki *et al.,* 1978
	99	circle	Synenki *et al.,* 1978
Beta vulgaris	1.3	circle	Powling, 1981
	1.4	circle	Powling, 1981
	1.45	circle	Powling, 1981
	1.5	circle	Powling, 1981
	7.3	circle	Powling, 1981
Nicotiana (cell culture)	10.1	oligomeric series	Dale, 1982
	28.8	oligomeric series	Dale, 1982
Phaseolus vulgaris (cell culture)	1.9	oligomeric series	Dale, 1982

Species	Length (kb)	Structure	Reference
Oenothera berteriana	6.3	circle	Brennicke and Blanz, 1982
	7.0	circle	Brennicke and Blanz, 1982
	8.2	circle	Brennicke and Blanz, 1982
	9.9	circle	Brennicke and Blanz, 1982
	13.5	circle	Brennicke and Blanz, 1982
Vicia faba	1.4	circle	Goblet *et al.*, 1983
	1.7	circle	Goblet *et al.*, 1983
(cms 350)	1.5	circle	Goblet *et al.*, 1983
	1.6	circle	Nikiforova and Negruk, 1983

Numbers in parentheses refer to the numerical relative abundance, for the molecules listed.

The earliest reports of molecular heterogeneity came from studies of closed circular DNA molecules observed in the electron microscope (Quetier and Vedel, 1977; Levings and Pring, 1978; Levings *et al.*, 1979). When DNA is purified from normal maize mitochondria, and circular molecules are recovered from CsCl gradients and examined in the electron microscope, a diversity of molecules is observed (Levings *et al.*, 1979). At least six different size classes of molecules were observed which differ in size and relative abundance (Table 2). The most abundant circular-class (48 %) was 68 kb, and the largest circular-class (136 kb) was present in low abundance (4 %). The distribution and relative abundance of the molecules did not fit that expected for a multimeric series. DNA purified from mitochondria of male-sterile cytoplasms, *cms-S* and *cms-T*, also showed heterogeneity of circular DNAs but the sizes and relative abundance of the circular molecules were different from that of normal mtDNA and from each other. It has been suggested that these circular molecules are derived from larger molecules by intramolecular recombination (Levings *et al.*, 1979). The diversity of sizes may reflect the distribution of recombination sites (repeated sequence elements) in the genome and variation in the number and location of such elements in the mtDNA of the male-sterile cytoplasms.

Small circular molecules, called minicircles, have been observed also (Levings *et al.*, 1979; Kemble and Bedbrook, 1980). These molecules are present as discrete size classes and are sufficiently abundant to be visible on agarose gels containing unrestricted DNA. Minicircles are also observed by electron microscopy of normal and male-sterile mtDNAs. A minicircle of 1.9 kb was observed in normal *(N)*, *cms-S*, *cms-T*, and *cms-C*. Smaller minicircles of 1.4 kb and 1.5 kb were found in *cms-C* (Kemble and Bedbrook, 1980). In mtDNA from cultured cells (Sparks and Dale, 1980; Dale, 1981; Dale, 1982) enriched for circular molecules, the DNA molecules followed the pattern of an oligomeric series. One molecular class of 1.5 kb showed a series of molecules with abundant monomers and decreasing abundance for larger multimers. Another group of circles was observed following a similar series with a monomer of 1.8 kb (Dale, 1982).

Discrete classes of large circular molecules have been described in soybean (Synenki *et al.*, 1978) ranging from 19 to 100 kb. In cucumber, wheat, potato and Virginia creeper (Quetier and Vedel, 1977) heterogeneity of large circular molecules was reported. A recent report on mtDNA from four species of Citrus (Fontarnau and Hernandez-Yago, 1982), demonstrated dispersion of large circular DNAs, but no discrete size classes. In cell cultures of bean and tobacco, a high percentage (20 to 40 %) of the mtDNA was isolated as supercoiled circles (Sparks and Dale, 1980). The molecules are heterogeneous, and contain some oligomeric series. In bean, the predominant molecule is 1.9 kb and is likely to be the monomeric circle of an abundant series of multimers (Dale, 1982). Tobacco mtDNA contains circular DNAs that are considerably larger. Circular molecules of 10.1, 20.2, and 28.8 kb were found (Dale, 1982). The restriction patterns of these larger molecules showed that the 10.1 and 20.2 molecules had a monomeric-dimeric relationship, but the 28.8 kb class of molecules contained additional restriction fragments not found in the 10.1 kb molecule. Both the small and large circular molecules from cultured cells hybridized to mtRNA (Dale, 1982) suggesting that the circles contain sequences (perhaps in part) of functional mitochondrial genes.

Minicircular DNAs have been found in the mtDNAs of *Vicia faba* and in some other legumes (Nikiforova and Negruk, 1983). A molecule of 1.6 kb was observed in all three tested cultivars. In *Vicia villosa,* a different pattern of minicircle sizes was observed. Similarly, *Medicago sativa,* showed a unique distribution of extrachromosomal mitochondrial molecules (Nikiforova and Negruk, 1983). A different study, which characterized small circular DNAs by electron microscopy, revealed plasmid-like DNAs in *Vicia faba* of 1.7 and 1.4 kb (Goblet *et al.*, 1983). A male-sterile cytoplasm was distinguished by the presence of an additional supercoil of 1.5 kb (Goblet *et al.*, 1983).

The mitochondrial genomes of sugarbeet *(Beta vulgaris)* possess a number of small supercoiled DNA molecules (Powling, 1981; Powling and Ellis, 1983). Several sizes of circles were detected ranging from 1.3 kb to 7.3 kb in size. Four different sizes of minicircles were distinguished between 1.3 kb and 1.5 kb in male-fertile lines in diverse combinations of two or three types while the male-sterile line examined contained only the 1.5 kb molecule. Minicircular DNAs have been cloned and tested by hybridization for homology with each other and with the mitochondrial chromosomal DNA (Powling and Ellis, 1983). Little if any homology was detected with the chromosomal DNA. Most minicircles do not cross hybridize with each other except for the 1.5 kb and the 1.45 kb molecules which have some sequences in common. These small DNA species also hybridize to molecules that appear to be members of the same oligomeric series.

A number of closed circular mtDNAs have been isolated from *Oenothera,* and characterized by electrophoresis and molecular hybridization (Brennicke and Blanz, 1982). The smallest five classes of circular molecules were analysed by restriction enzyme digestion. The linear sizes of these molecules were (in kb) 6.3, 7.0, 8.2, 9.9, and 13.5. Very little, if any, homology between the different circular molecules was detected by hybridization or by comparison of restriction sites.

The most intriguing supernumerary mtDNA molecules are the linear molecules first discovered in the cytoplasmic male-sterile cytoplasm *cms-S* (Pring *et al.,* 1977). These molecules are sufficiently abundant to be observed as discrete bands on agarose gels containing unrestricted mtDNA. Two major plasmid-like molecules are observed that are 6397 bp (S-1) and 5453 bp (S-2) long (Levings and Sederoff, 1983; Paillard *et al.,* 1985). A smaller linear molecule, about 2.3 kb long, called S-3, is found in the S type male-sterile cytoplasm. This molecule has not been sequenced, but restriction analysis indicates that it is related to S-2, and may have arisen from S-2 by a large internal deletion (Koncz *et al.,* 1981). Free plasmid-like linear DNAs have not been demonstrated in the nucleus or the chloroplasts in visible amounts. Plasmid-like molecules are maternally inherited as expected for a mitochondrially associated element (Conde *et al.,* 1979).

In a survey of 93 races of Latin American maize (Weissinger *et al.,* 1982; Weissinger *et al.,* 1983), plasmid-like elements were discovered in 18 races. One of these, Conico Norteno, contained molecules that are similar in size and abundance to S-1 and S-2. In the 17 other races, similar pairs of elements were found. One of these, R-2 is nearly indistinguishable from S-2 but a companion molecule R-1, is present that is larger than S-1 although similar in abundance. R-1 is about 1000 nucleotides longer than S-1 and contains a segment of about 2600 bp that is unique to R-1 (Levings *et al.,* 1983).

The teosintes, studied because of their close relationship to maize, also contain plasmid-like DNAs. Of 31 accessions thus far examined, one species, a diploid perennial *(Zea diploperennis)* contains molecules which resemble R-1 and R-2 (Timothy *et al.,* 1983). These molecules are designated D-1 and D-2. In a more distant relative, *Sorghum bicolor,* linear duplex molecules ~5.7 kb and ~5.3 kb have been found (Pring *et al.,* 1982; Dixon and Leaver, 1982). These molecules are found in the mitochondria from a male-sterile cytoplasm (IS 1112 C). Hybridization experiments indicate that the plasmid-like DNAs of sorghum are quite different from those of maize and teosinte. Interestingly, all of the races of South American maize that contain R-1 and R-2, and the teosintes containing D-1 and D-2, are male fertile. This result raises a question about the nature of the relationship between these plasmid-like DNAs and male sterility. In the *cms-S* mitochondria, there is a striking correlation between the presence of plasmid-like DNAs and male sterility (Levings *et al.,* 1980; Laughnan *et al.,* 1981). Revertants to male fertility almost always lose or greatly reduce the amounts of the plasmid-like DNAs. It is well established that nuclear genes (restorers) are able to suppress the phenotype of male sterility. Therefore, fertile races of maize or teosinte that carry plasmid-like DNAs may also contain restorer genes.

An extrachromosomal linear plasmid-like species of DNA has also been found in the mitochondria of *Brassica campestris* and *Brassica napus*

(Palmer *et al.,* 1983). This molecule is 11.3 kb long and is associated with cytoplasmic male sterility in some lines of *B. napus.* This plasmid-like DNA shows no homology to the mitochondrial chromosomal DNA of *Brassica* or to the S-1 and S-2 plasmid-like DNAs from maize. Of the 20 accessions of *Brassica* and *Raphanus* examined, less than half contained the plasmid-like DNA. The amounts of this linear molecule were highly variable, from 1/10 to 10 times that of the main mitochondrial genome (Palmer *et al.,* 1983).

II. Structure of Plasmid-like DNAs

The S-1 and S-2 DNAs of the maize cytoplasm *cms-S* have been examined more thoroughly than any of the other plasmid-like DNAs. These two are the only plasmid-like DNAs where the complete nucleotide sequences have been determined. S-1 is 6397 nucleotides long while S-2 is 5453 (Levings and Sederoff, 1983; Paillard *et al.,* 1985). The molar G+C contents of S-1 and S-2 are 39.4 and 37.5 % respectively; this is lower than that of the maize mtDNA, 47 % (Shah and Levings, 1974) and may suggest an exogenous origin for these DNA species.

S-1 and S-2 contain common sequences that are highly conserved, perhaps to maintain essential functions. For example, both molecules are terminated by exact 208 base pair inverted repeats that may be important in their replication. This possibility, which will be considered later, is suggested by investigations of their DNA termini and by analogy with adenovirus and *Bacillus* phage DNAs. A second large region of sequence homology begins at one end of the S-1 and S-2 molecules and runs for 1462 bp, including the 208 bp terminal inverted repeats. The 1462-base-pair homology is perfect except for two nucleotide differences. These minor differences may not be functionally significant because they do not occur in open-reading frames. The explanation for the 1462 bp of homology may lie in the origin of S-1 molecule. Previously, we proposed that S-1 arose by a DNA recombinational event between R-1 and R-2 (Levings *et al.,* 1983). This speculation is consistent with results obtained from restriction mapping, hybridization and nucleotide sequencing studies. The proposed recombinant molecule would contain sequences common to R-1 and R-2; this is in agreement with the structure of S-1 and nicely accounts for the 1462 base pairs of homology between S-1 and S-2. It was noted earlier that R-2 and S-2 are indistinguishable except for a difference in a single Bgl I site that is missing in S-2 (P. Sisco, unpublished).

S-1, S-2 and S-3 DNA molecules have proteins covalently attached to their 5′ termini that may be involved in priming DNA replication (Kemble and Thompson, 1982). Recent investigations have shown that the R-1 and R-2 plasmid-like DNAs have similar proteins linked to their ends (Meints *et al.,* in preparation). It is not known if the linked proteins are identical among the various DNA species since they have not been characterized with regard to attachment sites or amino acid constitution.

Among adenoviruses (Carusi, 1977; Rekosh *et al.*, 1977) and *Bacillus* phages (Salas *et al.*, 1978; Harding *et al.*, 1978; Ito, 1978; Yehle, 1978; Yoshikawa and Ito, 1981) similar DNA-protein associations have been demonstrated in which the DNA terminal proteins are thought to play a role in DNA replication. These viral DNAs initiate replication at or close to either termini and proceed by a mechanism called strand displacement (Lechner and Kelly, 1977; Ariga and Shimojo, 1977; Sussenbach and Kuijk, 1977; Inciarte *et al.*, 1980; Harding and Ito, 1980). It is believed that the protein covalently linked to the 5′ ends of the linear DNA strand serves as a primer for DNA synthesis in adenovirus and *Bacillus* phage (Rekosh *et al.*, 1977; Lechner and Kelly, 1977; Inciarte *et al.*, 1980; Harding and Ito, 1980; Winnacker, 1978). These viral DNAs are also structurally similar to the plasmid-like DNAs in that the linear molecules are terminated by inverted repeats. The *Bacillus* phage DNAs contain short terminal inverted repeats, which are 6 to 8 nucleotides long and the adenovirus DNAs have terminal inverted repeats, which are approximately 100 nucleotides in length. The 208 bp inverted repeats of S-1 and S-2 have a high degree of terminal homology with the inverted repeats of the *Bacillus* phages but not with adenoviruses (Levings and Sederoff, 1983). The terminal sequences of the *Bacillus* phages, adenovirus, S-1 and S-2 DNAs are rich in A-T pairs. This is of interest because origins of replication are rich in A+T regions presumably because local melting of DNA is required (Messer *et al.*, 1978; Moore *et al.*, 1978; Hobom *et al.*, 1978).

Electron microscopy studies have provided additional insight into the organization and replication of the DNA-protein complexes, S-1, S-2, R-1 and R-2 (Meints *et al.*, in preparation). When these complexes are digested with proteases, they are visualized as linear DNA molecules. However, if the molecules are not protease digested, the DNA-protein complexes remain intact and they are observed as circular molecules. Presumably the 5′ linked proteins bind to each other to generate the circular configuration. About 1% of the molecules examined appear to be replicative intermediates. Seemingly, the intermediates are replicating by the same strand displacement mechanism responsible for DNA replication in adenovirus and *Bacillus* phage. In the S and R plasmids, replication appears to initiate at or near either or both DNA termini.

Normally, the S-1 and S-2 elements are present in equimolar amounts and are about fivefold more abundant than other mtDNAs. However, in certain nuclear backgrounds differences are observed in the relative quantities of S-1 and S-2 DNAs (Laughnan and Gabay-Laughnan, 1983). For example, the S-1 plasmid is fivefold more prevalent than S-2 in the inbred line M825. In contrast, the situation is reversed in the inbred 38-11 where the amount of S-2 exceeds that of S-1 by at least threefold. Elegant backcrossing studies have provided further evidence that S-1 and S-2 content is influenced by nuclear background. It is not known how nuclear gene(s) are able to differentially control plasmid abundance. Although it is tempting to suggest differential replication as the mechanism, it is also possible that differences in degradation may explain the distinctions.

Even though functional genes have not yet been assigned to the S-1 and S-2 DNA molecules, the occurrence of open-reading frames strongly suggests the possibility of protein-encoding genes. S-2 DNA contains two large open-reading frames, which were identified by computer analysis using the universal code (Levings and Sederoff, 1983). These unidentified reading frames are 3513 and 1017 nucleotides long and occur on different DNA strands. Similar analysis has identified three open-reading frames of 2787, 1017 and 768 nucleotides on the S-1 molecule (Paillard *et al.*, 1985). The 1017 nucleotide-long reading frames of S-1 and S-2 are identical and occur in the 1462 bp homologous region of the two molecules. Preliminary studies have identified mitochondrial transcripts that map to the S-1 and S-2 DNAs (Traynor and Levings, unpublished). These transcripts coincide with open-reading frames with regard to strand and approximate length.

In vitro protein synthesis of maize mitochondria suggests that polypeptides are encoded by the S-1 and S-2 DNA species (Forde and Leaver, 1980). When translational products of mitochondria from *cms-S* and normal cytoplasms of maize were compared, eight additional polypeptides were detected in *cms-S* (Forde and Leaver, 1980). Seven of the additional polypeptides were of higher molecular weight than any of the translational products of *N, cms-T* and *cms-C*. The eight extra polypeptides of *cms-S* were unaffected by the presence of genes which restore male fertility to plants with S cytoplasm. It is possible that the additional polypeptides associated with the mitochondria *cms-S* are gene products of the S-1 and S-2 DNA species.

Numerous investigations have demonstrated that the mtDNAs of maize have integrated sequences which share homology with the R and S plasmids (Spruill *et al.*, 1980, 1981; Lonsdale *et al.*, 1981; Levings *et al.*, 1983; Lonsdale *et al.*, 1983 a, b; McNay *et al.*, 1984). For instance, mtDNA from normal maize cytoplasm contains two regions with extensive plasmid homology that are located adjacent to the three kb direct repeats. One region includes about 91 % of the R-1 sequence, while the second contains 94 % of R-2 or S-2. In both cases, it is the terminal sequences that are missing.

The amount of integrated plasmid-like sequence is variable among the mtDNAs of the male-sterile cytoplasm. The mtDNAs of *cms-C* and *cms-T* contain only small amounts of homology with the plasmid-like DNA. In sharp contrast, the *cms-S* mtDNA possesses large amounts of S-homologous sequence. The origin of the integrated sequences remains a mystery.

Interestingly, the S-1 DNA species contains a segment of chloroplast DNA. Sequence analysis has shown that about the latter one-third of the photogene 32 (*psb*A) is present in S-1. The *psb*A gene codes for the Mr 32,000 thylakoid membrane protein of chloroplasts. Two regions of 90 % nucleotide homology have been located between S-1 at positions 3711—3962 and 3599—3611 and the spinach gene, *psb*A. The S-1 sequences with homology to the *psb*A gene are also found in R-1 DNA and in the mtDNA of several maize cytoplasms. The occurrence of chloroplast

DNA sequences in the mtDNA of higher plants is well documented (Stern and Lonsdale, 1982; Stern and Palmer, 1984).

Previous sections have described and characterized the plasmid-like DNAs of the mitochondria. It is clear that these DNA species are distinct from the other mtDNAs. The major distinction is that the plasmid-like DNAs are packaged as DNA-protein complexes. More specifically, they are linear DNAs with protein attached at their 5′ termini that appear to replicate their DNAs by a mechanism peculiar to certain viruses. Although very little is known about mtDNA replication in higher plants, it is evident that the circular mtDNAs of plants must replicate by mechanisms more conventional to other mitochondrial genomes.

III. Reversion to Fertility in *cms-S*

In contrast to the other male-sterile cytoplasms of maize, *cms-S* is unstable and spontaneously mutates to the male-fertile phenotype (Laughnan and Gabay, 1975a, b; Laughnan and Gabay-Laughnan, 1983). Genetic analysis has revealed two distinct kinds of changes, reversions due to alterations of cytoplasmic factors and reversions due to newly arisen nuclear restorer genes. The overwhelming majority of male-fertile revertants are due to cytoplasmic changes, which are maternally inherited through subsequent generations. The cytoplasmic reversion phenomenon is influenced by nuclear gene(s). This is evident by the fact that in some nuclear backgrounds the reversion frequency is high (10%) but in others it does not occur. Laughnan and Gabay (1975b) proposed a male-fertility element with episomal characteristics to explain the spontaneous revertants.

Additional studies of cytoplasmic revertants showed that reversion from the male-sterile to the male-fertile plant type is accompanied by the disappearance of the free forms of S-1 and S-2 DNA species. Reversion is also associated with rearrangements in the mtDNA that most often involve sequences homologous with the S elements. These investigations were carried out with seven independent revertants of *cms-Vg* in the M825 nuclear background (*cms-Vg* is a member of the S group of male-sterile cytoplasms). The loss of plasmid-like DNA is not always complete since in one revertant trace amounts of S-1 and S-2 were detected by hybridization techniques. In any event, these findings led to the speculation that S-1 and S-2 may carry factors responsible for male fertility and behave like transposable elements. This view was supported by the variable nature of the rearrangements since five of the seven independent revertants examined were uniquely different. In addition, the S elements carry terminal inverted repeats, which are a common feature of transposable elements. Kemble and Mans (1983) re-examined the cytoplasmic reversion events and also suggested that S-2 sequences are transposed coincident with the reversion to male fertility.

To more closely examine the reversion phenomenon, we cloned and partially sequenced a fragment containing a rearrangement of mtDNA from a revertant, 369 (C. Braun *et al.*, unpublished). Nucleotide sequencing and restriction mapping indicated that most of the S-2 sequence is present in the fragment. Only the final 202 bp of the terminal inverted repeat, usually found in S-2, are absent. Because transpositional events often cause short duplication of host DNAs adjacent to the insertional site, we also examined the mtDNA sequences next to the S-2 sequence. A two nucleotide duplication was found, which is smaller than customarily encountered with transposition and has a high probability of occurring by chance. Since analysis of the border regions revealed a nucleotide sequence pattern that is atypical of transposition, it seems unlikely that the rearrangement arose by a transposition mechanism. Although this investigation was limited to a single cloned fragment, it casts doubt on the speculation that transpositional events are responsible for the mtDNA rearrangements associated with reversion in the S cytoplasms.

Even though specific genes have not been identified on the maize plasmids, there is ample evidence indicating that these DNA species do, in fact, carry genetic information. Similarities with certain viruses, adenovirus and *Bacillus* phages, suggest they may contain genes required for their replication. These viruses carry genes that participate in their DNA replication such as a polymerase, a DNA binding protein and the protein covalently linked to the 5′ termini. It seems appropriate that genes with corresponding functions should occur in the maize elements. Moreover, there is evidence indicating that the information encoded by maize plasmids is not essential to normal mitochondrial function (P. Sisco, unpublished). Hybridization studies have shown that S-2 sequences are totally absent from the mitochondria of certain maize cytoplasms, e. g. one of the *cms-S* revertants. Since these cytoplasmic types produce normal maize plants, it is doubtful that the S-2 sequences encode gene products essential to the maize mitochondria. These findings support the view that the plasmid encoded genes play a role in their own persistence. Finally, it is well established that S-1 and S-2 elements are associated with male sterility and instability; perhaps they also contain factors influencing these traits.

IV. A. The Diversity Paradox for Maize Mitochondrial DNA

Restriction fragment patterns of maize mtDNAs show an exceptional amount of diversity in comparisons of normal mtDNA and the male-sterile cytoplasms (Levings and Pring, 1976; Pring and Levings, 1978; Borck and Walbot, 1982). Visual comparisons of the patterns reveal that about one third of the fragments vary in pairwise comparisons between normal and male-sterile cytoplasms. More rigorous comparisons using cloned probes and molecular hybridization demonstrate that this view is correct (Spruill *et al.*, 1981). The high level of diversity suggests a rapid rate of evolution for mtDNA in maize, in contrast with chloroplast genomes that have much less divergence (Pring and Levings, 1978; Timothy *et al.*, 1979). A paradox arises when the maize genotypes are compared with those of the related species, the teosintes. Some of the teosintes, such as Central Plateau, a Mexican teosinte, have restriction patterns that are very close to those of normal mtDNA (Timothy *et al.,* 1979). Why a distinct group of teosintes should be so similar in this way to normal maize while different cytoplasmic genotypes within the species show greater diversity is paradoxical.

There are several possible explanations for this situation. It is well established that male-sterile cytoplasms can be generated from wide crosses within a species or from crosses between closely related species. For example, crosses between *Zea perennis* (a tetraploid perennial teosinte) and normal maize followed by repeated backcrosses, results in the transfer of the *Zea perennis* cytoplasm into the maize nuclear background (Laughnan and Gabay-Laughnan, 1983; J. Kermicle, personal communication). Since organelles are not transmitted through pollen (Conde *et al.,* 1979), no exchange would occur between different mitochondrial genotypes. These plants are cytoplasmic male steriles in a nonrestoring nuclear background. Therefore, the male-sterile cytoplasms may be derived from extinct or undiscovered species of *Zea.* None of the extant teosinte species have restriction patterns like those of the C, S, or T types of male-sterile cytoplasms.

Alternatively, the male-sterile genotypes could arise as "catastrophic" events that produce widespread rearrangement of the mitochondrial genome. Such events could occur spontaneously perhaps in response to genetic stress. It has been suggested that transposable elements in the nucleus of plant cells respond to stress, and that the consequences vary depending on the nature of the challenge (McClintock, 1978; Freeling, 1984). Therefore, it is of interest to explore the possible role of the plasmid-like DNAs in the evolution and divergence of the maize mitochondrial genome. While the origin of the different cytoplasmic genotypes and the diversity of supernumerary molecules cannot yet be explained, we have begun to learn about the evolutionary mechanisms that occur in maize mitochondria and to place limits on the kinds and extent of the processes that occur.

IV. B. Evolutionary Mechanisms in Maize mtDNA

Evolution of mtDNA may take place by structural variation and/or point mutation. Structural variation includes rearrangements and gross variations that occur within and between DNA molecules. Point mutations are substitution mutations and length mutations that occur at a single site. The distinction between a small length mutation and an insertion-deletion event that would be considered structural is, of course, arbitrary. Evidence indicates that the amount of point mutation that occurs in plant mitochondria is not unusually high; in fact it would appear low compared to animal mitochondria. Recent sequence information for the ribosomal genes of maize (Chao *et al.,* 1984) and wheat (Spencer *et al.,* 1984) indicates a relatively low rate of point mutation. The estimated base substitution rate for the plant mitochondrial small subunit ribosomal RNA genes using wheat and maize for comparison, is 0.17 substitutions per nucleotide per billion years (Chao *et al.,* 1984). The equivalent calculation for animal mitochondrial ribosomal genes, comparing rat, mouse, human, and bovine sequences estimates 2.6 to 2.9 substitution/base/billion years. In animal mitochondria, substitution mutations account for the major fraction of the restriction fragment diversity (Brown *et al.,* 1982).

While substitution mutations are higher for animal mtDNA than for plants, deletion-insertion events (length mutations) show less difference. Based on wheat and maize small subunit ribosomal RNA genes, the rate of length mutation is 0.37 events per nucleotide per billion years; the equivalent values for mammals range between 0.48 and 0.67.

Since the available evidence suggests a low point mutation rate for plant mitochondria, restriction fragment variation between normal maize mtDNA and that of the male steriles and the teosintes may be due to significant amounts of structural rearrangement. Lonsdale *et al.* (1983 a, b) using cosmid mapping data for normal and male-sterile cytoplasms, has observed several specific sites of rearrangement. Earlier work had suggested that rearrangements were frequent in the evolution of mtDNA in maize and teosinte (Spruill *et al.,* 1981; Sederoff *et al.,* 1981). On the basis of these data, it is reasonable to propose that the observed restriction fragment diversity which occurs in maize and the teosintes, is due largely to rearrangements rather than point mutations.

At least some rearrangement events are due to relatively recent transpositional events that have inserted chloroplast genes into the mtDNA of higher plants (Stern and Palmer, 1984). In maize, a 12 kb segment of the chloroplast genome containing a segment of the inverted repeat has been identified in mtDNA (Stern and Lonsdale, 1982). The restriction patterns of chloroplast DNAs in maize and the teosintes are relatively conserved, suggesting that transfer and exchange of DNA between chloroplast and mitochondria may not be reciprocal. It is interesting to note that the genome sizes of chloroplast DNAs are highly conserved compared to mitochondrial genomes in higher plants (Wallace, 1982).

In view of the extensive exchange that has taken place between mitochondrial and chloroplast genomes, we have inquired whether or not exchange has taken place between specific genes. If, for example, exchange has taken place between the ribosomal RNA genes of the chloroplast and mitochondria, the evolution of these genes would not be independent. This possibility can, in part, be tested by comparison of maize mitochondrial sequences with chloroplast genes from maize and with the corresponding genes from other chloroplasts, tobacco, Chlamydomonas, and Euglena. The results suggest that the evolution of these genes has been independent, because the maize mitochondrial gene is equally divergent from the maize chloroplast gene and the chloroplast genes of the other species.

The mapping studies of the mitochondrial genomes of maize (Lonsdale *et al.,* 1983 a, b) and *Brassica* (Palmer and Shields, 1984) provide a reasonable model to explain molecular heterogeneity of plant mitochondrial

DNA. Recombination at repeated sequences could generate multiple size classes of circular molecules. This model for molecular heterogeneity resembles the mechanisms thought to generate supernumerary molecules from the mitochondrial genomes of several fungi (Wright *et al.*, 1982; Kuntzel *et al.*, 1982). Distributions of circular molecules found in petite strains of yeast are striking in their similarity to the distribution of molecules found in mtDNA of higher plants. Although some petite mtDNAs consist of a simple oligomeric series, the pattern in plant cells appears to be more like petite populations that have arisen from multiple events. Therefore, recombination or rearrangement are plausible mechanisms for the origin of supernumerary circular molecules in the mitochondria. Alternatively, a DNA molecule from the nucleus or the chloroplast might enter a mitochondrion and become an autonomously replicating component of the mitochondrial genome. At present, the requirements for replication of the molecules in the mitochondrial genome are not known.

The linear plasmid-like DNAs appear so different in structure and in their mode of replication that alternative origins must be considered. The structure of these molecules strongly resembles that of known viruses suggesting an exogenous origin of these elements. The relationship of the plasmid-like DNAs with male sterility as well as the presence of integrated forms associated with normal and revertant genotypes remains to be explained.

If plasmid-like DNAs are not derived from the ancestral mitochondrial genome, then the integrated copies of plasmid-like DNA represent relatively recent events. In normal mtDNA the integrated forms of plasmid-like DNAs, R-1 and R-2, are not found as intact copies (Levings *et al.*, 1983). The integrated form of R-1 is missing 0.7 kb. One of the terminal inverted repeats and an adjacent segment of 0.5 kb has been deleted. Similarly, the terminal inverted repeat and an adjacent 0.1 kb is missing from the integrated form of R-2. This result suggests that the mechanism of insertion may not involve the integration of complete copies of these elements. Alternatively, the interpretation of these putative integration events could be confounded by additional recombinational events that have involved sequences near the boundaries of the inverted repeats. Detailed comparisons of the structure of integrated copies of the plasmid-like DNAs in normal, sterile and revertant lines of maize should help resolve these questions.

Recently, plasmid-like DNAs were implicated in rearrangements with DNA sequences in the nucleus and with chloroplast DNA sequences. When sequences cloned from S-1 were hybridized to nuclear DNA, four bands of hybridization were detected (Kemble *et al.*, 1983). One fragment of 10.2 kb may contain a complete copy of S-1, while two other fragments are missing the homology region. The fourth, a 4.7 kb fragment, contains only the central region of the S-1 molecule. The central region of S-1 contains a segment of a chloroplast gene for a thylakoid membrane protein (Ronald *et al.*, submitted for publication). The fourth band is likely to be due to trace amounts of chloroplast DNA contamination in the nuclear

DNA preparations and consequent hybridization with the homologous segment in S-1. No hybridization to nuclear DNA was detected when sequences unique to S-2 were used as probes.

In order to explain the properties of cytoplasmic and nuclear reversion to fertility in *cms-S,* Laughnan and Gabay (1973; 1975 a, b) postulated the existence of a male-fertility element and suggested that it might integrate into the nucleus and function as a nuclear restorer. The fertility element hypothesis was strengthened by the discovery of the plasmid-like DNAs in the *cms-S* cytoplasm (Pring *et al.,* 1977) and by evidence that rearrangement of these DNAs was associated with cytoplasmic reversion (Levings *et al.,* 1980). The fertility element hypothesis predicts that S-1 and/or S-2 sequences would be transposed into the nuclear DNA of nuclear revertants. While nuclear sequences with homology for S-1 were found, it appears that the same fragments are associated with all revertants and parental lines examined, indicating that these DNA fragments were not due to recent transpositional events associated with reversion (Kemble *et al.,* 1983).

V. References

Ariga, H., and Shimojo, H., 1977: Initiation and termination sites of adenovirus 2 DNA replication. Virology **78,** 415—424.

Augustyniak, H., Borsuk, P., Hirschler, I., Stepien, P. P., and Bartnik, E., 1983: Mitochondrial DNA from lupine: restriction analysis and cloning of fragments coding for tRNA. Gene **22,** 69—74.

Belliard, G., Vedel, F., and Pelletier, G., 1979: Mitochondrial recombination in cytoplasmic hybrids of *Nicotiana tabacum* by protoplast fusion. Nature (London) **281,** 401—403.

Bendich, A. J., 1982: Plant mitochondrial DNA: The last frontier. In: Slonimski, P., Borst, P., and Attardi, G. (eds.), Mitochondrial Genes, pp. 477—481. Cold Spring Harbor Laboratory, Cold Spring Harbor, N. Y.

Borck, K. S., and Walbot, V., 1982: Comparison of the restriction endonuclease digestion patterns of mitochondrial DNA from normal and male-sterile cytoplasms of *Zea Mays* L. Genetics **102,** 109—128.

Borst, P., Hoeijmakers, J. H. J., and Hajduk, S. L., 1981: Structure, function and evolution of kinetoplast DNA. Parasitology **82,** 81—93.

Braun, C. J., Sisco, P. H., Sederoff, R. R., and Levings, C. S. III: Nucleotide sequences of inverted repeats and flanking regions of five integrated segments of plasmid-like DNAs from maize mitochondria. Manuscript in preparation.

Brennicke, A., 1980: Mitochondrial DNA from *Oenothera berteriana.* Plant Physiol. **65,** 1207—1210.

Brennicke, A., and Blanz, P., 1982: Circular mitochondrial DNA species from *Oenothera* with unique sequences. Mol. Gen. Genet. **187,** 461—466.

Brown, W. M., Prager, E. M., Wang, A., and Wilson, A. C., 1982: Mitochondrial DNA sequences of primates: Tempo and mode of evolution. J. Mol. Evol. **18,** 225—239.

Carusi, E. A., 1977: Evidence for blocked 5′-termini in human adenovirus DNA. Virology **76,** 380—394.

Chao, S., Sederoff, R. R., and Levings, C. S. III, 1984: Nucleotide sequence and evolution of the 18S ribosomal RNA gene in maize mitochondria. Nucleic Acids Res.: In press.

Conde, M. F., Pring, D. R., and Levings, C. S. III, 1979: Maternal inheritance of organelle DNA's in *Zea mays* - *Zea perennis* reciprocal crosses. J. of Heredity **70,** 2—4.

Clark-Walker, G. D., and Sriprakash, K. S., 1982: Size diversity and sequence rearrangements in mitochondrial DNAs from yeast. In: Slonimski, P., Borst, P., and Attardi, G. (eds.), Mitochondrial Genes, pp. 349—354. Cold Spring Harbor Laboratory, Cold Spring Harbor, N. Y.

Dale, R. M. K., 1981: Sequence homology among different size classes of plant mtDNAs. Proc. Natl. Acad. Sci., U.S.A. **78,** 4454—4457.

Dale, R. M. K., 1982: Structure of plant mitochondrial DNAs. In: Slonimski, P., Borst, P., and Attardi, G. (eds.), Mitochondrial Genes, pp. 471—476. Cold Spring Harbor Laboratory, Cold Spring Harbor, N. Y.

Dixon, L. K., and Leaver, C. J., 1982: Mitochondrial gene expression and cytoplasmic male sterility in sorghum. Plant Mol. Biol. **1,** 89—102.

Fontarnau, A., and Hernandez-Yago, J., 1982: Characterization of mitochondrial DNA in *Citrus*. Plant Physiol. **70,** 1678—1682.

Forde, B. G., and Leaver, C. J., 1980: Nuclear and cytoplasmic genes controlling synthesis of variant mitochondrial polypeptides in male-sterile maize. Proc. Natl. Acad. Sci., U.S.A. **77,** 418—422.

Freeling, M., 1984: Plant transposable elements and insertion sequences. Ann. Rev. Plant Physiol. **35,** 277—298.

Goblet, J.-P., Boutry, M., Duc, G., Briquet, M., 1983: Mitochondrial plasmid-like molecules in fertile and sterile *Vicia faba* L. Plant Mol. Biol. **2,** 305—309.

Harding, N. E., Ito, J., and David, G. S., 1978: Identification of the protein firmly bound to the ends of bacteriophage ∅ 29 DNA. Virology **84,** 279—292.

Harding, N. E., and Ito, J., 1980: DNA replication of bacteriophage ∅ 29: Characterization of the intermediates and location of the termini of replication. J. of Virology **104,** 323—338.

Hobom, G., Grosschedl, R., Lusky, M., Scherer, G., Schwarz, E., and Kossel, H., 1978: Functional analysis of the replicator structure of lambdoid bacteriophage DNAs. Cold Spring Harbor Symposia on Quantitative Biology **43,** 165—178.

Inciarte, M. R., Salas, M., and Sogo, J. M., 1980: Structure of replicating DNA molecules of *Bacillus subtilis* bacteriophage. J. of Virology **4,** 187—199.

Ito, J., 1978: Bacteriophage ∅ 29 terminal protein: Its association with the 5′ termini of the ∅ 29 genome. Virology **28,** 895—904.

Kemble, R. J., and Bedbrook, J. R., 1980: Low molecular weight circular and linear DNA in mitochondria from normal and male-sterile *Zea mays* cytoplasm. Nature (London) **284,** 565—566.

Kemble, R. J., and Mans, R. J., 1983: Examination of the mitochondrial genomes of revertant progeny from S *cms* maize with cloned S-1 and S-2 hybridization probes. J. Mol. Appl. Genet. **2,** 161—171.

Kemble, R. J., and Thompson, R. D., 1982: S-1 and S-2, the linear mitochondrial DNAs present in a male sterile line of maize possess terminally attached proteins. Nucleic Acids Res. **10,** 8181—8190.

Kemble, R. J., Mans, R. J., Gabay-Laughnan, S., and Laughnan, J. R., 1982: Sequences homologous to episomal mitochondrial DNAs in the maize nuclear genome. Nature (London) **304,** 744—747.

Koncz, C., Janos, S., Udvardy, A., Racsmany, M., and Dudits, D., 1981: Cloning of mtDNA fragments homologous to mitochondrial S-2 plasmid-like DNA in maize. Mol. Gen. Genet. **183,** 449—458.

Kuntzel, H., Kochel, H. G., Lazarus, C. M., and Lunsdorf, H., 1982: Mitochondrial genes in *Aspergillus*. In: Slonimski, P., Borst, P., and Attardi, G., (eds.), Mitochondrial Genes, pp. 391—403. Cold Spring Harbor Laboratory, Cold Spring Harbor, N. Y.

Laughnan, J. R., and Gabay, S. J., 1973: Mutation leading to nuclear restoration of fertility in S male-sterile cytoplasm in maize. Theor. Appl. Genet. **43,** 109—116.

Laughnan, J. R., and Gabay, S. J., 1975a: Nuclear and cytoplasmic mutations to fertility in S male-sterile maize. In: Walden, D. B. (ed.), International Maize Symposium: Genetics and Breeding, p. 427. New York: Wiley.

Laughnan, J. R., and Gabay, S. J., 1975b: An episomal basis for instability of S male sterility in maize and some implications for plant breeding. In: Birky, C. W., Perlman, P. S., and Beyers, T. J. (eds.), Genetics and Biogenesis of Mitochondria and Chloroplasts, pp. 340—349. Columbus: Ohio State Univ. Press.

Laughnan, J. R., and Gabay-Laughnan, S. J., 1983: Cytoplasmic male sterility in maize. Ann. Rev. Genet. **17,** 27—48.

Laughnan, J. R., Gabay-Laughnan, S. J., and Carlson, J. E., 1981: Characteristics of *cms-S* reversion to male fertility in maize. Stadler Genet. Symp. **13,** 93—114.

Lechner, R. L., and Kelly, Jr., T. J., 1977: The structure of replicating adenovirus 2 DNA molecules. Cell **12,** 1007—1020.

Levings, C. S. III, and Pring, D. R., 1976: Restriction endonuclease analysis of mitochondrial DNA from normal and Texas cytoplasmic male-sterile maize. Science **193,** 158—160.

Levings, C. S. III, and Pring, D. R., 1978: The mitochondrial genome of higher plants. Stadler Genet. Symp. **10,** 77—94.

Levings, C. S. III, and Sederoff, R. R., 1983: Nucleotide sequence of the S-2 mitochondrial DNA from the S cytoplasm of maize. Proc. Natl. Acad. Sci., U.S.A. **80,** 4055—4059.

Levings, C. S. III, Sederoff, R. R., Hu, W. W. L., and Timothy, D. H., 1983: Relationships among plasmid-like DNAs of the maize mitochondria. In: Structure and Function of Plant Genomes, pp. 363—374. New York: Plenum.

Levings, C. S. III, Shah, D. M., Hu, W. W. L., Pring, D. R., and Timothy, D. H., 1979: Molecular heterogeneity among mitochondrial DNAs from different maize cytoplasms. In: Cummings, D. J., Fox, C. F., Borst, P., Dawid, I. G., and Weissman, S. M., (eds.), Extrachromosomal DNA, pp. 63—73. ICN-UCLA Symposia on Molecular and Cellular Biology, New York: Academic Press.

Levings, C. S. III, Kim, B. D., Pring, D. L., Conde, M. F., Mans, R. J., Laughnan, J. R., and Gabay-Laughnan, S. J., 1980: Cytoplasmic reversion of *cms-S* in maize: Association with a transpositional event. Science **209,** 1021—1023.

Lockhart, L., and Levings, C. S. III. Unpublished results.

Lonsdale, D. M., Thompson, R. D., and Hodge, T. P., 1981: The integrated forms of the S-1 and S-2 DNA elements of maize male sterile mitochondrial DNA are flanked by a large repeated sequence. Nucleic Acids Res. **9,** 3657.

Lonsdale, D. M., Hodge, T. P., Fauron, C. M.-R., and Flavell, R. B., 1983a: A predicted structure for the mitochondrial genome from the fertile cytoplasm of maize. In: Goldberg, R. B. (ed.), Plant Molecular Biology (UCLA Symposia on Molecular and Cellular Biology, New Series, Vol. 12), pp. 445—456. New York: Alan R. Liss, Inc.

Lonsdale, D. M., Fauron, C. M.-R., Hodge, T. P., Pring, D. R., and Stern, D. B., 1983b: Structural alterations in the mitochondrial genome of maize associated with cytoplasmic male sterility. In: Chater, K. F., Cullis, C. A., Hopwood, D. A., Johnson, A. W. B., and Woolhouse, H. W. (eds.), Genetic Rearrangement, pp. 183—206. London: Croom Helm.

McClintock, B., 1978: Mechanisms that rapidly reorganize the genome. Stadler Genet. Symp. **10,** 25—48.
McNay, J. W., Chourey, P. S., and Pring, D. R., 1984: Molecular analysis of genomic stability of mitochondrial DNA in tissue cultured cells of maize. Theor. Appl. Genet. **67,** 433—437.
Meints, R. H., Schuster, A., Hu, W. W. L., Timothy, D. H., and Levings, C. S. III: The structural nature of cytoplasmic male sterility associated plasmids in maize mitochondria. Manuscript in preparation.
Messer, W., Meijer, M., Bergmans, H. E. N., Hansen, F. G., von Meyenburg, K., Beck, E., and Schaller, H., 1978: Origin of replication, oriC, of the *Escherichia coli* K 12 chromosome: Nucleotide sequence. Cold Spring Harbor Symposia on Quantitative Biology **43,** 139—145.
Moore, D. D., Denniston-Thompson, K., Kruger, K. E., Furth, M. E., Williams, B. G., Daniels, D. L., and Blattner, F. R., 1978: Dissection and comparative anatomy of the origins of replication of lambdoid phages. Cold Spring Harbor Symposia on Quantitative Biology **43,** 155—163.
Nikiforova, I. D., and Negruk, V. I., 1983: Comparative electrophoretic analysis of plasmid-like mitochondrial DNAs in *Vicia faba* and in some other legumes. Planta **157,** 81—84.
Paillard, M., Sederoff, R. R., and Levings, C. S. III, 1985: Nucleotide sequence of the S-1 mitochondrial DNA from the S cytoplasm of maize. EMBO J.
Palmer, J. D., and Shields, C. R., 1984: Tripartite structure of the *Brassica campestris* mitochondrial genome. Nature (London) **307,** 437—441.
Palmer, J. D., Shields, C. R., Cohen, D. B., and Orton, T. J., 1983: An unusual mitochondrial DNA plasmid in the genus *Brassica.* Nature (London) **301,** 725—728.
Powling, A., 1981: Species of small DNA molecules found in mitochondria from sugar-beet with normal and male-sterile cytoplasms. Mol. Gen. Genet. **183,** 82—84.
Powling, A., and Ellis, T. H. N., 1983: Studies on the organelle genomes of sugarbeet with male-fertile and male-sterile cytoplasms. Theor. and Appl. Genet. **65,** 323—328.
Pring, D. R., and Levings, C. S. III, 1978: Heterogeneity of maize cytoplasmic genomes among male-sterile cytoplasms. Genetics **89,** 121—136.
Pring, D. R., Conde, M. F., and Schertz, K. F., 1982: Organelle genome diversity on sorghum: Male-sterile cytoplasms. Crop Sci. **22,** 414—421.
Pring, D. R., Levings, C. S. III, Hu, W. W. L., and Timothy, D. H., 1977: Unique DNA associated with mitochondria in the "S" type cytoplasm of male-sterile maize. Proc. Natl. Acad. Sci., U.S.A. **74,** 2904—2908.
Quetier, F., and Vedel, F., 1977: Heterogeneous population of mitochondrial DNA molecules in higher plants. Nature (London) **268,** 365—368.
Rekosh, D. M. K., Russell, W. C., Bellet, A. J. D., Robinson, A. J., 1977: Identification of a protein linked to the ends of adenovirus DNA. Cell **11,** 283—295.
Ronald, P., Bedinger, P., Rivin, C., Walbot, V., Bland, M., Levings, C. III, Sederoff, R.: Maize mitochondrial plasmid S-1 sequences share homology with chloroplast gene *psb*A (photogene 32). Submitted for publication.
Salas, M., Mellado, R. P., and Vinuela, E., 1978: Characterization of a protein covalently linked to the 5′ termini of the DNA of *Bacillus subtilis* phage ∅ 29. J. Mol. Biol. **119,** 269—291.
Sederoff, R. R., 1984: Structural variation in mitochondrial DNA. In: Advances in Genetics **22,** 1—108.
Sederoff, R. R., Levings, C. S. III, Timothy, D. H., and Hu, W. W. L., 1981: Evo-

lution of DNA sequence organization in mitochondrial genomes of *Zea*. Proc. Natl. Acad. Sci., U.S.A. **78,** 5953—5957.

Shah, D. M., and Levings, C. S. III, 1974: Mitochondrial DNA from maize hybrids with normal and Texas cytoplasms. Crop Science **14,** 852—853.

Sparks, R. B., Jr., and Dale, R. M. K., 1980: Characterization of 3 H-labeled supercoiled mitochondrial DNA from tobacco suspension culture cells. Mol. Gen. Genet. **180,** 351—355.

Spencer, D. F., Schnare, M. N., and Gray, M. W., 1984: Pronounced structural similarities between the small subunit ribosomal RNA genes of wheat mitochondria and *Escherichia coli*. Proc. Natl. Acad. Sci., U.S.A. **81,** 493—497.

Spruill, W. M., Levings, C. S., III, and Sederoff, R. R., 1981: Organization of mitochondrial DNA in normal and Texas male sterile cytoplasms of maize. Dev. Genet. **2,** 319—336.

Stern, D. B., and Palmer, J. D., 1984: Extensive and widespread homologies between mitochondrial DNA and chloroplast DNA in plants. Proc. Natl. Acad. Sci., U.S.A. **81,** 1946—1950.

Stern, D. B., and Lonsdale, D. M., 1982: Mitochondrial and chloroplast genomes of maize have a 12kb DNA sequence in common. Nature (London) **299,** 698—702.

Sussenbach, J. S., and Kuijk, M. G., 1977: Studies on the mechanism of replication of adenovirus DNA: V. The location of termini of replication. Virology **77,** 149—157.

Synenki, R. M., Levings, C. S. III, and Shah, D. M., 1978: Physicochemical characterization of mitochondrial DNA from soybean. Plant Physiol. **61,** 460—464.

Timothy, D. H., Levings, C. S. III, Hu, W. W. L., and Goodman, M. M., 1983: Plasmid-like DNAs in diploperennial teosinte. Maydica **28,** 139—149.

Timothy, D. H., Levings, C. S. III, Pring, D. R., Conde, M. F., and Kermicle, J. L., 1979: Organelle DNA variation and systematic relationships in the genus *Zea:* Teosinte. Proc. Natl. Acad. Sci., U.S.A. **76,** 4220—4224.

Wallace, D. C., 1982: Structure and evolution of organelle genomes. Microbiol. Rev. **46,** 208—240.

Ward, B. L., Anderson, R. S., and Bendich, A. J., 1981: The size of the mitochondrial genome is large and variable in a family of plants (Curcurbitaceae). Cell **25,** 793—803.

Weissinger, A. K., Timothy, D. H., Levings, C. S. III, and Goodman, M. M., 1983: Patterns of mitochondrial DNA variation in indigenous maize races of Latin America. Genetics **104,** 365—379.

Weissinger, A. K., Timothy, D. H., Levings, C. S. III, Hu, W. W. L., and Goodman, M. M., 1982: Unique plasmid-like mitochondrial DNAs from indigenous maize races of Latin America. Proc. Natl. Acad. Sci., U.S.A. **79,** 1.

Winnacker, E.-L., 1978: Adenovirus DNA: Structure and function of a novel replicon. Cell **14,** 761—773.

Wright, R. M., Horrum, M. A., and Cummings, D. J., 1982: Are mitochondrial structural genes selectively amplified during senescence in *Podospora anserina?* Cell **29,** 505—515.

Yehle, C. O., 1978: Genome-linked protein associated with the 5′ termini of bacteriophage ∅ 29 DNA. Virology **27,** 776—783.

Yoshikawa, H., and Ito, J., 1981: Terminal proteins and short inverted terminal repeats of the small *Bacillus* bacteriophage genomes. Proc. Natl. Acad. Sci., U.S.A. **78,** 2596—2600.

Chapter 7

Plant Mitochondrial DNA: Unusual Variation on a Common Theme

Arnold J. Bendich

Departments of Botany and Genetics, University of Washington, Seattle, WA 98195, U.S.A.

Contents

I. Introduction

There are two unusual features of the plant mitochondrial genome that seemingly distinguish it from the mitochondrial genome in all other eukaryotes. First, the size of the genome is very large — embarassingly so — and can be extremely variable even among closely related species. And second, the circular DNA molecules obtained from plant mitochondria are not easily related to the size of the genome as inferred from reassociation kinetics measurements or from restriction enzyme analyses. The circles are heterogeneous in size, do not approach the size of the genome inferred from the other methods, and do not amount to more than a few per cent of the molecules obtained from mitochondria of intact plant tissues, although they can be obtained in much greater yield from cultured cells. The unusual properties of the circles have led to the speculation that the mitochondrial genome in plants is organized differently than it is in other eukaryotes with the genes borne on different linkage groups of chromosomes, as

Table 1. Mitochondrial Genome Sizes, Translation Products and Number of Protein-Coding Genes

Organism	Genome size (Md)	Translation products*	Genes+
Metazoans[a]	10—11		6+ 8 URFs
Achlya[b]	33		
Yeasts[c]			
Torulopsis glabarata	13		
Brettanomyces custersianus	19		
Saccharomyces cerevisiae	50		10+ ≧8 URFs
Brettanomyces custersii	67		
Chlamydomomas[d]	10		
Chlorella[e]	53		
Maize[f]	320	15—20	
Pea[f]	240	18—20	
Vicia faba[g]	190	15—20	
Brassica[h]			
juncea	120		
campestris	130		
oleracea	140		
napus	140	about 20	
Cucurbits[f]			
watermelon	220	about 20	
zucchini	560	about 20	
cucumber	1000	about 20	
muskmelon	1600	about 20	

Genome size in megadaltons (Md): [a] Altman and Katz (1976); [b] Hudspeth *et al.* (1983); [c] Clark-Walker *et al.* (1981), McArthur and Clark-Walker (1983); [d] Grant and Chiang (1980); [e] Bayen and Rode (1973); [f] Ward *et al.* (1981); [g] Bendich (1982); [h] *B. campestris* is 144 (Palmer and Shields 1984) or 120 Md (Lebacq and Vedel, 1981). *B. oleracea* from Chétrit *et al.* (1984). Other *Brassica* species from Lebacq and Vedel (1981).

* The number of polypeptide bands on one-dimensional SDS-polyacrylamide gels after ^{35}S-methionine incorporation by isolated mitochondria from maize (Forde and Leaver, 1980; Boutry *et al.*, 1984), *V. faba* (Boutry and Briquet, 1982), *B. napus* (Vedel *et al.*, 1982), cucurbits (Stern and Newton, 1985) and pea (Leaver, personal communication). Hack and Leaver (1984) report 15—20 for cucumber.

+ Protein-coding genes from Borst *et al.* (1984). URF means unassigned reading frame.

is found in the nucleus (discussed in Bendich, 1982). The large size of the genome has led to the suggestion that there may be more — perhaps many more — genes in plant mtDNA than in mtDNA of other eukaryotes (Leaver and Gray, 1982; Grivell, 1983; Levings, 1983; Hack and Leaver, 1983). It will be the purpose of this article to briefly review the information concerning genome size and DNA circularity for plant mitochondria and to ask whether the limited data for plants do in fact point to a fundamental genetic difference between the mitochondria of plants and other eukaryotes.

In Table 1 we list mitochondrial genome sizes selected to illustrate the range of variation observed. Several points deserve emphasis. First, even the smallest plant genome is larger than the largest non-plant genome (the case of kinetoplast DNA may be an exception as discussed later). Second, closely related plant species may exhibit little variability (*Brassica* species) or a large variability (cucurbit species). Third, the variability among yeasts is nearly as great as among the cucurbits. It is, however, likely that the cucurbits are more "closely related" than are these yeasts since the criteria upon which such judgements are made are at present more limited for protists than for higher plants.

II. The "Extra" DNA in Plant Mitochondria

There are three explanations for the "extra" DNA in plant mitochondria. The large sequence complexity (used here as the measure of genome size) could represent additional genes or particular sequences necessary to mitochondrial function, or DNA that is useful to the plant in a manner not dependent on its sequence, or DNA that is of no use in mitochondrial function (selfish or ignorant DNA). The first of these alternatives implies mtDNA in plants actually encodes many more genes that contribute to mitochondrial function than it does in other organisms. This alternative, however, appears untenable for explaining the range of genome size among cucurbit species. Still, it is possible that the large plant genomes encode *some* additional genes. To date one gene not found on mtDNA in other organisms has been assigned to the plant mitochondrial genome. The α subunit of F_1-ATPase has been shown to be a translation product in mitochondria of maize (Hack and Leaver, 1983), *V. faba* (Boutry *et al.*, 1983) and six other angiosperm species (unpublished results mentioned in these two papers), although its location on a mtDNA fragment has not yet been reported.

A. The Number of Translation Products Does Not Vary with Genome Size

After incorporation of ^{35}S-methionine by isolated mitochondria, the number of translation products estimated by counting polypeptide bands on one-dimensional SDS-polyacrylamide gels has been used to approach the issue of gene number in plant mitochondria. As can be seen in Table 1, this assay produces the same number, some 15 to 20 bands, for plants ranging 10-fold in genome size. This same number of bands was obtained for every one of at least 13 additional angiosperm species (Dixon and Leaver, 1982; Boutry *et al.*, 1984; Leaver, personal communication). Taken as face value these data would suggest that irrespective of genome size, the number of translation products (and perhaps genes) is constant among higher plants. There are, however, major difficulties in equating translation products with genes. When two-dimensional gel electrophoresis is employed with maize, the 18—20 bands are resolved into 30—50 spots. But as Hack and Leaver (1983) note, some of these spots could be due to premature termination, charge modification or proteolytic cleavage. In addition, the bands or spots only represent those polypeptides present in high enough quantities to be detectable under the conditions used and potentially important low level gene products would be missed. Thus the use of translation products to estimate gene numbers could produce either over estimates or underestimates and this method is not presently adequate for counting genes. Nevertheless, it is interesting that the 15—20 translation products is close to the number of protein coding (including URFs) genes known to be present in mitochondria of animals and baker's yeast (Table 1).

B. The Sequence Complexity of Mitochondrial RNA Is Large

It is generally believed that most DNA sequences that function as genes are transcribed. The fraction of a genome able to hybridize with RNA at saturation (high R_0t values) might therefore provide an estimate of its genic content. Using this simple notion we measured the sequence complexity of transcribed DNA by monitoring the fraction of nick-translated mtDNA protected from digestion by S1 nuclease after hybridization with total mitochondrial RNA (Ward and Bendich, unpublished results). The kinetics of hybridization indicated the presence of 2 or 3 abundancy classes of RNA; and at saturation between 25 and 35% (watermelon) and 15% (muskmelon) of mtDNA was in RNA/DNA hybrids. Thus in shoots from etiolated seedlings 110—150 Md and 480 Md of the watermelon and muskmelon genomes, respectively, are transcribed, if we assume only one of the two DNA strands could encode a useful gene product. This result is difficult to interpret if one considers transcribed DNA to represent genic DNA since more mtDNA is transcribed in muskmelon than exists in the watermelon genome. The result is, however, not as puzzling in the light of those for human and yeast mitochondria. Although the H strand of HeLa cell mtDNA contains most of the genes, it is the L strand (which encodes only one URF and 8 of 22 tRNAs) that is much more intensively transcribed (Cantatore and Attardi, 1980; Attardi *et al.*, 1982). For baker's yeast there was "no difference in the kinetics of hybridization or in the hybridization saturation level" using RNA from aerobically or anaerobically grown cells and almost a full single stand equivalent was transcribed into RNA (Jakovcic *et al.*, 1979). These authors also note that "the turning off of mitochondrial protein synthesis in anaerobiosis may be independent of mitochondrial transcription". In addition, Benne *et al.* (1983) found abundant, polyadenylated transcripts for five URFs in the gene-bearing maxicircle of *Trypanosoma brucei* mtDNA that were unlikely to encode proteins because of non-homology with mitochondrial proteins and URFs of other organisms and because they have no apparent translation start codon. In view of the foregoing results we certainly cannot take the large sequence complexity of cucurbit mitochondrial RNA to indicate a large genic content. We do not yet understand the relationship between transcription and gene expression in plant mitochondria. It is possible that gene expression is effected entirely at a posttranscriptional level, although we are aware of no data on this issue for plants. We must conclude, as we did for the translation products approach, that comparison of saturation hybridization levels is not an adequate method for counting or comparing the number of mitochondrial genes.

C. Are There More Mitochondrial Genes in Plants than in Other Organisms?

The presently available evidence is certainly meager but does not indicate that plants are very different from other organisms in the number of mitochondrial genes. The prospects for counting all the mitochondrial genes in the near future are not bright. The direct approach of determining the nucleotide sequence of the entire genome might provide a compelling answer, but it also might not. What if many unusual URFs of the type mentioned above for *T. brucei* (Benne *et al.*, 1983) were found? In the absence of the ability to use the powerful genetic tools to test the function of a DNA sequence in a plant cell, it is virtually impossible to obtain definitive answers. What is needed is a means to replace a sequence in question with an altered version and to assess the modified genome for mitochondrial function in the plant cell. We need mutants and a transformation system for plant mitochondria.

D. Does Mitochondrial DNA Have a Sequence-Independent Function?

The second alternative for explaining the large genome in plant mitochondria is that bulk DNA performs a sequence-independent function. Such a function would, in principle, be served by nearly any sequence that is reiterated or by DNA that we classify as "non-repeated" because it appears in reassociation kinetics and restriction analyses to have a high sequence complexity. A group of properties called "nucleotypic" has been correlated in many instances with the amount of DNA in the nucleus (Bennett *et al.*, 1982). It has been proposed that the nucleotype affects many cellular and organismic properties including the volume of the nucleus and the cell, the duration of the mitotic cycle and the rate of development, and the ecological and geographical distribution of species (Cavalier-Smith, 1978; Bennett *et al.*, 1982). In seeking a possible "mitotypic" analog, Bendich and Gauriloff (1984) measured the volume of the chondriome (the total mitochondrial inventory of a cell irrespective of the number of individual mitochondria) but found no difference among the cucurbits listed in Table 1 despite the 7-fold range in both genome size and amount of mtDNA per cell. Thus there was no correlation between organellar volume and DNA content for mitochondria, as sometimes is found for the nucleus. We are aware of no other information on the question of whether putative non-coding DNA affects mitochondrial function in plants. At present the cucurbits would be the group in which to conduct such work since they would be expected to carry the same number of mitochondrial genes. In the future a useful experiment will be to eliminate large amounts of DNA from the muskmelon genome and compare mitochondrial function in plants carrying the reduced and normal-sized genomes.

E. Is Mitochondrial DNA Selfish or Ignorant?

The third alternative explanation for the large plant mitochondrial genome is that there exists a large and variable amount of extra DNA that is not useful to the plant either because of its sequence or as sequence-independent bulk polynucleotide. Instead, the extra DNA is somewhat deleterious (selfish DNA) or is neither harmful nor helpful for the plant (ignorant DNA) (Dover and others, 1980). In either case the plants have been unable to rid themselves of these sequences that are non-functional for the host plants. Animals (and evidently *Chlamydomonas*) have evolved a means for completely suppressing such extra DNA, but plants tolerate it even to the point in cucumber and muskmelon where it would account for 99 % of the mitochondrial genome.

There is one class of well-studied sequences that may be representative of extra mtDNA in fungi. "Optional" introns confer no obvious advantage upon the fungi in which they are found since by definition they are present in some wild type strains and absent in others. The mitochondrial genome of *Neurospora crassa* varies from 60—73 kb (40—48 Md) depending on the particular wild type strain. Most or all of this variation is due to optional introns. Different strains carry zero to four of these introns in the gene for cytochrome oxidase subunit I (COX I) (Collins and Lambowitz, 1983) which accounts for most of the variation; additional variation may be attributed to 2 versus 3 introns in the cytochrome b gene (Burke *et al.*, 1984). Among wild type strains of baker's yeast the genome varies from 68—78 kb (45—52 Md), again mainly due to optional introns. The "short" strains have 2 introns in cytochrome b and 7 in COX I, whereas in "long" strains these numbers are 5 and 9; for *Aspergillus nidulans* these two genes have 1 and 3 introns, respectively. Burke *et al.* (1984) summarize these data and also the variable locations of introns within the cytochrome b gene. By crossing wild type and petite strains of baker's yeast Labouesse and Slonimski (1983) have produced strains with several combinations of cytochrome b introns ranging from zero to five and have concluded that "all *cob-box* introns are optional for expression of this gene". No introns have been found in the cytochrome b (Dawson *et al.*, 1984) or COX I

(Leaver, personal communication) genes of maize. The gene for subunit II of COX (COX II) of maize (Fox and Leaver, 1981) and its close relative *Teosinte diploperennis* (Levings, personal communication) have 794 base pair introns that are identical in sequence, whereas in wheat this gene is interrupted at the same position by an intron that contains an insert relative to the maize intron which increases its overall length to 1.5 times that of maize (Bonen *et al.*, 1984). The COX II gene has no intron in the dicot *Oenothera berteriana* (Hiesel and Brennicke, 1983) nor in baker's yeast. No other introns have been reported for plant mitochondrial genes.

From the limited data available it is not yet clear whether optional introns are responsible for either the large size of the plant mitochondrial genome or the variability among species. The genome in maize may be somewhat larger than in *Oenothera* [restriction analysis indicated at least 130 Md, although lack of band resolution on the 1 % agarose gels (Brennicke, 1980) could have resulted in a 2-fold underestimate], but the larger intron is found in the smaller wheat genome (230 Md on 1 % agarose gels; Bonen and Gray, 1980). No introns have been found in the genes for the small rRNA of maize (Levings, personal communication; Dale, personal communication) or small (Spencer *et al.*, 1984) and large rRNA of wheat (Spencer and Gray, personal communication). It would require a large number of large introns to account for the "extra" mtDNA in plants. This seems unlikely, but it will remain a speculative judgement until we have more information.

F. Interorganellar DNA

Another class of sequences likely to be included in the category of extra mtDNA are those that have homology to chloroplast DNA (ctDNA). Despite an initial suggestion to the contrary (Stern and Lonsdale, 1982), these chloroplast-homologous sequences are considered unimportant for mitochondrial function (Stern *et al.*, 1983; Lonsdale *et al.*, 1983 a). In maize about 14 kb of mtDNA evidently originated in the chloroplast and includes altered versions of the chloroplast genes for 16 S rRNA and the large subunit of ribulosebisphosphate carboxylase. Stern *et al.* (1983) observed no correlation between the size of the cucurbit mitochondrial genome and the number or intensity of fragments that hybridized to total spinach ctDNA; and the greatest amount of mtDNA sequence homology to a particular ctDNA fragment was observed in the smallest genome (watermelon) and the least homology in the largest (muskmelon). Thus it should not be inferred that very much of the extra DNA in the mitochondrial genome is due to import of ctDNA.

G. Why Is the Mitochondrial Genome so Large?

At this point we find no convincing evidence to support any of the above explanations (more genes, DNA serving a sequence-independent function, optional introns, itinerant DNA from other organelles) for the large and variable plant mitochondrial genome, although we have very little information on optional introns and none on the possibility of nuclear DNA migrating to the mitochondrion. Before retreating to the position that we simply have no plausible explanation, we should recall the enigmatic DNA of the kinetoplast, a region of the mitochondrion in a group of parasitic, flagellated protozoans (reviewed by Englund, 1981). About 90% of this mtDNA consists of non-coding minicircles for which no function is known. In *Trypanosoma equiperdum* the approximately 10,000 1 kb minicircles are each equivalent in sequence, whereas in *T. brucei* there are 300 different types of minicircles. Except for the fact that the approximately 50 gene-bearing maxicircles (about 20 kb) are usually considered separately from the minicircles, the genome size would be 21 kb for *equiperdum* and 320 kb for *brucei*. As discussed by Ward *et al.* (1981), for dyskinetoplastic mutants of *equiperdum* and *brucei* in which repiratory ability is lost, the mtDNA is so altered from that of the wild type that the two no longer exhibit homology in DNA hybridization assays. [Note that Sloof *et al.* (1983) suggest that dyskinetoplastic strains do not contain mtDNA. The following, however, support the notion that the putative mtDNA is in fact mtDNA: (1) it has the same buoyant density as wild type mtDNA; (2) it represents the same fraction of total cell DNA as does wild type mtDNA, a situation also found for petite and wild type mtDNA in yeast; and (3) material identified as DNA was found in mitochondria of dyskinetoplastic *T. equiperdum* by fluorescence and electron microscopy (Hajduk, 1979)].

On what criteria do we base a judgement of "non-homology" or "unique sequence"? Two "different" types of minicircles of *T. brucei* were sequenced and found to contain many of the same short (10—15 base pairs) sequence elements albeit in scrambled arrangement in one with respect to the other. As Chen and Donelson (1980) conclude, "the ubiquity of small homologous elements throughout the two sequences suggests that massive recombination may occur among different minicircle sequences to generate a randomized version of a unique ancestral minicircle sequence". The primary sequence of members of two families of repetitive sequences (the operational definition of a family is that its members do not reassociate with members of another family) in sea urchin nuclear DNA revealed that the same five sequences of 10—14 base pairs in one are spaced differently in the other (Posakony *et al.*, 1981). Thus amplification followed by recombinative scrambling of a sequence on a fine scale can lead to greatly increased sequence complexity when DNA hybridization and restriction analysis are the techniques used to estimate genome size. As discussed by Ward *et al.* (1981), the base composition of mtDNA is virtually constant among plants, including *Chlamydomonas* (Ryan *et al.*, 1978), and the amplification and reshuffling of the same bits and pieces of non-coding mtDNA may generate the bewilderingly large and variable genome we observe. Support for this idea could come from searching sequence data from regions ouside of known genes for short, scrambled repeats.

Many molecular biologists find it difficult to accept the idea that some DNA may have no function whatsoever for its host organism. At the moment, however, this alternative seems as reasonable as any for the large plant mitochondrial genome. The ultimate test will come when we are able to delete large amounts of putative non-coding DNA and compare plants carrying the reduced and normal-sized genomes. Finally, we note that the problem of large variability of sequence complexity also exists for the nucleus and is called the "C-value paradox" (Gall, 1981). We suspect that the ultimate explanations for the problems will be the same for the mitochondrion and nucleus.

III. Circular Mitochondrial DNA

A. A Brief History

Since the first report of large circular plant mtDNA molecules (Kolodner and Tewari, 1972) most investigators have accepted the idea that the steady state disposition of DNA in mitochondria of living cells is circular. This supposition stemmed in part from earlier observations that mtDNA from animals is readily obtained as unit genome circular molecules, and in part from the close agreement between the prominent 30 μm circle class and the 74 Md kinetic complexity reported for pea mtDNA (Kolodner and Tewari, 1972). It was subsequently shown, however, that the 74 Md value should have been corrected to 126 Md and the kinetic complexity is 220 Md (Ward *et al.,* 1981). When restriction enzyme analysis indicated DNA molecular weights considerably larger than the largest mtDNA circles, it was still assumed that the circles were a representative subset of all the extracted mtDNA molecules and hence the genome size would be obtained by summing the lengths of circles in different size classes. It had been generally assumed that the failure to obtain a high yield of circular DNA was ascribable to physical and/or enzymatic degradation during extraction procedures. This failure will be considered in some detail below, but first we should review the properties of the circles that have been observed for mitochondria obtained both from native plant tissues and from cultured cells.

B. Circles in Native Plant Tissue

When maize leaf mtDNA is analyzed by gel elctrophoresis, a 1.9 kb circle is found (Kemble and Bedbrook, 1980; Kemble *et al.,* 1980). This plasmid does not have sequence homology with the main mitochondrial genome and is not detected in some other races and strains of maize (Lonsdale *et al.,* 1983 b). The properties of variable presence among strains and little or no homology with the main mitochondrial genome are shared by circular mitochondrial plasmids of *Neurospora crassa* which, in addition, are found tandemly arranged in monomer, dimer, etc. oligomeric series (Stohl *et al.,* 1982; Collins and Lambowitz, 1983). In what may be an analogous phenomenon, sugarbeet root mtDNA contains small (1.3 to 1.5 kb) circular molecules present as oligomeric series (see Fig. 1) that do not hybridize with high molecular weight mtDNA and are found in some but not other strains (Powling and Ellis, 1983). A linear plasmid of 11 kb is undetectable in some *Brassica* species and varies in others from about 1/10 to 10 times that of the main mitochondrial genome (Palmer *et al.*, 1983). It is possible that higher oligomers are also present since the plasmid also hybridizes to larger discrete bands in gels of unrestricted mtDNA. The well-studied linear plasmids S1 and S2 similarly are found in only some maize strains and hybridize to larger molecules proposed to be concatamers used in plasmid replication (Kemble and Mans, 1983). Small, plasmid-like circular mtDNA molecules were also found in *Vicia faba* leaves and seedlings by Nikoforova and Negruk (1983).

In addition to those mtDNA molecules considered as plasmids or plasmid-like, other circles from native tissue have been identified but not subjected to hybridization analysis. Small circular mtDNA molecules were

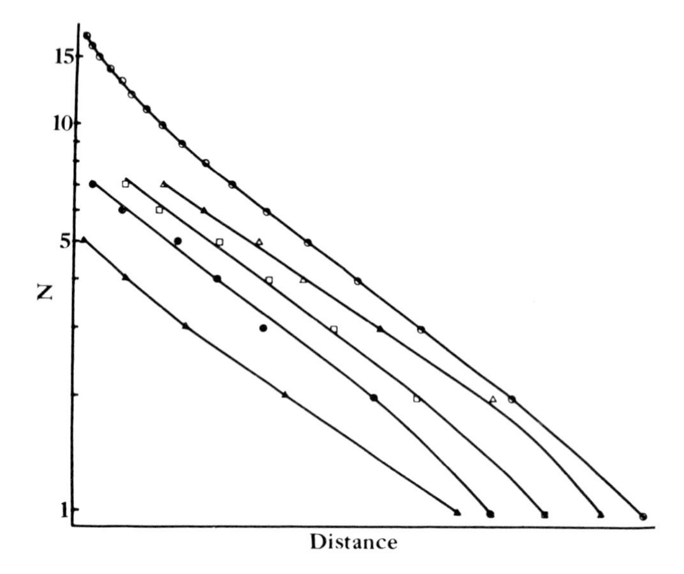

Fig. 1. Oligomeric Series for Supercoiled Mitochondrial DNA Molecules. The oligomer number, N, is plotted against the distance of migration in a 1.2 % agarose gel for sugarbeet (▲) (Fig. 3c of Powling and Ellis, 1983) or 0.7 % agrose gels for bean (○) (Fig. 1 of Dale *et al.* 1981) and maize (Fig. 7a of Dale, 1981). The smallest band was taken as the monomer and was, in kb: 1.4 (sugarbeet), 1.9 (bean), and 1.5, 1.8 and about 2 (maize). Each successively larger band (except for four faint bands that were clearly not in the oligomeric series for bean) was taken as the 2-mer, 3-mer, etc. For maize the first series (△) contains bands (numbered from smallest to largest) 1, 4, 7, etc.; the second (□) contains bands 2, 5, 8, etc.; the third (●) contains bands 3, 6, 9, etc. For sugarbeet S1 nuclease was used to relax supercoiled molecules before electrophoresis, whereas the bean and maize circles were in supercoiled form. For purposes of clarity the curves are arbitrarily separated laterally. The slopes of the curves cannot be compared meaningfully for the different plant species

found for pea (Kolodner and Tewari, 1972), *Citrus* (Fontarnau and Hernández-Yago, 1982) and pearl millet (Kim *et al.*, 1982) but no information was provided as regards strain distribution, oligomers, or possible homology with larger molecules. Circular forms of the mtDNA molecules obtained from soybean hypocotyls were grouped into seven size classes ranging from 6 to 30 μm (Synenki *et al.*, 1978). In order to enrich for circles, maize leaf mtDNA from the expected position of supercoiled molecules on CsCl-ethidium bromide gradients was analyzed in the electron microscope (Levings *et al.*, 1979). When normal (N, male fertile) plants were examined, size classes of about 15 and 30 μm, and 21 and 41 μm were found (along with molecules 4—10 μm) that were suggested to represent monomer/dimer relationships. When cytoplasmic male sterile (S and T cytoplasms) strains were examined, there were no similarities in circle size distribution among strain pairs for N, S or T, although one possible monomer/dimer size class relationship could be found for each of S and T along with molecules ranging from about 5 to 50 μm.

C. Circles in Cultured Cells

In contrast to the situation in whole plant tissues where circles account for a very low percentage of extracted molecules, the proportion of circular molecules is much higher in mtDNA prepared from suspension culture cells (Dale *et al.*, 1981; Brennicke and Blanz, 1982; Lifshitz *et al.*, 1982; Kool *et al.*, 1984). The ability to obtain greater amounts of covalently closed circular DNA has permitted analysis of its size distribution using gel electrophoresis. When the supercoiled fraction of bean mtDNA was examined, there was a striking pattern of regularly spaced bands beginning with a monomeric size of 1.9 kb and extending to at least the 17-mer of an oligomeric series (Fig. 1). When the monomer was used as a hybridization probe, each of the oligomers hybridized as if it contained mostly or only the sequences of the monomer (Dale, 1981). The complex size distribution of supercoiled mtDNA molecules from cultured maize cells appeared in sharp contrast to the simple pattern for bean (Dale, 1981). The maize pattern is, however, actually composed mostly or entirely of three different olgomeric series (Fig. 1); the monomer size is 1.5, 1.8 and about 2 kb for each series (Dale, 1981). When the 1.5 and 1.8 kb molecules were used as probes in separate hybrization experiments, "the two molecules hybridize to many of the same molecules, and each one hybridizes to a few size classes not recognized or recognized only slightly by the other size class" (Dale, 1981). Thus it appears that every (or nearly every) size class in each of the 3 oligomeric series contains a common sequence, although this sequence accounts for only a part of each molecule. One obvious, but speculative, possibility is that this common sequence is used as a replication origin. The supercoiled mtDNA from cultured tobacco cells also contains a head-to-tail oligomeric series reaching to at least the 3-mer, with the monomer length of 10 kb (Dale, 1981; Dale *et al.*, 1983). In addition, a supercoiled molecule of about 30 kb is present that shares about 120 base pairs of homology (heteroduplex assay in the electron microscope) with the 10 kb monomer molecule and has an entirely different restriction map from the 3-mer. Curiously, the size of the 30 kb molecule and the 3-mer are so closely similar that they appear as a single band on the gel, with each contributing about half the fluorescence. Judging from the florescence intensity, the three oligomers plus the open circular form of the 3-mer plus the 30 kb molecule account for about half the supercoiled DNA mass (if the series extends to more than three, these would account for even more than half) and collectively represent only about 40 kb of sequence complexity. These considerations lead us to what appears to be a paradoxical situation. The restriction patterns were almost identical in both the supercoiled and the open circle plus linear fractions from CsCl-ethidium bromide gradients of tobacco (the same Wisconsin 38 cell line of *Nicotiana tabacum*) mtDNA (Sparks and Dale, 1980). Furthermore there were no extremely prevalent restriction fragments in the complex pattern of fragments; the many fragments summed to about 240 kb (the genome size would surely be larger than this since no special attempts were described for resolving bands on the 1% agarose gels). We would expect very bright bands summing 10 kb (and other prominent bands to 30 kb) for both supercoiled and open circle plus linear fractions. Why was this expectation not realized? We have no simple explanation for these puzzling data. We note, however, that for a suspension culture of the NT-1 cell line of *N. tabacum* the Sal I restriction pattern for the supercoiled fraction of mtDNA was different from that for total mtDNA (Grayburn and Bendich, unpublished results).

Judging again from fluorescence intensity, it is obvious that more than half the DNA in the supercoiled mtDNA from maize cells is present in two sizes, 1.5 and 1.8 kb, yet the maize restriction patterns sum to hundreds of kb with no particularly bright bands (Dale, 1981) (reassociation kinetics yield a 480 kb genome, Table 1). Unfortunately, no restriction patterns were reported by Dale (1981) for the supercoiled fraction alone, but it is nearly inconcievable that this fraction could yield patterns similar to those shown for total mtDNA. It is possible that the entire sequence complexity of some 500 kb is represented in the 1.5 plus 1.8 kb circle classes, if as a consequence of establishing the callus culture, there were a scrambling of noncoding sequences on a fine scale and a partitioning of these onto small circles (recall the 300 types of 1 kb *T. brucei* minicircles dicussed above). An even more confusing situation is found for the simple oligomeric series of bean circles (Dale, 1981, and Fig. 1) since the restriction patterns imply hundreds of kb but the Hha I restriction fragments of both the 1.9 kb monomeric circle and the 3.8 kb dimer (and the 3-mer and 4-mer, as well) sum to 1.9 kb. For bean the sequence scrambling explanation simply cannot work and the only explanation consistent with the data is that the supercoiled fraction represents only a small part of the sequence complexity of the genome, rather than all of it as Dale implies.

Dale *et al.* (1981) also reported the electrophoretic separation of mtDNA supercoils from suspension cultures of *Nicotiana rustica, Solanum dulcimera* and *Datura stramonium,* but no oligomeric series were evident and no restriction data were presented. Gel analysis of *Oenothera* mtDNA revealed that the five smallest supercoiled size classes ranged from 6.3 to 13.5 kb (Brennicke and Blanz, 1982). Restriction analysis showed each size class to contain a single sequence type that shared no fragments with any of the others. When the 6.3 kb circle was used as a probe, some sequence homology was detected with circles "of much larger size", leaving open the possibilities of an oligomeric series or a common sequence that could be used in replication. Electron microscopic analysis of circular mtDNA from suspension cultures of *Petunia hybrida* revealed a continous range of sizes from 3 to 90 kb with no indication of an oligomeric series (Kool *et al.,* 1984).

The observation that provides our deepest insight concerning the functional significance of circular molecules from tissue cultures was made by Dale *et al.* (1981, 1983). When cultures established at different times from the same cultivar of tobacco (Wisconsin 38) were compared, the size class distributions for supercoiled mtDNA molecules were "totally different". The 10 kb size class was absent in the newer culture, although when used as probe, its homology was detected on different restriction fragments in the two cultures. The circle size distribution for the older culture was unchanged over several years and the restriction patterns for total mtDNA from the two cultures "looked identical". Dale *et al.* (1983) concluded that "it is relatively unimportant how plant mt genomes are arranged physically as long as essential sequence information is present". We agree with this statement but not with the implication that the 10.1 kb of sequence complexity is necessarily essential for mitochondrial function. It could be part of the "extra" DNA that may serve no function for the host plant. In a functional sense it could be analogous to the supercoiled mitochondrial plasmids found in some but not other strains of *Neurospora crassa* (Stohl *et al.,* 1982; Collins and Lambowitz, 1983) or the "optional" introns in fungal mitochondria. The cellular trauma of moving from plant to suspension culture may shock the mitochondrial genome so that certain sequences become rearranged, excised, amplified and as a consequence of replication or recombination, circularized. The process may be analogous to that when yeast petites are generated. The fact that many different sequences may undergo this transition and persist at high copy number suggests the phenomenon is an aberration that is of no use to the plant cell in which it occurs, and may illustrate the level of tolerance of useless or somewhat harmful DNA. Evidently such aberration is more tightly controlled in intact plant tissue. It is also possible that an individual plant is a mosaic in that the circular mtDNA molecules are not the same in each cell, just as the different petite colonies derived from a culture of cloned, grande yeast cells do not all contain mtDNA from the same region of the wild type genome. When a piece of plant tissue is placed into culture, some cells could outgrow others in a gradual cloning process of both cells and, incidentally, a particular mtDNA circle population that finally stabilizes. It would be informative to begin several tissue cultures from an individual plant and monitor changes in the distribution of circles for each as the cells are passaged in culture.

It is not implied that because a sequence is present in the high copy number circles it is not essential for mitochondrial function. The sequences found in high copy but excised from the mitochondrial genome of *N. crassa* include rDNA (Manella *et al.*, 1979). It is implied, however, that the additional dosage of those sequences that are genic is at best not deleterious to mitochondrial function and that the parental sequences in normal linkage arrangements are entirely sufficient for wild type function. The fact that the circular DNAs hybridize with mitochondrial RNA in maize, bean and tobacco does not imply, as Dale (1981) does suggest, that they contribute to mitochondrial function (see the above discussion on mitochondrial transcription).

D. What Do the Circles Represent?

We can now summarize the data on circles that have been reported for plant mtDNA. A satisfactory conclusion is difficult to reach if one expects that the circles obtained are either typically representative of the entire genome or represent multiple copies of certain sequences whose transcripts or encoded products are needed at elevated levels. If instead, one considers the circles as examples of irrepressible sequence replication that make little or no positive contribution to mitochondrial function, then their puzzling properties become understandable: the completely different size distributions for two cell cultures from the same tobacco cultivar, the completely different size distributions for maize suspension cells compared to leaf tissue (N, S or T types; these are entirely different among themselves), the variable presence of certain plasmid-like circles (and linear molecules too) among strains, and the low sequence complexity of circles for several cell cultures. (Recall from the discussion above that the restriction patterns for the supercoiled and non-supercoiled fractions were reported to be the same for tobacco and the same result was noted for *Petunia* by Lifshitz *et al.* (1982) and Kool *et al.* (1984), although this seems unlikely for bean, maize, sugarbeet and *Oenothera,* and is not the case for a different tobacco cultivar). Production of the circles thus far characterized may be considered as incidental events that accompany the process of DNA replication which may not be as stringently controlled in mitochondria as in the other organelles. Recombination events may be prerequisite to circle production, as discussed below, and may also be responsible for scrambling sequences to the point of "non-homology" in the case of some mitochondrial plasmids. A concatemeric array of monomer units is a consequence of the DNA replication process in phages T7, P22, T4, T1 (Watson, 1972), some bacterial plasmids (James *et al.,* 1982; Hakkaart *et al.,* 1984) the maxicircle component of kinetoplast (mitochondrial) DNA of *Crithidia fasciculata* (Hajduk *et al.,* 1984), and probably plasmids S1 and S2 of maize. That circles are more prevalent when extracted from suspension cultures than whole plant tissues may merely reflect the fact that many cells in culture divide and thus are replicating mtDNA, whereas very few cells in leaves and seedlings are meristematic and thus may not be replicating their DNA.

E. Evidence for a Circular Mitochondrial Genome

None of the circles yet detected microscopically or as bands on gels are large enough to represent the entire plant mitochondrial genome. [Note that the very large rosette structure shown for *Citrus* mtDNA (Fontarnau and Hernández-Yago, 1982) and *Shaerocarpus* chloroplast DNA (Herrmann *et al.*, 1980) appear similar to the rosettes shown to be artifacts by Leon and Macaya (1983).] Nevertheless, restriction fragment mapping has yielded circular maps for the mtDNA of *Brassica campestris* (Palmer and Shields, 1984), *B. oleracea* (Chétrit *et al.*, 1984), maize (Lonsdale *et al.*, 1983b) and wheat (Quetier, personnal communication). For *B. campestris* blot hybridization revealed several examples of small repeated sequences in addition to one major repeat of 2 kb. By using densitometer analysis certain restriction fragments that included the 2 kb repeat were found in substoichiometric levels relative to others. The simplest model proposed for the genome was one molecule of 218 kb containing a pair of 2 kb repeats in direct orientation plus circles of 135 and 83 kb each containing one copy of the 2 kb repeat. The two smaller circles were "postulated to interconvert with the master chromosome via a co-integration-resolution pathway mediated by reciprocal recombination with the 2-kb repeat". From the densitometer analysis it was surmised that "there are approximately two molecules of the two smaller circles for each molecule of the master chromosome". Although no molecules the size of any of the three circles were detected, the tripartite model for the mitochondrial genome is quite attractive. A similar tripartite model for maize has also been proposed (Lonsdale *et al.*, 1983b). It is ironic that among the myriad of circular plant mtDNA molecules described through the years, these hypothetical circles are the most likely to represent the entire genome. Here, as was the case with baker's yeast, it is not likely that genome-sized circles can be isolated intact for the reasons discussed below.

1. Site-Specific Versus General Recombination

The type of recombination proposed as responsible for fragmenting the master circle into the two smaller circles in the case of *B. campestris* and maize is a *site-specific* event, analogous to the integration — excision of lambda phage in *E. coli* and the inversion (flipping) of part of the "2 micron" plasmid in yeast that results from a crossover between its two repeats of opposite orientation. [An intramolecular crossover between direct repeats in a circle leads to two smaller circles (a "cutout"), whereas one involving inverted repeats causes an inversion or "flipping" of the region between the repeats. The *FLP* gene product of the 2 micron plasmid promotes either cutouts or flipping depending on whether its target sequence is in the direct or inverted orientation. It works on supercoils, relaxed circles and linear DNA (Vetter *et al.*, 1983).] The site-specific type of recombination was proposed to account for the different linkage arrangements of rRNA genes in wheat mtDNA (Falconet *et al.*, 1984). A *site-specific* recombination occurs at a particular sequence found in each of the two DNA regions undergoing the exchange because protein(s) mediating the event recognize that particular sequence. A *general* recombination (in the *RecA* system of *E. coli*, for example) occurs between homologous DNAs of many different primary sequences. One of the proteins required for sequence recognition in site-specific recombination is typically encoded on the DNA carrying the recombining sequence, and Palmer and Shields (1984) speculate that such a protein is linked to the 2 kb repeat in *B. campestris* mtDNA.

A major reason for proposing that the mtDNA recombination is site-specific in each of the three plants was that recombination was inferred to occur for one repeat sequence and not for different low copy repeats that were detected by hybridization and mapping. It may well be that site-specific recombination is responsible for these observations but we suspect that it is general recombination mediated by nuclear-encoded enzymes that is more likely operating here for two reasons. The first, and weaker reason is that if mtDNA circles that show no homology by hybridization tests (in *Oenothera* and sugarbeet, for example) are initially excised as cutouts (followed by amplification), then a different site-specific system would be necessary for each type of circle. There might, however, be a common sequence too short to be detected by hybridization. [The *FLP* recognition site is this type of short sequence (Jayaram and Broach, 1983).] Falconet *et al.* (1984) have argued for site-specificity to account for the inference that recombination is "highly specific, occurring only at a limited number of sites". The second reason is a stronger one, if we accept the analogy between recombination in yeast to generate petites and that in plants to generate circles. Sequence analysis has show a common structural motif in mtDNA from 32 petite mutants of baker's yeast (de Zamaroczy *et al.*, 1983). In each case the "excision sequences" within which crossing over generated the cutouts were "perfect direct repeats, often flanked on one or both sides by regions of patchy homology". Some of the excision sequences contained only A-T pairs, some contained mostly G-C pairs, and many shared no sequence similarity et all. General recombination at homologous sequences rather than site-specific recombination is clearly operating here. The frequency with which different petites arise is highly variable and "it is very likely that the most stable (longest and/or richest in GC) excision sequences are associated with the highest rates of excision" (de Zamaroczy *et al.*, 1983). If we apply these ideas to plant mtDNA, the fact that some repeats are used preferentially for cutouts and flipping may be explained by their preferential use as excision sequences in a general recombination system with the enzymes probably encoded by nuclear DNA. [Marotta *et al.* (1982) suggest a nuclear gene governs the excision process in baker's yeast.]

The apparent stability of the plant mitochondrial genome through many generations in plants and in cultured cells may not be due to site-specificity limiting recombination to certain sites, as suggested by Falconet *et al.* (1984), but rather to selection of those cells that adequately perform respiratory functions among others that cannot because of instability of the mitochondrial genome. The frequency of spontaneous petite mutants among yeast species, exclusive of baker's yeast, ranges from 0.003 to 0.17 % per generation; in strains of baker's yeast it ranges from 0.6 to 9 % (Clark-Walker *et al.*, 1981). The plant mitochondrial genome may well be unstable due to general recombination, but the cells with genomes rendered defective might be outgrown and we would therefore have no way to assess genome stability. Reassociation kinetics analysis revealed the presence of mtDNA sequences reiterated about 10—50 times and representing 5—10 % of the genome in pea and the cucurbits listed in Table 1 (Ward *et al.*, 1981). Since the repeated sequences were not evident as very prominent bands in restriction profiles, it is inferred that they are short and interspersed with each other or with non-repeated sequences. They could serve as sites for general recombination.

2. Circular Molecules Are Not Common in Mitochondrial DNA from Whole Plant Tissue

We now come to the question of why circular molecules are so rare in mtDNA extracted from whole plant tissue. One reason, as mentioned above, is that circular forms may be a consequence of DNA replication, and mtDNA replication occurs only in the small percentage of cells that are meristematic in the whole plant tissues examined. In this interpretation the genome would exist in linear molecules within the non-dividing cells. This possibility will be discussed below, but first we consider another possibility, that the genome is circular *in vivo* and the failure to observe the circles is a consequence of degradation of the molecules. Physical degradation due to shear forces cannot be the reason since <1 % of the molecules extracted from *Chlamydomonas* mitochondria are circular (Ryan *et al.*, 1978), yet this genome at 9.7 Md is the smallest reported for any eukaryote (Grant and Chiang, 1980) and much larger bacterial plasmid DNA molecules are routinely obtained in supercoiled form. It should be noted, too, that no unit circular genomes could be found in purified mtDNA from *Saccharomyces cerevisiae* (Hollenberg *et al.*, 1970; Locker *et al.*, 1974; Lozewska and Slonimski, 1976) or supercoiled mtDNA from *Achlya* (Hudspeth *et al.*, 1983) despite their modest genome sizes (Table 1). Genome-sized circles for *S. cerevisiae* and *S. carlsbergensis* were found very rarely and in relaxed form in purified mtDNA by Christiansen and Christiansen (1976) and in supercoiled form in mitochondrial lysates produced by osmotic shock by Hollenberg *et al.* (1970).

If shear degradation isn't responsible, perhaps nucleases are responsible for the paucity of circles in extracted mtDNA. Larger circles should be more susceptible to enzymatic (or physical) degradation than smaller ones. Thus the average length of circular molecules would be expected to be less than that of linear molecules. It is unfortunate that very few reports in which contour lengths of plant mtDNA circles are presented also contain length data for the more numerous linear molecules. For pea (Kolodner and Tewari, 1972) and *Citrus* (Fontarnau and Hernández-Yago, 1982), however, the average size of the circles is greater than that of the linear forms. For *Chlamydomonas* the average size of circular and linear molecules is about the same (Ryan *et al.*, 1978), but essentially all the linears have unique ends and the mtDNA restriction maps are linear (Grant and Chiang, 1980). Thus for *Chlamydomonas* the linears are not the result of random breakage of the circles. Either the linears result from a double stranded break at a single and unique point on the circle while in the mitochondrion or during extraction, as suggested by Grant and Chiang (1980), or this mitochondrial genome is actually linear as is the case in *Tetrahymena* (Suyama and Miura, 1968) and *Paramecium* (Cummings and Pritchard, 1982), and the rare circles are replicating forms or represent regions of the genome excised by recombination between direct repeats. It should be noted that the rare *Chlamydomonas* mtDNA circles that were measured ranged from 4.0 to 5.2 μm and "did not fall into a very narrow range as circular DNA molecules usually do" (Ryan *et al.*, 1978).

If the linear mtDNA molecules obtained from higher plants were not simply breakdown products of the circles that have thus far been characterized, they could be degradation products of genome-sized circles that have not yet been isolated but are common in mitochondria. It is also possible that DNA within mitochondria of living plant cells is continuously undergoing rearrangement at sites of directly repeating sequences so that genome-sized circular molecules are very rare or non-existent. In this case the steady would be a population of molecules in which deletions of regions between direct repeats generate smaller cutouts which can recombine via the same repeated sequences, as postulated for *Brassica campestris* (Palmer and Shields, 1984) and maize (Lonsdale *et al.*, 1983b). If "flipping" or inversion occurs between inverted repeats, as has been inferred from restriction mapping of *Achlya* mtDNA, the process might not be expected to prevent the detection of genome-sized circles, but yet no band is found at the position of supercoiled DNA in CsCl-ethidium bromide gradients (Hudspeth *et al.*, 1983). Thus despite circular restriction maps for mtDNA of maize and *Achlya*, not only are no genome-sized molecules found in purified mtDNA, but very few circles of any size are found. (No electron microscopic analysis has been reported for *Achlya* or *B. campestris*, but we may assume that the latter will yield no more circular mtDNA than is typical from native plant tissue of any species.) The very act of recombination, which involves double-stranded cleavage, might actually prevent the isolation of circular molecules. The many identical copies of the mitochondrial genome per cell are probably accessible to one another because of continuous mitochondrial fusion and fragmentation in baker's yeast (Stevens, 1981) and algae and plants (Bendich and Gauriloff, 1984, and references therein), and could lead to intermolecular strand exchange. This would not result in new combinations detectable by either restriction or genetic analyses but could, along with cutouts and flipping, contribute to a steady state of broken-stranded intermediates, given a high frequency of strand exchange. We have no such data for plants, but for baker's yeast the rate of recombination is very high (Dujon, 1981; de Zamaroczy *et al.*, 1983) and the inability to demonstrate semi-conservative mtDNA replication in density-shift experiments was attributed to frequent strand exchange (Williamson and Fennell, 1974).

3. Circles and mtDNA Replication

Perhaps the plant mitochondrial genome is not circular, as has been more or less universally assumed, but all the mtDNA sequence complexity is actually arranged in linear form within mitochondria of whole plant tissues. In cultured plant cells the genome might also be linear with the additional presence of many circular molecules, frequently in oligomeric array, representing only a subset of the entire sequence complexity. The first suggestion that the plant mitochondrial genome might not be circular was made by Quetier and Vedel (1980) who commented that "the possibility cannot be ruled out that" mtDNA "occurs predominantly as linear molecules" and "circularization takes place only in specific conditions, i. e. for replication". How can we reconcile a linear molecule with clearly circular restriction maps? If the termini of a linear molecule were repetitious, then restriction fragments containing the left-hand terminus plus contiguous sequences would hybridize with fragments containing the right-hand terminus plus contiguous sequences, providing that the fragments were longer than the repeated sequence(s) within the termini. For example, in the linear array a-b-c-d-e-f-a, a fragment containing sequences a-b would hybridize also to one containing f-a and thus a circular map would be obtained.

A relevant and interesting observation was made by Hajduk *et al.* (1984) for the maxicircle component of *Crithidia fasciculata* kinetoplast DNA. Recently replicated maxicircle DNA was found as linear molecules 43 kb in length with repeated termini of 6 kb. Most of the circular maxicircles were found interlocked with the network of thousands of circular mtDNA molecules and were only 37 kb in length. The replication of maxicircles was found to occur on the network by a rolling circle mechanism in which the terminal repetition was thought to be produced. The 6 kb of terminally repeated DNA was subsequently eliminated upon reincorporation of the maxicircle into the network. Electron microscopic analysis of pearl millet mtDNA revealed "lasso-like" structures of a circle with a tail that might represent rolling circle replication forms, although the lengths of these molecules would be much less than the full genome (Kim *et al.*, 1982a). The linear mitochondrial plasmids S1 and S2 of maize contain inverted terminal repeats of 200 base pairs (Kim, 1982b) and their proposed replication forms are thought to be concatemeric (Kemble and Mans, 1983). Another example relevant to our discussion is the linear form of the proviral DNA of retroviruses (Varmus, 1982). Here the genome is a linear molecule with long terminal repeats (LTRs) that would yield a circular restriction map using suitable sized fragments for mapping. The phage T7 provides yet another example (Watson, 1972).

Two major pieces of evidence for the tripartite circular model of the mitochondrial genome in *B. campestris* and maize were stubstoichiometric levels of certain restriction fragments containing a direct repeat, and the repeat flanked by four unique sequences in paired combinations. Recombination between direct repeats in a linear molecule would also lead to the underrepresentation of the fragments and the four paired combinations, but in this case the products of recombination are one excised circle plus one linear molecule deleted for the region between the repeats.

The data leading to a circular restriction map for the mitochondrial genome of both *B. campestris* (Palmer, personal communication) and *Achlya* (Grossman, personal communication) are not compatible with a linear molecule bearing terminal repeats, as exists for phage T7. There is one way in which a linear molecule would be compatible with the data: a circularly permuted molecule. The genome of phage T4 obtained from virions exists as a collection of molecules with sequences a-b-c-d-e-f-a, b-c-d-e-f-a-b, c-d-e-f-a-b-c, etc. The terminal redundancy is a consequence of the replication mechanism (Watson, 1972). The circular restriction maps for plants, *Achlya* and baker's yeast would all be consistent with circularly permuted linear DNA molecules.

The last refuge of the investigator unable to obtain nucleic acids of anticipated size is to invoke nuclease activity. Nucleases may well be responsible for this inability in the case of mtDNA and there is as yet no evidence for circular permutation of this genome. Such evidence would be difficult to obtain since the prediction based on the T4 precedent would be unit genome-sized linear molecules terminally repetitious for any sequence. *Brassica* species offer the best hope for obtaining unbroken linear molecules the length of the entire plant genome, and for the smaller genomes of *Achlya* and baker's yeast one would expect to isolate such linear molecules without difficulty. Though some genome-sized linear molecules are found, they are not common in mtDNA from baker's yeast (Hollenberg *et al.*, 1970; Christiansen and Christiansen, 1976).

4. Is the Genome Really Circular?

Is the plant mitochondrial genome really circular? The restriction maps would seem to indicate circularity for plants, *Achlya* and baker's yeast. The only alternative is circularly permuted linear molecules, and this is unprecedented among cellular DNAs. Nevertheless, two questions cause us to accept the circular model only with some reservation. Why are genome-sized circles almost never found in purified mtDNA from baker's yeast? For the smallest mitochondrial genome, in *Chlamydomonas*, why are numerous genome-sized circles not found? Certainly nuclease activity during extraction may be the answer. A steady state of broken-stranded recombination intermediates tenously held in recombination apparatuses could also be the answer. Genome-sized circles by observation would be preferable to the circles by inference, but we are not likely to see them soon.

5. *Is Circularity of the Genome Important for Mitochondrial Function?*

We now come to the question of whether circularity of this genome is useful for its replication and/or expression. Despite the special theoretical problems attendant to the replication of linear but not circular DNA molecules (Watson, 1972), chromosomes and many genomes are linear, including the mitochondrial genomes of *Paramecium* and *Tetrahymena*. Restriction mapping suggests a circular mitochondrial genome for normal (male fertile) maize, but a predominantly linear form for type S cytoplasmic male sterile (cms) maize Schardl *et al.*, 1984). Aside from their apparent failure to function properly during anther maturation, mitochondria with their linear genome function adequately in all other stages of maize development since there is no other obvious phenotypic difference between normal and cms-S plants. Furthermore, cms-S plants restored to fertility by nuclear restorer genes still have the S-type cytoplasm and the linear mitochondrial genome (data cited by Schardl *et al.*, 1984). Therefore, "linearization alone is not sufficient to cause male sterility" (Schardl *et al.*, 1984).

Having huffed and puffed through the properties of circular mtDNA molecules, we must now conclude that circularity may not be necessary or even useful for mitochondrial function.

IV. Summary and Conclusions

Two questions are addressed in this article. Why is the mitochondrial genome so much larger in plants than in other eukaryotes? And what is the relationship between the circular mtDNA molecules that have been observed and the entire genome?

A. The "Extra" DNA

The very large sequence complexity (genome size) for plants compared to other organisms could represent more genes, DNA that functions in a sequence-independent manner, or DNA that is not useful for mitochondrial function (selfish or ignorant DNA). The number of mitochondrial translation products is about the same among plants with widely different genome sizes and is roughly the same as the number of protein-coding genes in the much smaller yeast and mammalian mitochondrial genomes. Although there are severe reservations in equating translation products with genes, the available data do not suggest that plants carry many more mitochondrial genes than do other organisms. Although a large fraction of the genome is represented in plant mitochondrial RNA, data from fungal and animal cells suggest that the extent of transcription may not be a good indicator of the number of expressed genes.

There is no evidence at present to indicate that much of the mtDNA serves a sequence-independent function, as has been postulated (but not demonstrated) for the very large and variable nuclear DNA content (the C-value paradox). It should be emphasized, however, that only one such function (organelle volume determined by its DNA content) has been investigated.

Among types of DNA in the selfish or ignorant category, optional introns (as in mtDNA of some but not other wild type strains of fungal species) and interorganellar DNA (chloroplast DNA in mitochondria, for example) may contribute to the size of the genome, but from the very limited evidence thus far do not explain much of the extra DNA in plant mitochondria. One possible explanation draws upon an analogy to the mtDNA of the kinetoplast in which recombinative scrambling of sequences on a fine scale can evidently convert "repetitive" DNA to "unique" DNA of high sequence complexity.

In order to test the idea that "extra" DNA is not necessary for mitochondrial function a transformation system is needed for producing modified versions of the mitochondrial genome that can be assayed in whole plants. The requirement for a transformation system in assessing the functional significance of mtDNA sequences is far more stringent for plants than other organisms because the genetic approach is, except for cytoplasmic male sterility, simply unavailable at present. Developing a transformation system is, consequently, our most important objective for further progress in the molecular biology of plant mitochondria.

B. Circles

Only small amounts of circular molecules have been obtained from whole plant tissues, whereas supercoiled molecules can sometimes be isolated in much larger amounts from tissue culture cells thus permitting more extensive characterization. The circles are frequently in an oligomeric series of a low-complexity unit, as in fungi (yeast petites, for example), but in two instances the supercoiled fraction was found to contain the sequence complexity of the entire genome. In no case has a single circle been found that approaches the size of the entire genome. The circles may be formed by excision of regions between directly repeating sequences by general rather than site-specific recombination (the same process that produces yeast petite mtDNA) and may persist at high copy number as a consequence of mtDNA replication. Their elevated levels in tissue culture cells may, therefore, merely reflect the higher percentage cells that divide (and replicate mtDNA) in culture than in whole plant tissues where meristematic cells are rare. The circles do not appear to represent sequences needed at high copy number for mitochondrial function since many different sequences can be found in the circles and different tissue cultures from the same tobacco species contain completely different size class distributions, as do different cultivars of maize which also differ from cultured maize cells. It is possible that different cells in a leaf, for example, contain different circle populations, as do individual petite colonies derived from a wild type yeast culture, and that cells with their particular circle distributions are gradually cloned in different tissue culture lines.

Restriction mapping and analysis for some plants suggests that the form of the genome within mitochondria is circular, although no genome-sized circles have been observed and circularly permuted linear molecules (as for phage T4) are also compatible with the data. In any case the conclusion that all the sequence complexity is on a single linkage group applies also to mitochondrial genomes of all other organisms. In fact, the very large and variable size of the genome and the seemingly perplexing circle distributions appear merely to be unusual variations on a common mitochondrial theme.

V. References

Altman, P. L., Katz, D. D., eds., 1976: In: Biological Handbooks. I. Cell Biology, pp. 217—218. Bethesda, Maryland: Federation of American Society of Experimental Biology.

Attardi, G., Cantatore, P., Chomyn, A., Crews, S., Gelfand, R., Merkel, C., Montoya, J., Ojala, D., 1982: A comprehensive view of mitochondrial gene expression in human cells. In: Slonimski, P., Borst, P., Attardi, G. (eds.), Mitochondrial Genes, pp. 51—71. Cold Spring Harbor, New York: Cold Spring Harbor Lab.

Bayen, M., Rode, A., 1973: The 1.700 DNA of *Chlorella pyrenoidosa:* heterogeneity and complexity. Plant Sci. Lett. **1,** 385—389.

Bendich, A. J., 1982: Plant mitochondrial DNA: the last frontier. In: Slonimski, P., Borst, P., Attardi, G. (eds.), Mitochondrial Genes, pp. 477—481. Cold Spring Harbor, New York: Cold Spring Harbor Lab.

Bendich, A. J., Gauriloff, L. P., 1984: Morphometric analysis of cucurbit mitochondria: the relationship between chondriome volume and DNA content. Protoplasma **119,** 1—7.

Benne, R., De Vries, B. F., Van den Berg, J., Klaver, B., 1983: The nucleotide sequence of a segment of *Trypanosoma brucei* mitochondrial maxi-circle DNA that contains the gene for the apoctyochrome b and some unusual unassigned reading frames. Nucleic Acids Res. **11,** 6925—6941.

Bennett, M. D., Smith, J. B., Smith, R. I. L., 1982: DNA amounts of angiosperms from the Antarctic and South Georgia. Environ. Exp. Bot. **22,** 307—318.

Bonen, L., Boer, P. H., Gray, M. W., 1984: The wheat cytochrome oxidase subunit II gene has an intron insert and three radical amino acid changes relative to maize. EMBO J. **3,** 2531—2536.

Bonen, L., Gray, M. W., 1980: Organization and expression of the mitochondrial genome of plants. I. The genes for wheat mitochondrial ribosomal and transfer RNA: evidence for an unusual arrangement. Nucleic Acids Res. **8,** 319—335.

Borst, P., Grivell, L. A., Groot, G. S. P., 1984: Organelle DNA. Trends in Biochem. Sciences **9,** 128—130.

Boutry, M., Briquet, M., 1982: Mitochondrial modifications associated with the cytoplasmic male sterility in faba beans. Eur. J. Biochem. **127,** 129—135.

Boutry, M. Briquet, M., Goffeau, A., 1983: The α subunit of a plant mitochondrial F_1-ATPase is translated in mitochondria. J. Biol. Chem. **258,** 8524—8526.

Boutry, M., Faber, A.-M., Charbonnier, M., Briquet, M., 1984: Microanalysis of plant mitochondrial protein synthesis products: detection of variant polypeptides associated with cytoplasmic male sterility. Plant Molec. Biol. **3,** 445—452.

Brennicke, A., 1980: Mitochondrial DNA from *Oenothera berteriana*. Plant Physiol. **65**, 1207—1210.

Brennicke, A., Blanz, P., 1982: Circular mitochondrial DNA species from *Oenothera* with unique sequences. Molec. Gen. Genet. **187**, 461—467.

Burke, J. M., Breitenberger, C., Heckman, J. E., Dujon, B., RajBandary, U. L., 1984: J. Biol. Chem. **259**, 504—511.

Cantatore, P., Attardi, G., 1980: Mapping of nascent light and heavy strand transcripts on the physical map of HeLa cell mitochondrial DNA. Nucleic Acids Res. **8**, 2605—2625.

Cavalier-Smith, T., 1978: Nuclear volume control by nucleoskeletal DNA, selection for cell volume and cell growth rate, and the solution to the DNA C-value paradox. J. Cell Sci. **34**, 247—278.

Chen, K. K., Donelson, J. E., 1980: Sequences of two kinetoplast DNA minicircles of *Trypanosoma brucei*. Proc. Natl. Acad. Sci., U.S.A **77**, 2445—2449.

Chétrit, P., Mathieu, C., Muller, J. P., Vedel, F., 1984: Curr. Genet. **8**, 413—421.

Christiansen, G., Christiansen, C., 1976: Comparison of the fine structure of mitochondrial DNA from Saccharomyces cerevisiae and S. carlsbergensis: electron microscopy of partially denatured molecules. Nucleic Acids Res. **3**, 465—476.

Clark-Walker, G. D., McArthur, C. R., Daley, D. J., 1981: Does mitochondrial DNA length influence the frequency of spontaneous petite mutants in yeasts? Curr. Genet. **4**, 7—12.

Collins, R. A., Lambowitz, A. M., 1983: Structural variations and optional introns in the mitochondrial DNAs of *Neurospora* strains isolated from nature. Plasmid **9**, 53—70.

Cummings, D. J., Pritchard, A. E., 1982: Replication mechanism of mitochondrial DNA from *Paramecium aurelia:* sequence of the cross-linked origin. In: Slonimski, P., Borst, P., Attardi, G. (eds.), Mitochondrial Genes, pp. 441—447. Cold Spring Harbor, New York: Cold Spring Harbor Lab.

Dale, R. M. K., 1981: Sequence homology among different size classes of plant mtDNAs. Proc. Natl. Acad. Sci., U.S.A. **78**, 4453—4457.

Dale, R. M. K., Duessing, J. H., Keene, D., 1981: Supercoiled mitochondrial DNAs from plant tissue culture cells. Nucleic Acids Res. **9**, 4583—4593.

Dale, R. M. K., Wu, M., Kiernan, M. C. C., 1983: Analysis of four tobacco mitochondrial DNA size classes. Nucleic Acids Res. **11**, 1673—1685.

Dawson, A. J., Jones, V. P., Leaver, C. J., 1984: The apocytochrome b gene in maize mitochondria does not contain introns and is preceded by a potential ribosome binding site. EMBO J. **3**, 2107—2113.

de Zamaroczy, M., Faugeron-Fonty, G., Bernardi, G., 1983: Excision sequences in the mitochondrial genome of yeast. Gene **21**, 193—202.

Dixon, L. K., Leaver, C. J., 1982: Mitochondrial gene expression and cytoplasmic male sterility in sorghum. Plant Molec. Biol. **1**, 89—102.

Dover, G., 1980: Ignorant DNA? Nature **285**, 618—620; Cavalier-Smith, T., Smith, T. F., Reid, R. A., Nature **285**, 617—620.

Dujon, B., 1981: Mitochondrial genetics and functions. In: Strathern, J. N., Jones, E. W., Broach, J. R. (eds.), The Molecular Biology of the Yeast Saccharomyces. Life Cycle and Inheritance, pp. 505—635. Cold Spring Harbor, New York: Cold Spring Harbor Lab.

Englund, P., 1981: Kinetoplast DNA. In: Levandowsky, M., Hunter, S. H. (eds.), Biochemistry and Physiology of Protozoa, 2nd ed., Volume 4, pp. 333—383. New York: Academic Press.

Falconet, D., Lejeune, B., Quetier, F., Gray, M. W., 1984: Evidence for homologous recombination between repeated sequences containing 18 S and 5 S ribosomal RNA genes in wheat mitochondrial DNA. EMBO J. **3,** 297—302.

Fontarnau, A., Hernández-Yago, J., 1982: Characterization of mitochondrial DNA in *Citrus.* Plant. Physiol. **70,** 1678—1682.

Forde, B. G., Leaver, C. J., 1980: Nuclear and cytoplasmic genes controlling synthesis of variant mitochondrial polypeptides in male-sterile maize. Proc. Natl. Acad. Sci., U.S.A. **77,** 418—422.

Fox, T. D., Leaver, C. J., 1981: The Zea mays mitochondrial gene coding cytochrome oxidase subunit II has an intervening sequence and does not contain TGA codons. Cell **26,** 315—323.

Gall, J. G., 1981: Chromosome structure and the C-value paradox. J. Cell Biol. **91,** 3 s—14 s.

Grant, D., Chiang, K. S., 1980: Physical mapping and characterization of *Chlamydomonas* mitochondrial DNA molecules: Their unique ends, sequence homogeneity and conservation. Plasmid **4,** 82—96.

Grivell, L. A., 1983: Mitochondrial DNA. Scientific Amer. **248,** 78—89.

Hack, E., Leaver, C. J., 1983: The α-subunit of the maize F_1-ATPase is synthesized in the mitochondrion. EMBO J. **2,** 1783—1789.

Hack, E., Leaver, C. J., 1984: Synthesis of a dicyclohexylcarbodiimide-binding proteolipid by cucumber *(Cucumis sativus* L.) mitochondria. Curr. Genet. **8,** 537—542.

Hajduk, S. L., 1979: Dyskinetoplasty in two species of trypanosomatids. J. Cell Sci. **35,** 185—202.

Hajduk, S. L., Klein, V. A., Englund, P. T., 1984: Replication of kinetoplast DNA maxicircles. Cell **36,** 483—492.

Hakkaart, M. J. J., van den Elzen, P. J. M., Veltkamp, E., Nijkamp, H. J. J., 1984: Maintenance of multicopy plasmid Clo DF13 in E. coli cells: evidence for site-specific recombination at parB. Cell **36,** 203—209.

Herrmann, R. G., Palta, H. K., Kowallik, K. V., 1980: Chloroplast DNA from three archegoniates. Planta **148,** 319—327.

Hiesel, R., Brennicke, A., 1983: Cytochrome oxidase subunit II gene in mitochondria of *Oenothera* has no intron. EMBO J. **2,** 2173—2178.

Hollenberg, C. P., Borst, P., Van Bruggen, E. F. H., 1970: Mitochondrial DNA. V. A 25-μ closed circular duplex DNA molecule in wild-type yeast mitochondria. Structure and genetic complexity. Biochim. Biophys. Acta **209,** 1—15.

Hudspeth, E. S., Shumard, D. S., Bradford, C. J. R., Grossman, L. I., 1983: Organization of *Achlya* mtDNA: a population with two orientations and a large inverted repeat containing the rRNA genes. Proc. Natl. Acad. Sci., U.S.A. **80,** 142—146.

Jakovcic, S., Hendler, F., Halbreich, A., Rabinowitz, M., 1979: Transcription of yeast mitochondrial deoxyribonucleic acid. Biochem. **18,** 3200—3205.

James, A. A., Morrison, P. T., Kolodner, R., 1983: Isolation of genetic elements that increase frequencies of plasmid recombination. Nature **303,** 256—259.

Jayaram, M., Broach, J. R., 1983: Yeast plasmid 2-μm circle promotes recombination within bacterial transposon Tn5. Proc. Natl. Acad. Sci., U.S.A. **80,** 7264—7268.

Kemble, R. J., Bedbrook, J. R., 1980: Low molecular weight circular and linear DNA in mitochondria from normal and male-sterile *Zea mays* cytoplasm. Nature **284,** 565—566.

Kemble, R. J., Gunn, R. E., Flavell, R. B., 1980: Classification of normal and male-sterile cytoplasms in maize. II. Electrophoretic analysis of DNA species in mitochondria. Genetics **95,** 451—458.

Kemble, R. J., Mans, R. J., 1983: Examination of the mitochondrial genome of revertant progeny from S *cms* maize with cloned S-1 and S-2 hybridization probes. J. Molec. Appl. Genet. **2,** 161—171.

Kim, B. D., Lee, K. J., DeBusk, A. G., 1982a: Linear and 'lasso-like' structures of mitochondrial DNA from *Pennisetum typhoides*. FEBS Letts. **147,** 231—234.

Kim, B. D., Mans, R. J., Conde, M. F., Pring, D. R., Levings, C. S. III, 1982b: Physical mapping of homologous segments of mitochondrial episomes from S male-sterile maize. Plasmid **7,** 1—14.

Kolodner, R., Tewari, K. K., 1972: Physicochemical characterization of mitochondrial DNA from pea leaves. Proc. Natl. Acad. Sci., U.S.A. **69,** 1830—1834.

Kool, A. J., de Haas, J. M., Mol, J. N. M., van Marrewijk, G. A. M., 1984: Isolation and physicochemical characterization of mitochondrial DNA from cultured cells of *Petunia hybrida*. Theor. Appl. Genet. **68,** in press.

Labouesse, M., Slonimski, P. P., 1983: Construction of novel cytochrome b genes in yeast mitochondria by subtraction or addition of introns. EMBO J. **2,** 269—276.

Lazowska, J., Slonimski, P. P., 1976: Electron microscopy analysis of circular repetitive DNA molecules from genetically characterized rho$^-$ mutants of *Saccharomyces cerevisiae*. Molec. Gen. Genet. **146,** 61—78.

Leaver, C. J., Gray, M. W., 1982: Mitochondrial organization and expression in higher plants. Ann. Rev. Plant Physiol. **33,** 373—402.

Lebacq, P., Vedel, F., 1981: Sal I restriction enzyme analysis of chloroplast and mitochondrial DNAs in the genus *Brassica*. Plant Sci. Lett. **23,** 1—9.

León, P., Macaya, G., 1983: Properties of DNA rosettes and their relevance to chromosome structure. Chromosoma **88,** 307—314.

Levings, C. S. III, 1983: The plant mitochondrial genome and its mutants. Cell **32,** 659—661.

Levings, C. S. III, Shah, D. M., Hu, W. W. L., Pring, D. R., Timothy, D. H., 1979: Molecular heterogeneity among mitochondrial DNAs from different maize cytoplasms. In: Cummings, D. J., Borst, P., David, I. B., Weissman, S. M., Fox, C. F. (eds.), Extrachromosomal DNA, ICN-UCLA Symposia on Molecular and Cellular Biology, Volume 15, pp. 63—73. New York: Academic Press.

Lifshitz, I., Shamay, I., Beckmann, J., 1982: Isolation of circular mitochondrial DNA from suspension culture cells of Petunia. Plant Molec. Biol. Newslett. **3,** 6—7.

Locker, J., Rabinowitz, M., Getz, G. S., 1974: Electron microscopic and renaturation kinetic analysis of mitochondrial DNA of cytoplasmic petite mutants of *Saccharomyces cerevisiae*. J. Molec. Biol. **88,** 489—507.

Lonsdale, D. M., Hodge, T. P., Fauron, M.-R., Flavell, R. B., 1983b: A predicted structure for the mitochondrial genome from the fertile cytoplasm of maize. In: Goldberg, R. B. (ed.), Plant Molecular Biology, ICN-UCLA Symposium on Molecular and Cellular Biology, New Series, Volume 12, pp. 445—456. New York: Alan R. Liss.

Lonsdale, D. M., Hodge, T. P., Howe, C. J., Stern, D. B., 1983a: Maize mitochondrial DNA contains a sequence homologous to the ribulose-1,5-bisphosphate carboxylase large subunit gene of chloroplast DNA. Cell **34,** 1007—1014.

Mannella, C., Goewert, R. R., Lambowitz, A. M., 1979: Characterization of variant Neurospora crassa mitochondrial DNAs which contain tandem reiterations. Cell **18,** 1197—1207.

Marotta, R., Colin, Y., Goursot, R., Bernardi, G., 1982: A region of extreme instability in the mitochondrial genome of yeast. EMBO J. **1,** 529—534.

McArthur, C. R., Clark-Walker, G. D., 1983: Mitochondrial DNA size diversity in the *Dekkera/Brettanomyces* yeasts. Curr. Genet. **7,** 29—35.

Nikiforova, I. D., Negruk, V. I., 1983: Comparative electrophoretical analysis of plasmid-like mitochondrial DNA in *Vicia faba* and in some other legumes. Planta **157,** 81—84.

Palmer, J. D., Shields, C. R., 1984: Tripartite structure of the *Brassica campestris* mitochondrial genome. Nature **307,** 437—440.

Palmer, J. D., Shields, C. R., Cohen, D. B., Orton, T. J., 1983: An unusual mitochondrial DNA plasmid in the genus *Brassica.* Nature **301,** 725—728.

Posakony, J. W., Scheller, R. H., Anderson, D. M., Britten, R. J., Davidson, E. H., 1981: Repetitive sequences of the sea urchin genome. III. Nucleotide sequences of cloned repeated elements. J. Molec. Biol. **149,** 41—67.

Powling, A., Ellis, T. H. N., 1983: Studies on the organelle genomes of sugarbeet with male-fertile and male-sterile cytoplasms. Theor. Appl. Genet. **65,** 323—328.

Quetier, F., Vedel, F., 1980: Physico-chemical and restriction endonuclease analysis of mitochondrial DNA from higher plants. In: Leaver, C. J. (ed.), Genome Organization and Expression in Plants, pp. 401—406. New York: Plenum.

Ryan, R., Grant, D., Chiang, K.-S., Swift, H., 1978: Isolation and characterization of mitochondrial DNA from *Chlamydomonas reinhardtii.* Proc. Natl. Acad. Sci., U.S.A. **75,** 3268—3272.

Schardl, C. L., Lonsdale, D. M., Pring, D. R., Rose, K. R., 1984: Linearization of maize mitochondrial chromosomes by recombination with linear episomes. Nature **310,** 292—296.

Sloof, P., Bos, J. L., Konings, A. F. J. M., Menke, H. H., Borst, P., Gutteridge, W. E., Leon, W., 1983: Characterization of satellite DNA in *Trypanosoma brucei* and *Trypanosoma cruzi.* J. Molec. Biol. **167,** 1—21.

Sparks, R. B., Jr., Dale, R. M. K., 1980: Characterization of ^{3}H-labeled supercoiled mitochondrial DNA from tobacco suspension culture cells. Mol. Gen. Genet. **180,** 351—355.

Spencer, D. F., Schnare, M. N., Gray, M. W., 1984: Pronounced structural similarities between the small subunit ribosomal RNA genes of wheat mitochondria and *Escherichia coli.* Proc. Natl. Acad. Sci., U.S.A. **81,** 493—497.

Stern, D. B., Lonsdale, D. M., 1982: Mitochondrial and chloroplast genomes of maize have a 12-kilobase DNA sequence in common. Nature **299,** 698—702.

Stern, D. B., Newton, K. J., 1985: Mitochondrial gene expression in *Cucurbitaceae:* conserved and variable features. Curr. Genet., in press.

Stern, D. B., Palmer, J. D., Thompson, W. F., Lonsdale, D. M., 1983: Mitochondrial DNA sequence evolution and homology to chloroplast DNA in angiosperms. In: Goldberg, R. B. (ed.), Plant Molecular Biology, ICN-UCLA Symposium on Molecular and Cellular Biology, New Series, Volume 12, pp. 467—477. New York: Alan R. Liss.

Stevens, B., 1981: Mitochondrial structure. In: Strathern, J. N., Jones, E. W., Broach, J. R. (eds.), The Molecular Biology of the Yeast Saccharomyces. Life Cycle and Inheritance, pp. 471—504. Cold Spring Harbor, New York: Cold Spring Harbor Lab.

Stohl, L. L., Collins, R. A., Cole, M. D., Lambowitz, A. M., 1982: Characterization of two new plasmid DNAs found in mitochondria of wild-type Neurospora intermidia strains. Nucleic Acids Res. **10,** 1439—1458.

Suyama, J., Miura, K., 1968: Size and structural variations of mitochondrial DNA. Proc. Natl. Acad. Sci., U.S.A. **60,** 235—242.

Synenki, R. M., Levings, C. S. III, Shah, D. M., 1978: Physicochemical characterization of mitochondrial DNA from soybeans. Plant Physiol. **61,** 460—464.

Varmus, H., 1982: Form and function of retroviral proviruses. Science **216,** 812—820.

Vedel, F., Mathieu, C., Lebacq, P., Ambard-Bretteville, F., Remy, R., Pelletier, G., 1982: Comparative macromolecular analysis of the cytoplasms of normal and cytoplasmic male sterile *Brassica napus*. Theor. Appl. Genet. **62,** 255—262.

Vetter, D., Andrews, B. J., Roberts-Beatty, L., Sadowski, P. D., 1983: Sitespecific recombination of yeast 2-μm DNA *in vitro*. Proc. Natl. Acad. Sci., U.S.A **80,** 7284—7288.

Ward, B. L., Anderson, R. S., Bendich, A. J., 1981: The mitochondrial genome is large and variable in a family of plants (Cucurbitaceae). Cell **25,** 793—803.

Watson, J. D., 1972: Origin of concatemeric T7 DNA. Nature **239,** 197—201.

Williamson, D. H., Fennel, D. J., 1974: Apparent dispersive replication of yeast mitochondrial DNA as revealed by density labelling experiments. Molec. Gen. Gent. **131,** 193—207.

Chapter 8

Repeated Sequences and Genome Change

R. B. Flavell

Plant Breeding Institute, Trumpington, Cambridge CB2 2LQ, England

With 6 Figures

Contents

I. Introduction

Repeated sequences, by their very nature of being present in many copies in the genome, appear to be prone to rapid change. Enzymes, especially those involved in recombination and replication, appear to mistake one copy for another and produce various new DNA structures. Other sorts of repeats, notably those in transposable elements, are recognised by enzymes, excised from the chromosomes and integrated elsewhere.

The amount of repeated DNA in plant genomes is high, especially in genomes containing more than 2 pg DNA, where it is in excess of 75 % of the total DNA (Flavell, 1980, 1982b; Thompson and Murray, 1981). These estimates were made using techniques which recognise repeats about 50 base pairs or longer. Many shorter repeats revealed by DNA sequencing are also present, and so the estimates are underestimates. Much of the repeated DNA does not code for proteins or play sequence-specific roles and so it can move, be amplified, deleted or replaced without apparent catastrophic effects on the species (Hinegardner, 1976; Flavell, 1982a). However, many genes are also reiterated and are not immune from the 'mutations' characteristic of repeats (reviewed in Ohta, 1983b). The high proportion of repeated DNA and the prevalence of multigene families leads one to suggest that plant genomes are potentially very unstable and therefore the genetic stability from generation to generation, i. e. over short time scales, must result from mechanisms that have evolved to suppress the potential instability. There would clearly be strong selection pressure for such mechanisms.

In discussion of repeated sequence flux it is necessary to discuss both the nature of the "mutations" (amplifications, deletions, transpositions, etc.) occurring in repeats of individuals and also the fate of the mutations in the population and species. Sometimes the mutation will be lethal or highly deleterious. If the mutation does not increase in frequency in the population it will have no significance in evolution. Mutations spread in a population and increase in frequency as a result of natural selection, random drift or various mechanisms such as unequal crossing over, transposition and biased gene conversion to which certain repeated sequences are particularly susceptible (Dover, 1982).

When arguments are proposed that repeats have enhanced probabilities of change in individuals and populations, it is, of course, important to quantify the extent of change within defined time frames. This is difficult to do for most repeats in plants, except for certain situations such as the excision of transposable elements from specific genes in a particular set of somatic or meiotic cells (Dellaporta, 1985). Such examples illustrate the most frequent of changes but most changes have to be evaluated over evolutionary time scales. Nevertheless, such comparisons are made easily and it is often straightforward to contrast repeats with, for example, conserved coding sequences and conclude that some repeats arise, evolve and turnover much more rapidly (Flavell, 1982a; Thompson and Murray, 1980).

The diversity of repeats in any one plant genome is enormous. It is therefore difficult to create general arguments about them. However, for the present discussion, it is convenient to divide them into categories relating to their arrangement and/or their function. There are families of repeats arranged in tandem arrays. There are others where the individual members are highly dispersed in the chromosomes. Transposable elements are included in this category. Then there are families whose members are clustered together at a locus but not in a tandem array. Many small multigene families are in this group.

The long-term stability of a repeat appears to depend, amongst other things, on its sequence, its functions, its arrangement with respect to its homologues and what changes are occurring to its relatives and other sequences in the genome. In this chapter some of the sorts of changes to which repeats are susceptible are discussed together with the consequences. The examples chosen represent only a few of those that would be included in an exhaustive review. Other examples are included elsewhere in this volume.

Amplification, Deletion and Transposition of Tandem Arrays

All species examined possess tandem arrays of identical, or nearly identical sequences (reviewed in Flavell, 1980, 1982b; Thompson and Murray, 1981). Often the sequences are very numerous. The length of DNA amplified, i. e. the size of the repeating unit, can vary from a few base pairs (Dennis *et. al.*, 1980; Deumling, 1981) to thousands of base pairs (Bedbrook *et al.*, 1980b). The genome of one species can usually be distinguished from that of a close relative by the presence of arrays of at least one specific repeat family. Often this highly amplified sequence is present in only one or relatively few copies in the close relative (Bedbrook *et al.*, 1980a, b; Jones and Flavell, 1982b; Gerlach and Peacock, 1980). Comparisons between related genomes, using restriction endonucleases, have revealed extensive evidence for the amplification of different forms in each species of sequences found in both species (Bedbrook *et al.*, 1980a, b; Jones and Flavell, 1982b; Flavell, 1982b; Flavell *et al.*, 1983). The different forms can be small mutational variants of one another or new 'compound' repeat units created by recombination between two or more different repeats (Bedbrook *et al.*, 1980b; Flavell, 1980).

The common occurrence of tandem arrays of repeats and the major repeat differences between related species imply that large-scale amplifi-

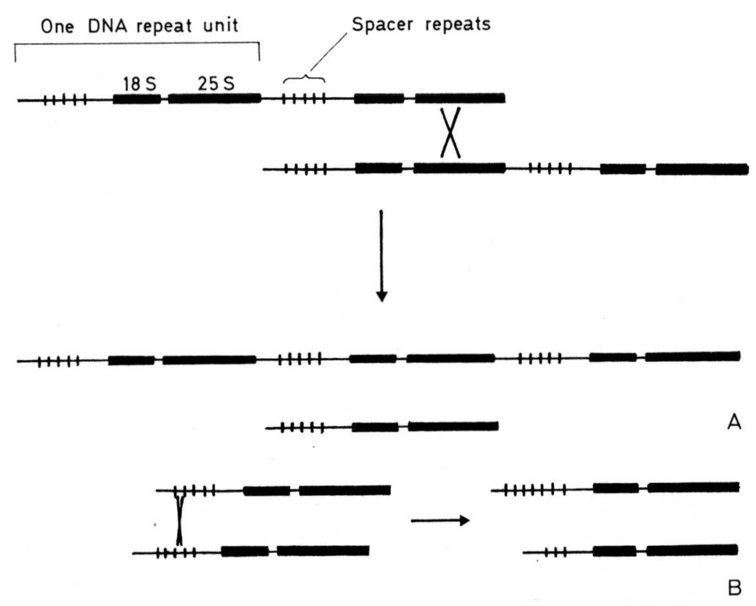

Fig. 1. Unequal crossing over in ribosomal RNA genes. Two members of a tandem array of repeat units containing 18 S, 5.8 S and 25 S ribosomal RNA genes are illustrated

A. Unequal crossing over between complete repeating units leads to chromosomes with more or fewer genes.

B. Unequal crossing over between the repeats in the spacer regions leads to genes with more or fewer spacer repeats.

cation events are very common kinds of mutations and some are also fixed rapidly in species. Estimates of such events in mouse cells in culture have suggested that 10 % of the cells may suffer such mutations! (Bostock and Tyler-Smith, 1982; Schimke, 1982.) Rapid amplification may occur by excision of DNA followed by replication on a rolling circle type model and reintegration (Hourcade *et al.*, 1973). Alternatively, aberrant replication may occur *in situ* leading to a localised tandem array (Schimke, 1982). Another model which can explain the amplification of sequences over a longer time-span involves recurrent unequal crossing over (Smith, 1976). Unequal crossing over is a process which creates a sequence duplication in one chromatid or chromosome and a corresponding deletion in the other as a consequence of, for example, inaccurate alignment in pairing and recombination in meiosis. The inaccuracy is provoked by the tandem repeat arrangement (see Figure 1 A). Such exchanges have been well documented in the ribosomal DNA arrays in yeast (Szostak and Wu, 1980; Petes, 1980). Because the process creates deletions as well as amplifications, it can account for the major differences between individuals within species for the number of copies of repeats in tandem array (see later).

Arrays of the same repeat are usually, but not always, found on more than one chromosome. This has been shown either by *in situ* hybridisation (e. g. Bedbrook *et al.*, 1980a; Jones and Flavell, 1982a, b; Deumling and Greilhuber, 1982; Gerlach and Peacock, 1980) or by hybridisation to a series of DNAs isolated from chromosome addition lines each containing only a single chromosome of a species (Bedbrook *et al.*, 1980a, b; Jones and Flavell, 1982a). These results show that arrays often become duplicated or divided and transposed to other chromosomes (homologous and non-homologous) and that the new variant chromosomes are fixed efficiently in populations. The transpositions of repeats may involve double crossovers between interacting chromosome segments or excision and integration of DNA fragments. Repeats in tandem array are presumably susceptible to deletions of circular segments of repeats by intrastrand recombination (Figure 2) and the circles could be reintegrated elsewhere. Several examples of plants carrying deletions from specific tandem arrays of ribosomal RNA genes have been found in a relatively small sample of inbred wheat lines (unpublished results) but whether the deletions occurred as in

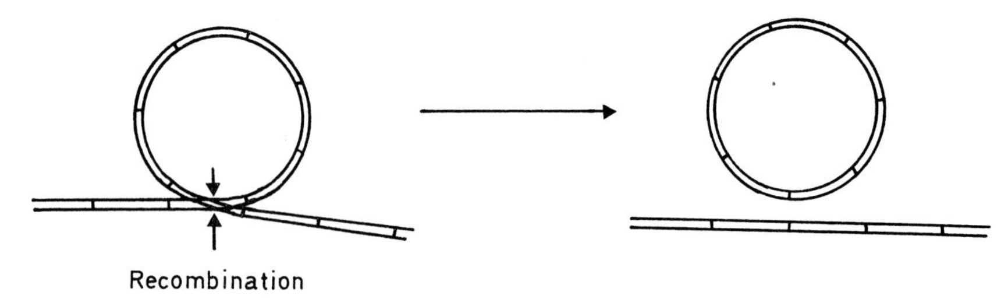

Fig. 2. Deletion of repeats from a tandem array by intrastrand recombination. An array of repeats is represented. Reciprocal recombination between members of the same array leads to deletion of a circular array of repeats

Figure 2 or by unequal crossing over is not known. The transfer of arrays of repeats between specific regions of chromosomes which frequently lie close together in the nucleus has been discussed elsewhere (Jones and Flavell, 1982a; Flavell, 1983). Transposition is likely to be very important in aiding the fixation of arrays in a species because a high rate of transfer to new chromosomes within individuals of an outbreeding population, substantially increases the proportion of individuals containing the repeat family. This is due to the segregation of homologous and non-homologous chromosomes at meiosis in each generation as shown schematically in Figure 3 (see Dover, 1982; Ohta and Dover, 1983). Dover (1982) has pointed out that in these circumstances, the population can change, with respect to the repeat family, in a 'cohesive' manner.

Why are certain arrays of sequences transposed at apparently high frequencies to other chromosomes? It could be due to some structural features of the amplified sequences which enhance the probability of such events, e. g. the frequent generation of circular fragments by intrastrand recombination (Figure 2). Alternatively, when an amplified array occupies a high proportion of the genome of an individual, e. g. 0.1 %, there is a similarly high probability that this sequence will be further amplified and/or transposed by random amplification and transposition events. Because such events greatly increase the probability of an array being spread through a population (see Figure 3), even in the absence of selection for the array, it is likely that many of the arrays found in plant genomes are the ones that have elevated frequencies of being duplicated and transposed. Those arrays which do not have these properties are likely to have been confined to a few individuals or lost from populations, unless of course, the arrays have been propagated in populations as a result of selection or genetic drift. Several discussions of the possible 'functions' of tandem arrays of non-coding repeats have appeared, e. g. (John and Miklos, 1979; Bostock, 1980; Jones and Flavell, 1982a; Miklos and Gill, 1982; Thompson and Murray, 1981) as contributions to the debate of the likelihood that such arrays have been spread and maintained by natural selection.

Repeats arranged in tandem arrays are particularly prone to molecular events which alter the number of repeats in the array as noted above. This is well illustrated by studies on the arrays of the genes encoding the 18S, 5.8S and 25S ribosomal RNAs but many other examples affecting non-coding arrays are known. The number of repeated rRNA genes differs considerably between individuals within all plant species studied, e. g. wheat (Flavell and Smith, 1974); pea (Cullis and Davies, 1975); maize (Ramirez and Sinclair, 1975). Genetic analyses in wheat have shown that the number of rRNA genes at a specific locus can vary up to ten fold (Flavell and Smith, 1974). This variation could arise from unequal crossing over which produces deletions as well as duplications and/or deletions as shown in Figures 1 and 2.

In many plant species (e. g. wheat, barley, *Vicia faba*) there is a short tandem array of repeats within the 'spacer' DNA (Appels and Dvorak,

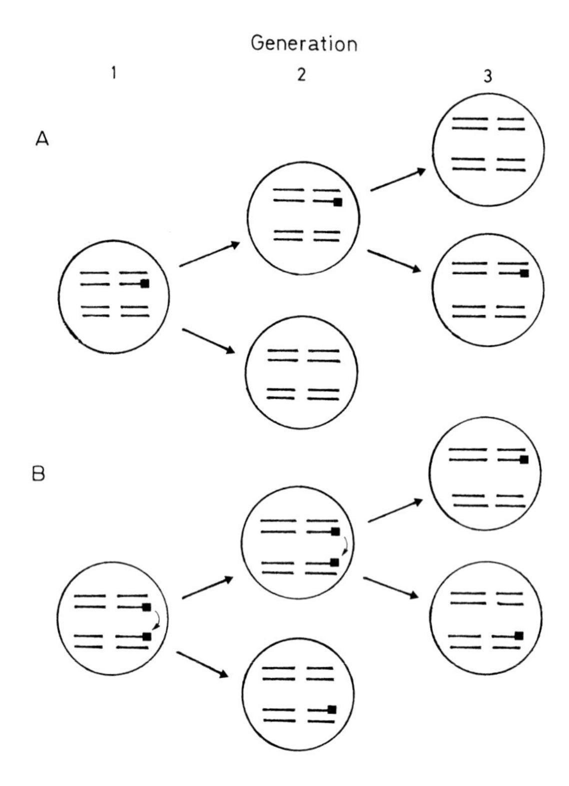

Fig. 3. The role of transposition in increasing the frequency of individuals in a population carrying a specific repeat. Representative individuals with two homologous pairs of chromosomes are shown. Members of each pair are localised together for ease of representation

A. Without transposition, selection or genetic drift the proportion of individuals carrying the array of repeats does not increase in the population.

B. With transposition in generations 1 and/or 2, the proportion of individuals carrying the repeat increases. Transposition to a non-homologous chromosome is illustrated by arrows. Many other transposition schemes, e. g. transposition to a homologue, are possible which similarly result in a net increase in the proportion of individuals with the repeat in the population.

1982; Yakura *et al.*, 1983; Yakura and Tanifuji, 1983) (Figure 1 B). Unequal crossing over between these repeats would lead to variation in the number of short spacer repeats and hence in the total length of the repeating gene unit (Figure 1 B). Such events must occur relatively frequently because there is considerable variation within species for the length of the repeated gene units (Figure 4) and this has been shown for wheat, barley and *Vicia faba* to be due to variation in the number of short spacer repeats (Appels and Dvorak, 1982; Flavell, unpublished; Yakura *et al.*, 1983).

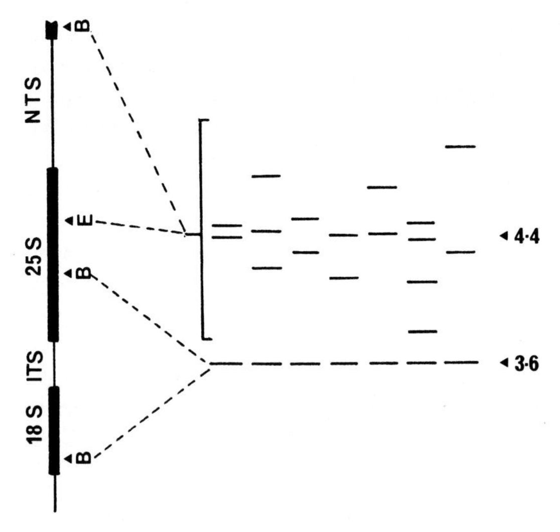

Fig. 4. Concerted evolution of rDNA repeat units. Total DNA from different wheat varieties was cleaved with EcoRI and BamHI, the fragments separated by electrophoresis, transferred to nitrocellulose and the rDNA revealed by hybridisation with labelled rDNA. The 3.6 kbp BamHI fragments are invariant between genes in all the varieties. The EcoRI-BamHI fragments differ between varieties due to the number of repeats in the spacer (NTS). However, each band in each variety represents thousands of genes. Therefore there is very little spacer length variation between the genes at a single locus. This illustrates concerted evolution

II. Concerted Evolution

Repeats in a tandem array are usually more similar to one another than would be expected if each evolved independently. They thus appear to evolve 'in concert' (Zimmer *et al.,* 1980; Kedes, 1979; Arnheim, 1983). The phenomenon is also well illustrated by genetic analyses of the rRNA genes in wheat (Flavell, 1983; Snape *et al.,* 1984; Appels and Dvorak, 1982). When the structure of all the rRNA genes of the species are considered together, the variation in repeat unit length due to spacer repeat variation, is very extensive. However, within a single locus most of the genes have the same number of spacer repeats or the locus is constructed of a small number of arrays each internally homogeneous. This is illustrated in Figure 4. Similar concerted evolution can be seen when rRNA genes of a locus are classified for restriction enzyme polymorphisms (unpublished results).

Tandemly arrayed repeats evolve together probably due to frequent unequal crossing over (Smith, 1976; Szostak and Wu, 1980; Petes, 1980) and/or gene conversion (Baltimore, 1981; Klein and Petes, 1981; Whitehouse, 1983) types of events where one repeat is converted to the sequence of another by an error in template choice during DNA replication or repair. This is illustrated in Figure 5. Gene conversion, in contrast to unequal crossing over, does not result in a change in the number of repeats. These processes operating at random within an array would reduce the accumulation of mutational variants and thus maintain a high degree of homogeneity within the array. They should be recognised as capable of causing a continual turnover or replacement of members of the array. Occasionally, however, a new variant will be used as template to convert pre-existing copies or a new variant will increase in frequency by unequal crossing over, at the expense of an old sequence. These stochastic processes also therefore provide the means of spreading a new variant through an array, and subsequently keeping this new variant homogeneous, even without selection for the variant. This source of genetic flux is important in the evolution of repeats and is probably responsible, for example, for the rapid changes in the spacer regions of rRNA genes during species divergence (Appels and Dvorak, 1982; Dover and Flavell, 1984).

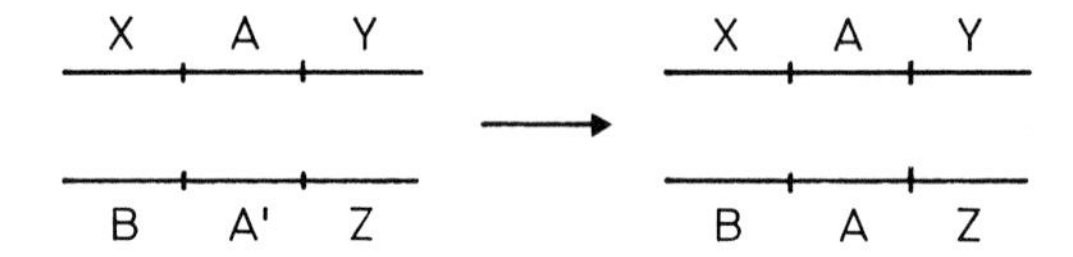

Fig. 5. Gene conversion. The allele A[1] is converted to the sequence of its related allele A. The process involves migration of a DNA strand from A to A[1] and replacement of the A[1] sequence by the A sequence. For simplicity, the drawing shows when both strands of the A[1] duplex have been converted. For hypotheses on the mechanisms of conversion see Whitehouse (1983)

The rapid evolution of the spacer DNA repeats is in contrast to the regions specifying the ribosomal RNAs. The latter are presumably conserved by selection. However, it is believed that the spacer DNA repeats also play an important function in regulating transcription because they are the binding sites for a regulatory protein required for transcription (Reeder *et al.*, 1983). How then can the functional DNA evolve so rapidly? It has been proposed (Dover and Flavell, 1984; Arnheim, 1983) that because the rRNA gene family contains many more genes than essential, new variants can replace old ones in tandem arrays without appreciable effect on the phenotype. These variants then allow the spread of mutations in the gene specifying the regulatory binding protein which are compatible with the new spacer DNA sequence. Fixation of this new variant protein gene together with the new rDNA repeats would replace the 'old' combination of protein gene and rDNA spacer repeats. The rapid change in the spacer repeat but not the region specifying the ribosomal RNAs is possible, it is postulated, because the former has to interact with only one protein and therefore needs to co-evolve with only one gene while the ribosomal RNAs interact with many proteins in the ribosome and must co-evolve with a very large number of genes — something that could happen only extremely slowly. In the context of this chapter, it is important to note that the rapid co-evolution of the genes occurs because of the ability to fix concertedly new variants into tandem arrays and because an excess of repeated genes is maintained in each individual.

Arrays of non-coding sequences on different chromosomes also evolve in concert and are maintained homogeneous (Bedbrook *et al.,* 1980a, b). This may be because gene conversion occurs not only between members of an array but also between sequences on different homologous and non-homologous chromosomes, albeit at a lower frequency than between members of the same array (Scherer and Davies, 1980). The process also explains how occasionally a new variant can be transferred to arrays on different chromosomes. Conversion of many repeats to a single form will be relatively rare due to chance alone but if the conversion in 'biased' in favour of a particular variant or selected for, then fixation could be relatively rapid.

'Biased' gene conversion is well known in fungi (Whitehouse, 1983). The magnitude of the bias in conversion is variable from gene to gene and species to species. In extreme cases one allele emerges one hundred times more frequently than its homologue. A gene lacking bases relative to its homologous alleles is often preferentially converted to the allele with an additional base(s). Also, the invading DNA strand (the one cleaved first?) which moves to the other allele may sometimes be preferred as template during subsequent DNA synthesis. Therefore any repeat variant with a higher probability of being cleaved or with additional bases might be more likely to be fixed by biased gene conversion. Gene conversion frequencies have not been measured for any specific loci in plants.

III. Transposable Elements and Dispersed Repeats

Transposable elements are segments of DNA which autonomously control their own excision from the chromosome and their integration into a new site. They have now been recognised in many plant species by their ability to cause mutations at high frequencies (Freeling, 1984; Dellaporta, 1985). It is presumed that they contain genes coding for proteins which catalyse and possibly regulate the transposition process. They are often capable of increasing in copy number in individuals. These phenomena have been demonstrated to be characteristic of Mu-I and other elements in maize, for example (Fincham and Sastry, 1974; Freeling, 1984).

Autonomous transposable elements are often related to a series of non-autonomous elements which, by themselves cannot transpose, but can respond to the presence of an autonomous element. The non-autonomous elements are probably mutant derivatives of autonomous elements (Federoff *et al.,* 1983).

Elements with the properties of duplication and transposition are likely to spread through populations rapidly (see Figure 3), even where their presence confers some selective disadvantage to the individuals carrying them (Hickey, 1982; Ohta, 1983a). The number of copies of active elements and their frequency of movement to create mutations is likely to be maintained in balance during evolution of a population to prevent the mutation frequency exceeding acceptable limits.

The transposable elements studied so far differ in the kinds of sequences they contain. Ac and its related Ds elements in maize, for example, contain sequences repeated many times in the maize genome (Federoff *et al.*, 1983; Freeling, 1984). Whether each copy of these sequences belongs to a potentially active or fossilised element or whether the active elements have, by chance, picked up representatives of these repeated sequence families (see Döring *et al.*, 1984) is not clear at present. Nevertheless, it is established that at least some of the dispersed repeated sequences in the maize genome are due to the presence and activity of transposable elements.

Transposable elements have inverted repeats at their termini and cause a duplication of a few bases upon insertion (Sutton *et al.*, 1984). Thus it will be interesting to see how many of the large number of dispersed repeats in plant genomes have these properties and so can be postulated to have been spread within individuals and populations by these processes (Flavell, 1984a, b). One example has already been published for a maize dispersed repeat (Shepherd *et al.*, 1984).

Short sequences containing inverted repeats are extremely common in all plant genomes (reviewed in Flavell, 1980, 1982). The wheat genome, for example, contains several million with inverted repeats long enough to be stable under highly stringent renaturation conditions. When shorter inverted repeats are included, the number is likely to increase substantially. Thus perhaps 10% or more of the wheat genome consists of short sequences containing inverted repeats (Flavell, 1984a).

This clearly supports, but by no means proves, that transposable elements and the mechanisms of transposition associated with them, have played major roles in moving repeats through genomes and populations during evolution, thereby giving plant chromosomes the structural feature of many dispersed repeats.

The failure to detect active elements in plants on this scale would be expected because such numbers would create enormous genetic loads. As indicated above, the number of different elements and frequency of each type tolerated in a population are likely to be limited by natural selection to "acceptable" levels even though active elements can still spread where the presence confers some selective disadvantage.

The hypothesis that many of the dispersed repeats in plant genomes have been propagated by or are similar to transposable elements is attractive because without such a mechanism it would be necessary to invoke their spreading as a result of selection or genetic drift. A plausible function for most dispersed repeat sequences has yet to be put forward. All the evidence is against them playing a sequence-dependent role and having been 'selected'. Some may have spread by genetic drift but again it is difficult to believe this is how so many spread through species so rapidly.

Where repeats are localised close to genes under selection, then the repeats may be fixed in population by being linked to the genes. This is likely to be a very important means of spreading new non-coding sequences through species.

If many dispersed repeats have not spread by mechanisms akin to those responsible for the genetically-defined transposable elements, then perhaps other mechanisms of excision and integration occur frequently in plants and some sequences are particularly susceptible to these mechanisms. Such a hypothesis would still give the dispersed repeats a higher flux frequency than other sequences and would accelerate their fixation in populations.

Dispersed repeats can also become involved in gene conversion processes, but at a lower frequency than members of a tandem array (Scherer and Davis, 1980). Thus it is possible for unlinked repeats to coevolve. If gene conversion were biased in favour of a particular variant a large population of repeats could be maintained more homogeneous than expected for random, separate evolution.

IV. Repeated DNA Flux and Species Divergence

So far I have pointed out that the presence of repeats in genomes confers potential instability and that the extent of variation involving repeats within species supports this view. Furthermore, the kinds of mutational changes to which repeats are susceptible can explain how the changes are spread through populations. It is this latter property that is important for evolutionary considerations because to have an effect, a mutation must spread. With this knowledge in mind, what would be expected to happen to the complement of repeats as populations diverged and distinct species separated?

We could expect new transposable elements to arise and spread, causing all sorts of rearrangements and new patterns of dispersed repeats. There would be considerable new variation within each population due to these elements and new families of dispersed repeats. Other transposable elements might be lost, either by genetic drift or selection against too high a frequency of mutation. It seems likely that when autonomous elements are deleted from or inactivated in the population, then many of the non-autonomous elements would also be lost. In the absence of autonomous elements to give rise to new non-autonomous elements, they could decrease by random deletion. Alternatively, they may often be lost because the genes conferring the ability to excise them from chromosomes are maintained in the population but those with the ability to reinsert them are lost. It is known that for some transposable elements the frequency of excision is considerably greater than the frequency of reinsertion. Thus with the turnover of transposable elements, the populations would diverge in genome structure considerably due to rearrangements and the appearance and disappearance of dispersed repeats.

If a repeat variant arose in one population that was preferred in the process of gene conversion then it might replace a significant fraction of the pre-existing copies and so lead to different variants being present in each population or species. The same result could be obtained by deletion of pre-existing copies and separate amplification and dispersal of a new variant. New tandem arrays of repeats would be expected to arise by large scale amplification and some might be expected to be spread through the populations. They may accumulate in addition to pre-existing ones either as a result of selection, drift or because of having a higher probability of being amplified or transposed or may replace the old ones. Different variants are likely to be fixed in related tandem arrays due to ongoing turnover within the arrays.

Where new repeats are added to the genomes of a population and they do not replace old variants, an increase in total DNA results. Some variation in total DNA content and chromosome size would often be tolerated but there is likely to be selection against genome growth above a certain size, since total DNA content limits rates of development, and life cycle times and also affects cell size (Bennett, 1972, 1973). Thus there is likely to be selection for deletions to maintain total DNA content and chromosome size. Where the DNA deleted is *not* the most recently evolved, then the non-coding repeated DNA complement of the species will turnover with time. Such models of repeated DNA turnover have been discussed elsewhere (Flavell, 1980, 1982 a, Thompson and Murray, 1980).

The rate of turnover will depend on many factors, including genome size. In larger genomes, there is a greater probability of amplifications or deletions occurring because more DNA has to be replicated etc. If more amplification events are tolerated in larger genomes because selection against small changes in genome size is less, then amplification rates will appear greater and the turnover rate will probably also be greater. Thompson and Murray (1980) and Preisler and Thompson (1981 a, b) have provided some experimental evidence consistent with the hypothesis that the amplification rate is greater in species with larger genomes.

This projection of how repeated DNA changes over time periods in which populations diverge substantially and new species arise, is strongly supported by numerous comparative interspecies studies, e. g. Osmunda (Stein *et al.*, 1979), Cichorieae (Bachmann and Price, 1977), Vicia (Straus, 1972).

As an example of one set of comparative analyses I draw attention to those on species in the genus *Lathyrus* (Narayan and Rees, 1976, 1977). There is a three fold variation in nuclear DNA content between species in the genus but all species have the same number of chromosomes. This variation is mostly but not entirely due to repetitive DNA. The variation is also highly correlated with the amount of heterochromatin, a conclusion which is consistent with heterochromatin consisting predominantly of repetitive DNA (Bedbrook *et al.*, 1980 a). The DNA that is amplified in the genus differs from species to species as illustrated by the percentage of repetitive DNA from one species that hybridises to the DNA of another (Table 1).

For example only 14 % of the repetitive DNA of *L. hirsutus* hybridised to the repetitive DNA of *L. clymenum* which has a genome size 67 % that of *L. hirsutus* while 44 % hybridised to the repetitive DNA of *L. articulatus* which has a genome size 61 % that of *L. hirsutus*.

Table 1. *Repeated Sequence DNA Homologies Between Different Lathyrus Species*

Repeated DNAs hybridised together	% *L. hirsutus* repeated DNA hybridised	ΔT_m of hybrids °C
L. hirsutus × *L. hirsutus* (20.3)	100	0.0
L. hirsutus × *L. tingitanus* (17.9)	50	1.25
L. hirsutus × *L. odoratus* (17.2)	62	2.25
L. hirsutus × *L. sphaericus* (14.2)	17	4.0
L. hirsutus × *L. clymenum* (13.8)	14	4.5
L. hirsutus × *L. articulatus* (12.5)	44	3.0
L. hirsutus × *L. angulatus* (9.2)	21	3.5

Data taken from Narayan and Rees (1977).

The % hybridisation values and stabilities (ΔT_m) of the DNA/DNA hybrids have been normalised to the values obtained for the self-hybridisation of *L. hirsutus* DNA. The DNA contents (pg) of each species are given in brackets.

When the sequences "common" to each species are compared by examining the thermal stabilities of the DNA/DNA hybrids formed in vitro, in each case the interspecies hybrid DNAs are less stable than the intraspecies hybrids (Table 1). This shows that in fact the "common" repeated sequences must frequently be different in different species. As the proportion of repeated DNA which can form interspecies duplexes declines in the comparisons the extent of similarity (ΔT_m) between "common" sequences also declines (Table 1). These data illustrated in Figure 6 show that in the evolution of this genus, many changes in repetitive DNA have occured and "turnover" of repeated sequences must be invoked to explain the extent of divergence.

In other research closely related species in the *Aegilops, Triticum* and *Secale* genera have been compared as well as other Gramineae species (barley, oats, rye, wheat) which have developed from a common ancestor (Rimpau *et al.*, 1978, 1980; Flavell *et al.*, 1977; Jones and Flavell, 1982b; Flavell *et al.*, 1979; Gerlach and Peacock, 1980). Each species can be distinguished from another by at least one major tandem array. Other species-specific, highly repeated sequences are dispersed and lie between sequences found in other, often distant, species. Clearly the linear organisation of related sequences in different species is very different. Fur-

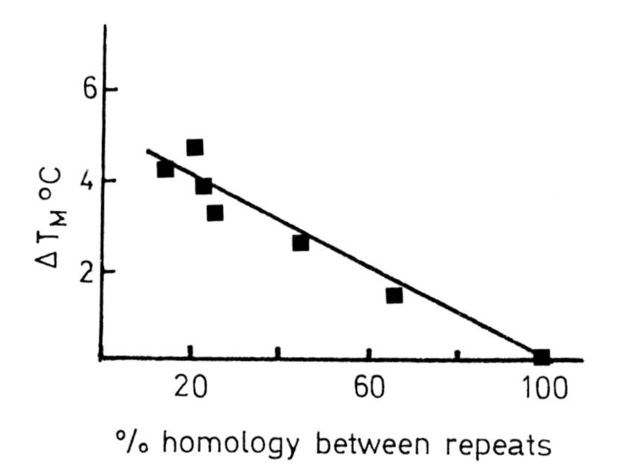

Fig. 6. Repeated sequence homologies between Lathyrus species. The percentages of *L. hirsutus* repeated DNA able to hybridise to the repeated DNAs of other *Lathyrus* species (data taken from Table 1) are plotted against the reductions in thermal stability (ΔT_m) of the interspecies repeated DNA hybrids relative to the intraspecies repeated DNA hybrids (data from Table 1). Results from Naryan and Rees (1977)

thermore, different variants of sequences found in related species are amplified in each species. The results clearly show that the products of amplification, deletion, turnover/replacement and transposition events are spread throughout populations at a rapid rate.

These changes produce major chromosomal structural differences between diverging species which are likely to reduce meiotic chromosome pairing between members of divergent populations and related species and are likely to reinforce, if not create, species barriers. This with some supporting data is discussed elsewhere (Flavell, 1982).

If all these changes occur in non-coding sequences or ones in which selection plays little role, what of the genes? Since many of the genes belong to multigene families, the DNA sequences are prone to the same 'mutations' but clearly selection will operate to eliminate many of the mutants. Gene conversion and unequal crossing over will occur between members of the gene family and these processes may contribute to maintaining homogeneity within the family (Ohta 1983b). However, where variation is not lethal, for example in the 3′ non-translated regions of mRNAs, then the genes might evolve rapidly and different species might have major differences in these regions but be relatively homogeneous within the family due to the turnover processes.

V. Concluding Remarks

The evidence is now becoming substantial that repeats are susceptible to rapid evolutionary change because of the way they are recognised by specific enzymes. The presence of repeats creates potential, perhaps inevitable, instability for a genome and a population or species. Why then are there so many repeats maintained in the genomes of higher organisms? Perhaps they are generated so frequently, that they are not easily eliminated. Where repeats have the capacity to spread efficiently, as do transposable elements, they can accumulate "selfishly" even though they may reduce fitness. Perhaps the associated higher rate of mutation is beneficial.

Repeated DNA clearly contributes to chromosome structure and in this plays an important role in the chromosome biology of each species. Perhaps it is selected in nature to fulfil roles relating to chromosome pairing, recombination, chromosome position in the nucleus and chromatin condensation (Flavell, 1982b) and these roles can be fulfilled without the repeats having very defined sequence characteristics.

I suggested in the Introduction that the potential instability associated with repeated DNA implied that mechanisms must have evolved to suppress or control the variation. Such mechanisms would presumably be under genetic and physiological control. Therefore, when major genetic or physiological stresses are on a species by the introduction of alien genes in wide crosses or when plant cells are put into tissue culture, variation involving repeated DNA is likely to be unleashed. It will therefore be interesting to discover to what extent somaclonal variation is due to such mutations (Larkin and Scowcroft, 1981) and whether the mixing of diverged genotypes provokes an eruption of new mutations. It is also interesting to consider how nature might make use of an increase in frequency of mutations when plants become ill-adapted to their environment. Perhaps following such a burst of mutations, a new genotype can be selected which confers better fitness to the environment and the thriving individuals will then found a new population.

VI. References

Appels, R., Dvorak, J., 1982: The wheat ribosomal DNA spacer region: Its structure and variation in populations and among species. Theoret. and Applied Genetics **63,** 337—348.

Arnheim, N., 1983: Concerted evolution of multigene families. In: Nei, M., and Koehn, R. K. (eds.): Evolution of Genes and Proteins, pp. 38—61. Sunderland, Mass., U.S.A.: Sinauer Associates.

Baltimore, D., 1981: Gene conversion: some implications for immunoglobulin genes. Cell **24,** 592—594.

Bachmann, K., Price, H. J., 1977: Repetitive DNA in Cichorieae (compositae). Chromosoma (Berl.) **61,** 267—275.

Bedbrook, J. R., Jones, J., O'Dell, M., Thompson, R. D., Flavell, R. B., 1980a: A molecular description of telomeric heterochromatin in *Secale* species. Cell **19,** 545—560.

Bedbrook, J. R., O'Dell, M., Flavell, R. B. 1980b: Amplification of rearranged sequences in cereal plants. Nature **288,** 133—137.

Bennett, M. D., 1972: Nuclear DNA content and minimum generation time in herbaceous plants. Proc. R. Soc. Lond. **B 181,** 109—135.

Bennett, M. D., 1973: Nuclear characters in plants. Brookhaven Symposia in Biology **25,** 344—366.

Bostock, C., 1980: A function for satellite DNA? Trends in Biochem. Sciences **5,** 117—119.

Bostock, C., Tyler-Smith, 1982: Changes to genomic DNA in methotrexate resistant cells. In: Dover, G. A., Flavell, R. B., (eds.): Genome Evolution, pp. 69—94. New York: Academic Press.

Cullis, C. A., Davies, D. R., 1975: Ribosomal DNA amounts in *Pisum sativum*. Genetics **81,** 485—492.

Dellaporta, S., 1985: This volume.

Dennis, E. S., Gerlach, W. L., Peacock, W. J., 1980: Identical polypyrimidine polypurine satellite DNAs in wheat and barley. Heredity **44,** 349—366.

Deumling, B., 1981: Sequence arrangement of a highly methylated satellite DNA of a plant, *Scilla:* a tandemly repeated inverted repeat. Proc. Nat. Acad. Sci., U.S.A. **78,** 338—342.

Deumling, B., Greilhuber, J., 1982: Characterization of heterochromatin in different species of the *Scilla siberica* group *(Liliaceae)* by *in situ* hybridization of satellite DNAs and fluorochrome banding. Chromosoma **84,** 535—555.

Doring, H. P., Tillmann, E., Starlinger, P., 1984: DNA sequence of the maize transposable element *Dissociation.* Nature **307,** 127—130.

Dover, G. A., 1982: Molecular drive: a cohesive mode of species evolution. Nature **299,** 111—117.

Dover, G. A., Flavell, R. B., 1984: Molecular Coevolution: DNA divergence and the maintenance of function. Cell **38,** 622—623.

Federoff, N., Wessler, S., Shure, M., 1983: Isolation of the transposable maize controlling elements *Ac* and *Ds.* Cell **35,** 235—242.

Fincham, J. R. S., Sastry, G. R. K., 1974: Controlling elements in maize. Annual Rev. Genetics **8,** 15—50.

Flavell, A. J., Ish-Horowicz, D., 1982: Extrachromosomal circular copies of the eukaryotic transposable element *copia* in cultured *Drosophila* cells. Nature **292,** 591—595.

Flavell, R. B., 1980: The molecular characterization and organization of plant chromosomal DNA sequences. Annual Rev. Plant Physiology **31,** 569—596.

Flavell, R., 1982a: Amplification deletion and rearrangement: major sources of variation during species divergence. In: Dover, G. A., Flavell, R. B. (eds): Genome Evolution, pp. 301—324. London: Academic Press.

Flavell, R. B., 1982b: Chromosomal DNA sequences and their organisation. In: Nucleic Acids and Proteins in Plants II, Encyclopedia of Plant Physiology, New series **14B,** 46—74.

Flavell, R. B., 1983: Repeated sequences and genome architecture. In: Ciferri, O., and Dure, L. (eds.): Structure and Function of Plant Genomes, pp. 1—14. New York: Plenum Press.

Flavell, R. B., 1984a: DNA transposition — a major contributor to plant chromosome structure. Bio Essays **1,** 21—22.

Flavell, R. B., 1984b: Transposable elements. Oxford Surveys of Plant Molec. and Cell Biology **1,** 207—210.

Flavell, R. B., Jones, J., Lonsdale, D., O'Dell, M., 1983: Higher plant genome structure and the dynamics of genome evolution. In: Downey, K., Voellmy, R. W., Ahmed, F., and Schultz, J. (eds.): Advances in gene technology, pp. 47—59. New York: Academic Press.

Flavell, R. B., O'Dell, M., Smith, D. B., 1979: Repeated sequence DNA comparisons between *Triticum* and *Aegilops* species. Heredity **42,** 309—322.

Flavell, R. B., Rimpau, J., Smith, D. B., 1977: Repeated sequence DNA relationships in four cereal genomes. Chromosoma (Berl.) **63,** 205—222.

Flavell, R. B., Smith, D. B., 1974: Variation in nucleolar organiser rRNA gene multiplicity in wheat and rye. Chromosoma (Berl.) **47,** 327—334.

Freeling, M., 1984: Plant transposable elements and insertion sequences. Ann. Rev. Plant Physiology **35,** 277—298.

Gerlach, W. L., Peacock, W. J., 1980: Chromosomal locations of highly repeated DNA sequences in wheat. Heredity **44,** 269—276.

Hickey, D. A., 1982: Selfish DNA: a sexually-transmitted nuclear parasite. Genetics **101,** 519—531.

Hinegardner, R., 1976: Evolution of genome size. In: Ayala, F. J. (ed.): Molecular Evolution, pp. 179—199. Sunderland, Mass., U.S.A.: Sinauer Associates Inc.

Hourcade, D., Dressler, D., Wolfson, J., 1973: The amplification of ribosomal RNA genes involving a rolling circle intermediate. Proc. Nat. Acad. Sci., U.S.A. **70,** 2926—2930.

Jones, J. D., Flavell, R. B., 1982a: The mapping of highly repeated DNA families and their relationship to C bands in chromosomes of *Secale cereale.* Chromosoma (Berl.) **86,** 595—612.

Jones, J. D., Flavell, R. B., 1982b: The structure amount and chromosomal localization of defined repeated DNA sequences in species of the genus Secale. Chromosoma (Berl.) **86,** 613—641.

John, B., Miklos, G. L. G., 1979: Functional aspects of heterochromatin and satellite DNA. Int. Rev. Cytol. **58,** 1—114.

Kedes, L. H. A., 1979: Histone genes and histone messengers. Ann. Rev. of Biochem. **48,** 837—870.

Klein, H. L., Petes, T. D., 1981: Intrachromosomal gene conversion in yeast. Nature **289,** 144—148.

Larkin, P. J., Scowcroft, W. R., 1981: Somaclonal variation — a novel source of variability from cell cultures for plant improvement. Theoretical and Applied Genetics **60,** 197—214.

Miklos, G. L., Gill, A. C., 1982: Nucleotide sequences of highly repeated DNAs; compilation and comments. Genet. Res. Camb. **39,** 1—30.

Narayan, R. K. J., Rees, H., 1977: Nuclear DNA divergence among *Lathyrus* species. Chromosoma (Berl.) **63,** 101—107.

Narayan, R. K. J., Rees, H., 1976: Nuclear DNA variation in *Lathyrus.* Chromosoma (Berl.) **54,** 141—154.

Ohta, T., 1983a: Theoretical study on the accumulation of selfish DNA. Genet. Research **41,** 1—15.

Ohta, T., 1983b: On the evolution of multigene families. Theoretical Population Biology **23,** 216—240.

Ohta, T., Dover, G. A., 1983: Population genetics of multigene families that are dispersed in two or more chromosomes. Proc. Natl. Acad. Sci., U.S.A. **80,** 4079—4083.

Petes, T. D., 1980: Unequal meiotic recombination within tandem arrays of yeast ribosomal DNA genes. Cell **19,** 765—774.

Preisler, R. S., Thompson, W. F., 1981a: Evolutionary sequence divergence within repeated DNA families of higher plant genomes. I. Analysis of reassociation kinetics. Journal Molecular Evolution **17,** 78—84.

Preisler, R. S., Thompson, W. F., 1981b: Evolutionary sequence divergence within repeated DNA families of higher plant genomes. II. Analysis of thermal denaturation. J. Molecular Evolution **17,** 85—93.

Ramirez, S. A., Sinclair, J. H., 1975: Intraspecific variation of ribosomal gene redundancy in *Zea mays.* Genetics **80,** 495—504.

Reeder, R. H., Roan, J. G., Dunaway, M., 1983: Spacer Regulation of Xenopus ribosomal gene transcription: competition in oocytes. Cell **35,** 449—456.

Rimpau, J., Smith, D. B., Flavell, R. B., 1978: Sequence organisation analysis of the wheat and rye genomes by interspecies DNA/DNA hybridization. J. Molec. Biol. **123,** 327—359.

Rimpau, J., Smith, D. B., Flavell, R. B., 1980: Sequence organization in barley and oats chromosomes revealed by interspecies DNA/DNA hybridization. Heredity **44,** 131—149.

Scherer, S., Davis, R. W., 1980: Recombination of dispersed repeated DNA sequences in yeast. Science **209,** 1380—1384.

Schimke, R., 1982: In: R. Schimke (ed.): Gene amplification, pp. 317—333. New York: Cold Spring Harbour Press.

Shepherd, N. S., Schwarz-Sommer, A., Velspalve, J. B., Gupta, M., Wienand, U., Saedler, H., 1984: Similarity of the Cin 1 repetitive family of *Zea mays* to eukaryotic transposable elements. Nature **307,** 185—187.

Smith, G. P., 1976: Evolution of repeated DNA sequences by unequal crossover. Science **191,** 528—535.

Snape, J. W., Flavell, R. B., O'Dell, M., Hughes, W. G., Payne, P. I., 1984: Intrachromosomal mapping of the nucleolar organiser region relative to three marker loci on chromosome IB of wheat *(Triticum aestivum)* Theoret. and Applied Genetics, in press.

Stein, D. B., Thompson, W. F., Belford, H. S., 1979: Studies on DNA sequences in the Osmundaceae. J. Mol. Evol. **13,** 215—232.

Straus, N., 1972: Reassociation of bean DNA. Carnegie Inst. Wash. Yearbook **71,** 257—259.

Sutton, W. D., Gerlach, W. L., Schwartz, D., Peacock, W. J., 1984: Molecular analysis of Ds controlling element mutations as the Adh-1 locus of maize. Science **223,** 1265—1268.

Szostak, J. W., Wu, R., 1980: Unequal crossing over in the ribosomal DNA of Saccharomyces cerevisiae. Nature (Lond.) **284,** 426—430.

Thompson, W. F., Murray, M. G., 1980: Sequence organisation in pea and mung bean DNA and a model for genome evolution. In: Davies, D. R., and Hopwood, D. A. (eds.): Fourth John Innes Symposium, pp. 31—45, Norwich, U.K., John Innes Institute.

Thompson, W. F., Murray, M. G., 1981: The nuclear genome: structure and function. In: Strumpf, P. K., and Conn, E. E. (eds.), pp. 10—81. Biochemistry of Plants, New York: Academic Press.

Whitehouse, H. L. K., 1983: Genetic recombination — understanding the mechanisms. New York: Wiley.

Yakura, K., Kato, A., Tanifuju, S., 1983: Structural organisation of ribosomal DNA in four *Trillium* species and *Paris* verticillate. Plant and Cell Physiol. **24,** 1231—1240.

Yakura, K., Tanifuji, S., 1983: Molecular cloning and restriction analysis of EcoRI fragments of *Vicia faba* rDNA. Plant and Cell Physiol. **24,** 1327—1330.

Zimmer, E. A., Martin, S. L., Beverley, S. M., Kan, Y. W., Wilson, A. C., 1980: Rapid duplication and loss of genes coding for the α chains of hemoglobin. Proc. Natl. Acad. Sci., U.S.A. **77,** 2158—2162.

Chapter 9

Sequence Variation and Stress

C. A. Cullis

John Innes Institute, Colney Lane, Norwich NR4 7UH, England

With 1 Figure

Contents

I. Introduction

Higher eukaryotes have very complex nuclear genomes. An understanding of the origins of this complexity is beginning to emerge with the recognition that the DNA of higher organisms is subject to a variety of sequence rearrangements including amplification, deletion, mutation and translocation both within and between chromosomes. The changes thus generated can spread through populations so that certain segments of the genome change rapidly during evolution. Comparisons of the genome of closely related species (see Flavell, 1982, this volume) describe the products of these processes rather than the actual series of events by which a particular arrangement was produced. Potential mechanisms for the production of sequence rearrangements include unequal crossing-over within tandem arrays, amplifications, deletions and rearrangements associated with the activation of transposable elements, and the amplification of portions of the genome containing a gene conferring a selectable advantage.

To study the mechanisms by which the genome is modified, and the interaction between these mechanisms and some form of stress applied to the organism, it is necessary for the events to occur in a small number of generations, and ideally within a single generation. Rapid nuclear DNA variation has been described in a number of different types of experimental material. Three of these examples in plants are: — (a) The environmental induction of heritable changes in flax (Cullis, 1983 a); (b) The generation of somaclonal variation (Larkin and Scowcroft, 1983; de Paepe, Prat and Huget, 1982) and (c) The formation of certain interspecific hybrids (Price *et al.*, 1983). Examples from all three of these systems will be considered and the following questions posed. What are the molecular mechanisms which generate rapid nuclear DNA variation? Are there distinct regions of the genome in which this variation occurs? Can specific subsets of the genome be defined by differing responses to stress? Finally, does rapid nuclear change have any evolutionary consequences?

II. Environmentally Induced DNA Changes in Flax

Heritable changes can be found in some flax varieties after they have been grown in different, characterised environments for a single generation (Durrant, 1962, 1971; Cullis, 1981 a). The stable lines produced (termed genotrophs) differ from each other and the original variety from which they were derived (Stormont Cirrus, termed P 1) in a number of characters. These include plant weight and height, the total nuclear DNA amount (as determined by Feulgen staining) (Evans, 1968), the number of genes coding for the 18 S and 25 S ribosomal RNAs (rDNA) (Cullis, 1976, 1979) the number of genes coding for the 5 S RNA (5 S DNA) (Goldsbrough *et al.*, 1981) and for a number of other cloned repetitive sequences (Cullis, 1984 a).

There is no obvious relationship between the defined aspects of the environmental conditions and the variation of the characters mentioned above. For this reason the environmental conditions have been termed "inducing environments" in the absence of any clear molecular mechanisms. However, since only a small number of phenotypic characters have been studied it is possible that only a minor portion of the variation is of adaptive significance with respect to the stresses of these inducing environments. The remainder of the variation may be neutral in the particular inducing environments and so any adaptive changes may appear insignificant among the wide range of variation observed.

III. Nuclear DNA Variation

The nuclear DNA amount for a number of genotrophs has been estimated by Feulgen cytophotometry (Evans, 1968; Joarder *et al.*, 1975). Highly significant differences have been observed between the genotrophs and the difference can be up to 15% of the total nuclear DNA.

The genome size of flax is 1.5 pg/2 C nucleus (Timmis and Ingle, 1973) and it has a diploid complement of 30 small chromosomes, all of which are approximately the same size. A comparison of the chromosomes from two genotrophs differing in nuclear DNA content by 15% of the total nuclear DNA amount showed no significant differences. Thus the sequences altered must be located on a large number of the chromosomes of the normal complement, as the localisation to a few sites should have produced significant chromosomal changes.

IV. Analysis of Nuclear DNA

The repeated and single copy sequences of the flax genome are arranged in a long period interspersion pattern (Cullis, 1981 b). The DNAs from twelve genotrophs have been characterised by renaturation kinetics using the hydroxylapatite method (Britten *et al.*, 1974) with DNA of an average single-stranded length of 300 base pairs (Cullis, 1983 b). The C_{ot} curves obtained were analysed by a non-linear regression computer program. Wide variation for both the rate constants and the proportions of the fitted curves were obtained when the DNAs from different genotrophs were analysed (Cullis, 1983 b). The DNAs from two lines, which showed large differences, were characterised further in mixed drive/trace experiments. The resulting C_{ot} curves confirmed that there were significant differences in the amount of both highly repetitive and intermediately repetitive sequences, with the high DNA line having more of these sequences than the low DNA line (Cullis, 1983 a). The characterisation of the DNA from the genotrophs by renaturation kinetics demonstrated that a wide range of sequences could be affected by the events leading to the formation of the genotrophs. However, such studies could only give information on relatively large changes within the genome and could not detect small changes, especially those occurring in the unique or low copy number sequences. For the determination of this type of variation specific cloned probes would be required.

The variation in the highly repetitive fraction of the genome has been extensively characterised using a number of cloned repetitive sequences (Cullis, 1984 a). The determinations have been made using a modified dot-blot procedure (Brown *et al.*, 1983). Together the repetitive families comprise some 30% of the flax genome and their contribution to the total DNA in 17 lines has been determined (Cullis, 1984 a). These lines include

12 genotrophs, one flax variety, one linseed variety, two *Linum bienne* accessions and one *Linum grandiflorum* Desf. accession. The values obtained from a subset of these lines is given in Table 1. The variation in the individual sets of sequences will be considered separately.

Table 1. The relative amounts of different repeated sequences in leaf DNA from flax genotrophs and related *Linum* species. L_6, L^H, C_1, C_2 and C_3 were all derived from P1 (Cullis, 1977, 1981). Liral Monarch was supplied by the Northern Ireland Agricultural Department and *Linum grandiflorum* Desf. was obtained from the Botanical Garden, Aachen. All the values are given with the value for P1 as standard. * Values are significantly different from P1 at the 1 % level. Data taken from Cullis, 1984a

Source of DNA	Probe							
							Satellite DNA	
	rDNA	5SDNA	pCL2	pCL8	pCL53	pBG87	pCL13	pCL21
P1	1.0	1.0	1.0	1.0	1.0	1.0	1.0	1.0
L_6	*0.68	*0.47	*0.86	0.83	*0.77	0.90	*0.59	0.87
L^H	1.04	1.08	*1.21	*1.28	*0.47	0.86	*0.76	1.10
C_1	1.0	*0.67	0.91	0.94	*0.65	0.84	*0.59	1.0
C_2	1.11	0.98	1.0	1.06	0.90	0.94	*0.66	1.03
C_3	*0.52	*0.72	*0.6	*0.81	*0.72	*0.77	*0.53	0.99
Liral Monarch	*0.77	*0.75	*0.73	*0.80	*0.41	*0.80	*0.54	0.87
Linum grandiflorum Desf.	*1.21	*1.33	*0.2	*0.25	0.83	*0.74	0.85	0.98
% of total genome in P1	1.5	3	4.2	2.6	2.3	2.5	15	

V. Ribosomal DNA Variation

The flax genes coding for the 18S and 25S ribosomal RNAs consist of a homogeneous set arranged in tandem arrays with a repeat length of 8.6 kilobases (Goldsbrough and Cullis, 1981). The number of rRNA genes varied nearly three fold between genotrophs but there was neither length heterogeneity nor restriction enzyme site polymorphisms in the genotroph rDNAs.

The rDNA variation has also been followed during the growth of plants under inducing conditions (Cullis and Charlton, 1981). It was shown, for P1, that the amount of rDNA varied during growth under inducing conditions and that the change observed could be transmitted to the progeny. Under the same set of growth conditions two of the stable genotrophs showed no variation in rDNA amount either during their growth or in their progeny. Thus, in this case, the change in rDNA depended on the interaction of a particular genotype (P1) with the environment in which it was grown.

VI. 5S DNA Variation

The 5S DNA of flax is arranged in tandem arrays of 350 to 370 base pair repeating sequence (Goldsbrough *et al.,* 1981). There is both length and sequence heterogeneity and the number of copies can vary more than two-fold. In the genotrophs with the highest number of copies this sequence can comprise about 3 % of the genome. It has been possible to distinguish a sub-set of the 5S genes which appears to be preferentially deleted when the 5S gene number is reduced (Goldsbrough *et al.,* 1981; Cullis, 1984b). This subset is one in which a TaqI site is missing and so is uncut with this enzyme. Another example is shown in Figure 1 where the 5S gene number has been reduced in tissue culture cells and again part of the same subset has been eliminated. Thus, within the 5S genes there is differentiation into two recognisable subsets, one of which is more frequently varied than the other. However there is evidence that both subsets can vary over a longer time scale than that required to generate the genotrophs (Cullis, 1984a).

VII. Satellite DNA

A light satellite DNA, comprising about 15 % of the genome, can be isolated from the flax genome on neutral cesium chloride gradients. Estimates from analytical neutral cesium chloride gradients have given little variation between genotrophs for the amount of satellite DNA (Cullis, 1975). The satellite DNA is complex and consists of diverged repeating units as shown by its renaturation characteristics (Cullis, 1981b; Ingle, Pearson and Sinclair, 1973). Two clones (pCL13 and pCL21, Table 1) which did not cross hybridise at T_m-10° have been used to characterise the satellite DNA fraction (Cullis, 1984a). One, pCL21, gave no detectable variation either between the genotrophs or the related Linum species, while the other, pCL13, varied about two-fold. Thus another subset of the genome can be separated into variable and constant components.

VIII. Other Repetitive Sequences

The variation for four other highly repetitive clones is given in Table 1. It can be seen that the extent of variation observed both between the genotrophs and between flax and related species depends on the probe. These four families together with the rDNA, 5S DNA and satellite DNA make up more than 30% of the total flax genome. It can be seen that variation can occur in most of these families with the induction of heritable changes and the pattern of variation shows similarities with that observed when different flax and linseed cultivars are compared with each other or related *Linum* species.

IX. Somaclonal Variation

Variants and mutant lines have often been found in plant tissue cultures and in the plants regenerated from these cultures (Larkin and Scowcroft, 1983). Gross karyotypic changes have frequently been observed in tissue cultured plant cells and these undoubtedly account for some of the variability. However in a number of cases phenotypic variation has been found in regenerated plants in the absence of karyotypic changes (Edallo *et al.*, 1981; Shepard *et al.*, 1980). The only molecular analysis of this mutational variation has been effected on microspore-derived doubled haploids of *Nicotiana tabacum* and *N. sylvestris* (de Paepe *et al.*, 1982; Dhillon *et al.*, 1983) where it has been suggested that specific DNA amplification and modification can occur during the culture process.

Table 2. The relative amounts of different repeated sequences in leaf and callus DNA from two genotrophs. S_6 was derived from P1 (Cullis, 1977). * Differences between leaf and callus DNA from same line significant at 1 % level. Data from Cullis, 1984b)

Source of DNA	Probe							
							Satellite DNA	
	rDNA	5SDNA	pCL2	pCL8	pCL53	pBG87	pCL13	pCL21
P1 leaf	1.0	1.0	1.0	1.0	1.0	1.0	1.0	1.0
P_1 callus	*0.58	0.87	*0.80	0.88	*0.59	0.98	*0.45	1.12
S_6	0.86	0.9	0.63	0.90	0.89	0.90	0.81	0.98
S_6 callus	0.75	0.9	0.63	0.94	0.95	0.80	0.76	0.98

DNA variation associated with the culture process has also been demonstrated in flax (Cullis, 1983b, 1984b). DNAs from leaf and callus tissue from the same lines have been compared using the same cloned probes previously described for the comparison of the genotroph genomes. The data for two pairs are given in Table 2, from which it can be seen that the two lines behave differently. The callus DNA derived from P1 was much more variable than that derived from S_6, in the latter case there being no significant differences between DNAs extracted from leaves and from callus tissue. Over the eight pairwise comparisons, P1 was most variable and differences were observed for all the probes except pCL21, which was also invariant in the genotroph comparisons. In addition when the 5S DNA was shown to vary between leaf and callus in one line it was also shown that the reduction could be accounted for by a deletion in the TaqI resistant fraction (Fig. 1; Cullis, 1984b). Thus the DNA variation observed in flax tissue culture follows that already described for the genotrophs. One additional sequence shown to vary in flax tissue culture is a cloned sequence which shows some similarities with a transposable element (Cullis, 1983b). This has been shown to show restriction fragment length polymorphisms in some lines when DNAs from leaf and callus were compared. The pattern obtained with DNA from regenerated plants closely

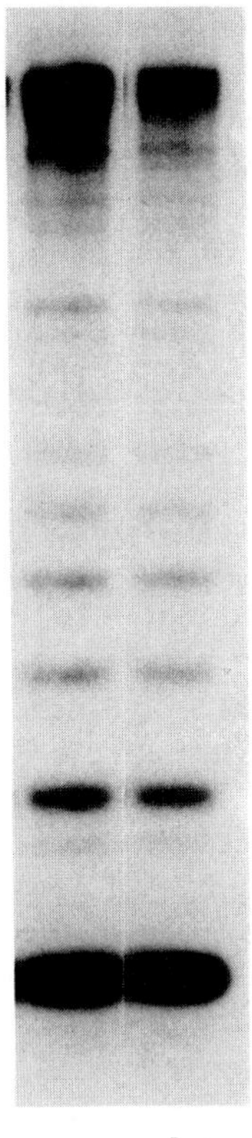

Fig. 1. DNA from leaves (a) and callus (b) from the same flax genotroph digested with the restriction enzyme TaqI, separated on a 1.5 % agarose gel, blotted onto nitrocellulose and hybridised with nick translated plasmid pBG 13 (a clone of the 5 S DNA, Goldsbrough *et al.*, 1981)

resembled that obtained for the callus DNA from which the plants were regenerated (Cullis, 1983 b).

In addition to the apparent nuclear mutations there is evidence for variation occurring in the organelle genomes during tissue culture. In *Zea mays* the mitochondrial DNA has been shown to undergo rearrangements during tissue culture with the generation of specific disease resistance and male fertility (Gengenbach *et al.*, 1981). In barley there is evidence for specific genome changes in the plastids of the many albino plants amongst regenerants from barley microspores (A. Day, personal communication). The relationship, if any, between the variation of the nuclear genome and the organelle genomes is unknown.

Potential mechanisms for the generation of somaclonal variation have been discussed previously (Larkin and Scowcroft, 1981). It is abundantly clear that tissue culture can result in chromosome deletions, translocations and other minor rearrangements (McCoy *et al.*, 1982; Ogihara, 1981). Since these modifications can occur at a cytologically observable level, they may also be occurring at a finer structure level. These finer structural rearrangements may include amplifications, deletions, translocations and the movement of transposable elements, all of which can alter subsequent gene expression. The demonstration as to which of these mechanisms predominate awaits further experimental data.

X. Instabilities in Hybrid Plants

McLintock (1978) suggested that wide species crosses were among the stresses that might trigger the reorganisation of parental genomes. Evidence for this comes from interspecific crosses in *Nicotiana* and in *Microseris* species. There is also evidence, from *Microseris,* maize and flax, that wide crosses may not be necessary and the reorganisation may be initiated in intraspecific crosses.

In *Microseris* species the inheritance of the nuclear 2C DNA content in interspecific and intraspecific hybrids has been determined (Price *et al.,* 1983). In the interspecific crosses between *M. douglasii* and *M. bigloveii* the 2C DNA value of the F1 progeny did not cluster around the parental mid point but instead encompassed nearly the entire range between the parental means. Five families of F2 progeny were characterised for their 2C DNA content and each had a mean value corresponding to that of the F1 plant from which they were derived, with no evidence of F2 segregation. This indicated that the F1 plants were not of identical DNA content. Two intraspecific crosses between *M. douglasii* biotypes, differing in nuclear DNA content by about 10%, have been analysed. The F2 progeny from one intraspecific cross showed no striking evidence for segregation for DNA content. From a second intraspecific cross, the mean DNA contents of the F2 progeny from two sister hybrids were significantly different.

Crosses between high DNA and low DNA genotrophs of the flax variety Stormont Cirrus to Liral Monarch, which had an intermediate DNA content, gave F1's with DNA contents similar to the Liral Monarch parent (Durrant, 1981). Here, as in the case of the *Microseris* species, there is an indication of a quantitative change in the DNA amount. However the DNA variation was observed in the F1 material so that these differences cannot be attributed to irregular segregation.

In *Nicotiana* species there is evidence for some hybrids being unstable in their DNA content. Gerstel and Burns (1966) observed chromosomes up to 15 times the normal length in some corolla, root tip and pollen mother cells from *N. tabacum* × *N. otophora* hybrids. The mechanism for the production of these megachromosomes is unknown but they may be the result of amplifications such as those which produce homogeneously staining regions in the chromosomes of some methotrexate resistant cell lines which have an amplified number of copies of the gene coding for dihydrofolate reductase (Dolnick *et al.,* 1975). An example of possible interaction and amplification in mammalian hybrids is the description of double minute chromosomes in cell lines independently derived from a *Mus musculus* × *M. caroli* fetus (Graves, 1984). This quantitative variation in interspecific hybrids does not always occur. In Lolium hybrids Hutchinson *et al.* (1979) showed evidence for the segregation of DNA amounts, in both backcross and F2 progeny. Specific sequence changes have been shown in crosses between flax genotrophs and in crosses between maize inbred lines. The rDNA in the F1 of crosses between flax genotrophs differed from the expected mid-parent value (Cullis, 1979). In crosses between maize inbreds a number of sequences including the rDNA and 5S DNA were shown to differ from their expected value in F1 hybrids (Rivin and Cullis, 1983). In this case whether or not a variation was observed was dependent both on the inbred lines used and on the particular DNA sequence investigated.

It is clear that changes in DNA amount can occur on the combination of two genomes after either an intraspecific or interspecific cross. The magnitude of such changes appears to depend on the parents involved in the cross but the genetical and biochemical bases for this interaction are presently unknown.

XI. Discussion

It is clear that parts of the plant genome can be rapidly modulated in response to stress applied to the genome. The three different types of stress considered here have been a nutrient stress applied to the whole plant (the generation of the flax genotrophs), the growing of cells in tissue culture and the combination of dissimilar genomes. In the one experimental system where all these three perturbations have been applied, namely the flax genotrophs, it appeared that sequences from the same subset of the genome were affected whenever variation was observed. Thus it is possible that the same mechanism was responsible for the changes in all three instances.

Another point to emerge is that not all species, nor all members of the same species, show a susceptibility of the genome to stress. Not all flax and linseed varieties can undergo environmentally induced heritable changes nor are there large DNA shifts in all interspecific hybrids. The basis for this variability in response is not clear but since it is restricted to some lines it should be possible to investigate the genetic basis of this modulating mechanism.

The mechanisms by which these DNA rearrangements occur are also presently unknown. Possible means of achieving variation include unequal recombination, amplification by unscheduled DNA synthesis and transposition. The latter introduces one further variable not considered here, namely the role of transposable elements in DNA variation. It is known that integration and excision of transposable elements can cause amplifications and deletions, but whether they can do so on the scale described here is unclear. It is possible that they may be one of the subsets of the DNA which can be activated by stress.

The plant genome, from the information described, appears to contain islands of instability in that only a subset of the DNA is affected in stress induced variation. In one case it has been shown that a particular sub-set of the same set of genes, namely the 5S RNA genes, can be differentially affected. How is this subset distinguishable within the genome? It appears unlikely that the differentiation can be on the basis of the primary nucleotide sequence so it is likely to reside in the structure of the chromatin. Perhaps there are particular sequences which, because of the chromatin structure of their particular chromosomal location, are more susceptible to variation. These subsets may vary with the cell type so the resultant genomic changes in response to applied stress may depend on the cell type being subjected to that stress.

In conclusion, the plant genome is variable and appears to contain islands of instability and constancy with respect to applied stress. However, the genetic basis for this response, the mechanisms by which the variation occurs and the role such variation may play in evolution are not at all clear.

XII. References

Britten, R. J., Graham, D. E., Neufeld, B. R., 1974: Analysis of repeating sequences by reassociation kinetics. In: Grossman, L., Moldave, K. (eds.), Methods in Enzymology, Vol. 29, pp. 363—405, New York: Academic Press.

Brown, P. C., Tlsty, T. D., Schimke, R. T., 1983: Enhancement of methotrexate resistance and dihydrofolate reductase gene amplification by treatment of mouse 3T6 cells with hydroxyurea. Mol. Cell Biol. **3**, 1097—1107.

Cullis, C. A., 1975: Environmentally induced DNA differences in flax. In: Markham, R., Davies, D. R., Hopwood, D. A., and Horne, R. W. (eds.): Modification ot the Information Content of Plant Cells. pp. 27—36. Amsterdam: North Holland.

Cullis, C. A., 1976: Environmentally induced changes in ribosomal RNA cistron number in flax. Heredity **36**, 73—79.

Cullis, C. A., 1977: Molecular aspects of the environmental induction of heritable changes in flax. Heredity **38**, 129—154.

Cullis, C. A., 1979: Quantitative variation of ribosomal RNA genes in flax genotrophs. Heredity **42**, 237—246.

Cullis, C. A., 1981 a: Environmental induction of heritable changes in flax: Defined environments inducing changes in rDNA and peroxidase isozyme band pattern. Heredity **47**, 87—94.

Cullis, C. A., 1981 b: DNA sequence organisation in the flax genome. Biochim. Biophys. Acta **652**, 1—15.

Cullis, C. A., 1983 a: Environmentally induced DNA changes in plants. CRC Critical Reviews in Plant Sciences **1**, 117—131.

Cullis, C. A., 1983 b: Variable DNA sequences in flax. In: Chater, K. F., Cullis, C. A., Hopwood, D. A., Johnston, A. W. B., Woolhouse, H. W. (eds.): Genetic Rearrangement, pp. 253—264. Croon Helm U. K.

Cullis, C. A., 1984 a: Rapidly varying DNA sequences in flax. Theoret. App. Genet. Submitted.

Cullis, C. A., 1984b: DNA variation in flax tissue culture. Theoret. App. Genet. submitted.

Cullis, C. A., Charlton, L. M., 1981: The induction of ribosomal DNA changes in flax. Plant Sci. Lett. **20,** 213—217.

De Paepe, R., Prat, D., and Huget, T., 1982: Heritable nuclear DNA changes in doubled haploid (D. H.) plants obtained by pollen culture of *Nicotiana sylvestris.* Plant Sci. Lett. **28,** 11—28.

Dhillon, S. S., Wernsman, E. A., Miksche, J. P., 1983: Evaluation of nuclear DNA content and heterochromatin changes in anther-derived dihaploids of tobacco *(Nicotiana tabacum).* C. V. Coker 139. Can. J. Genet. Cytol. **25,** 169—173.

Dolnick, B. J., Berenson, R. J., Bertino, J. R., Kaufman, R. J., Nunberg, J. H., Schimke, R. T., 1979: Correlation of dihydrofolate reductase elevation with gene amplification in a homogeneously staining chromosomal region in C5178Y cells. J. Cell Biol. **83,** 394—402.

Durrant, A., 1962: The environmental induction of heritable changes in *Linum.* Heredity **17,** 27—61.

Durrant, A., 1971: Induction and growth of flax genotrophs. Heredity **27,** 277—298.

Durrant, A., 1981: Unstable genotypes. Phil. Trans. R. Soc. London, B. **292,** 467—474.

Edallo, S., Zucchinali, C., Perezin, M., Salamini, F., 1981: Chromosomal variation and frequency of spontaneous mutation associated with *in vitro* culture and plant regeneration in maize. Maydica **26,** 39—56.

Evans, G. M., 1968: Nuclear changes in flax. Heredity **23,** 25—38.

Flavell, R. B., 1982: Sequence amplification, deletion and rearrangement: major sources of variation during species divergence. In: Flavell, R. B., Dover, G. A. (eds.): Genome Evolution, pp. 301—323, New York: Academic Press,.

Gegenbach, B. G., Conelly, D. R., Pring, D. R., Conde, M. F., 1981: Mitochondrial DNA variation in maize plants regenerated during tissue culture selection. Theor. Appl. Genet. **59,** 161—167.

Gerstel, D. U., Burns, J. A., 1976: Enlarged euchromatic chromosomes ("megachromosomes") in hybrids between *Nicotiana tabacum* and *N. plumbaginifolia.* Genetica **46,** 139—153.

Goldsbrough, P. B., Cullis, C. A., 1981: Characterisation of the genes for ribosomal RNA in flax. Nuc. Acid. Res. **9,** 1301—1309.

Goldsbrough, P. B., Ellis, T. H. N., Cullis, C. A., 1981: Organisation of the 5S RNA genes in flax. Nuc. Acid. Res. **9,** 5895—5904.

Graves, J. A. M., 1984: Gene amplification in a mouse embryo? Double minutes in cell lines independently derived from a *Mus musculus* × *M. caroli* fetus. Chromosoma (Berl.) **89,** 138—142.

Hutchinson, J., Rees, H., Seal, A. G., 1979: An assay of the activity of supplementary DNA in Lolium. Heredity **43,** 411—421.

Ingle, J., Pearson, G. G., Sinclair, J., 1973: Species destribution and properties of nuclear satellite DNA in higher plants. Nature New Biol. **242,** 193—197.

Joarder, I. O., Al-Saheal, Y., Begum, J., Durrant, A., 1975: Environments inducing changes in the amount of DNA in flax. Heredity **34,** 247—253.

Larkin, P. J., Scowcroft, W. R., 1981: Somaclonal variation — A novel source of variability from cell cultures for plant improvement. Theor. Appl. Genet. **60,** 197—214.

Larkin, P. J., Scowcroft, W. R., 1983: Somaclonal variation and crop improvement. In: Kosuge, T., Meredity, C. P., Hollander, A., (eds.): Genetic Engineering in Plants, pp. 289—314. New York: Plenum Press,.

McClintock, B., 1978: Mechanisms that rapidly reorganize the genome. Stadler Symposium. **10,** 25—47.

McCoy, T. J., Phillips, R. L., Rives, H. W., 1982: Cytogenetic analysis of plants regenerated from oat *(Avena sativa)* tissue cultures; high frequency of partial chromosome loss. Can. J. Genet. Cytol. **24,** 37—50.

Ogihara, Y., 1981: Tissue culture in *Haworthia.* Part 4: Genetic characterisation of plants regenerated from callus. Theoret. Appl. Genet. **60,** 353—363.

Price, H. J., Chambers, K. L., Bachman, K., Riggs, J., 1983: Inheritance of nuclear 2C DNA content variation in intraspecific and interspecific hybrids of Microseris (Asteraceae). Amer. J. Bot. **70,** 1133—1138.

Rivin, C. J., Cullis, C. A., 1983: Modulation of repetitive DNA in the maize genome. Genetics **104,** 859—860.

Shepard, J. F., Bidney, D., Shahin, E., 1980: Potato protoplasts in crop improvement. Science **208,** 17—24.

Timmis, J. N., Ingle, J., 1973: Environmentally induced changes in rRNA gene redundancy. Nature New Biol. **244,** 235—236.

Chapter 10

The Activation of Maize Controlling Elements

S. L. Dellaporta and P. S. Chomet

Cold Spring Harbor Laboratory, Cold Spring Harbor, NY 11724, U.S.A.

With 18 Figures

Contents

I. Introduction

A major contribution to the dynamic flux of the plant genome is the presence of discrete mobile genetic elements that can cause high rates of genetic instability including spontaneous unstable mutations and chromosome rearrangements. These mobile elements, first identified in maize, have since been discovered in a wide variety of prokaryotic and eukaryotic organisms. They may well be ubiquitous but simply not yet identified in other organisms.

There is genetic and molecular evidence that these elements are quiescent components of the plant genome that can be activated following disruption to chromosomes. In this cryptic state, these elements are genetically undetectable. Molecular data have shown that sequences homologous to most maize transposons are indeed present regardless of whether these elements are found in a detectable state. Once activated, mobility of these elements is responsible for spontaneous unstable mutations and chromosome restructuring that contribute to heritable altered progeny. This instability may play a significant role in genome reorganization and provide accelerated rates of diversification in plant populations.

This chapter presents an brief overview of the genetic behavior of selected maize transposable elements. We shall not attempt to make an exhaustive survey through the vast literature on plant transposable elements. The reader is referred to recent reviews by Fedoroff (1983), Freeling (1984), Nevers *et al.* (1984), and Doring and Starlinger (1984) for a more complete reference to plant transposable elements. We will focus our discussion on the transposable element systems of maize, particularly the *Ac/Ds* system, the consequences of transposon instability, and their relationship to the rest of the maize genome.

II. Unstable Mutations in Maize

A. Genetic Loci

The vast collection of unstable mutations caused by insertion elements, and the identification of transposable element systems in maize far outnumbers the identification of these elements in other eukaryotes. This is because of several factors: i) the array of unstable mutations readily visible in the maize endosperm; ii) somatic mutations associated with mobility of transposable element can be isolated in germinal progeny, since plants do not segregate germ line tissue early in development. If an event occurs in meristematic cells during plant development that develop into gametophytic tissue (pollen and ovules), then sectors of mutant tissue will eventually lead to mutant progeny. (iii) Cultural practices such as inbreeding and selection may propagate these types of instabilities; and (iv) the concentrated genetic research on maize controlling elements has greatly contributed to the identification of these elements.

Mutations affecting either anthocyanin production in the aleurone and pericarp cells, or the quality and quantity of starch and protein synthesis in the endosperm have received the most attention, since these are highly visible and developmentally dispensable. The loci discussed in this review are described in detail in a review by Coe and Neuffer (1977) of maize genetics for additional information. Generally, the genes involved in anthocyanin biosynthesis have been especially useful in the genetic study of unstable mutations because alterations in the pigmentation pattern of aleurone and plant tissues can reflect subtle changes in the patterns of anthocyanin gene expression caused by controlling element regulation. Cloning of abundantly expressed genes during endosperm development such as sucrose synthetase (*shrunken* locus) (Burr and Burr, 1981; Chaleff *et al.,* 1981; Geiser *et al.,* 1982) and UDP-glucose transferase (*waxy* locus) (Shure *et al.,* 1983) began the work on molecular biology of maize transposable elements. Both loci contain several transposon insertions that were soon characterized once these gene probes became available.

B. General Considerations

Many transposable element systems in maize have been described that act as regulators of gene action, often giving rise to variegated patterns of gene expression and chromosomal rearrangements that are determined by: i) the genetic properties of the element; ii) the structural gene being controlled; and iii) perhaps by environmental and cellular factors during the development of the organism. What's more, a single element has been shown to be capable of simultaneous regulation of independent genes acting at different times during plant development (McClintock, 1956a). Hence, the term *controlling element* was used by McClintock to describe these mobile elements and will be used throughout this discussion to emphasize the regulatory nature of these systems.

Mutations caused by controlling elements can be somatically unstable; that is, during plant development, sectors of mutant tissue containing altered levels of gene action are responsible for the variegated or unstable patterns of gene expression observed with these types of mutations. The term *mutable* locus was used by McClintock (1947) to describe these unstable genes, since germinal tissue derived from mutant somatic sectors contains mutated alleles that can express a range of stable levels of gene action (Fig. 1). These derivative alleles no longer respond to controlling element activity, as if the element has left the locus restoring various stable levels of gene action. Molecular evidence shows that variable levels of gene expression of these events probably reflect the precision of excision of the element from the locus as discussed below. The patterns of somatic variegation of mutable alleles has been the subject of intensive genetic research although little is known concerning molecular factors responsible for generating patterned gene expression. These patterns of instability appear to be developmentally regulated. That is, the induction of somatic variegation (i. e. transposition) is under the genetic control of the element resident at a gene locus and this regulation appears to be under both temporal and spatial control during plant development.

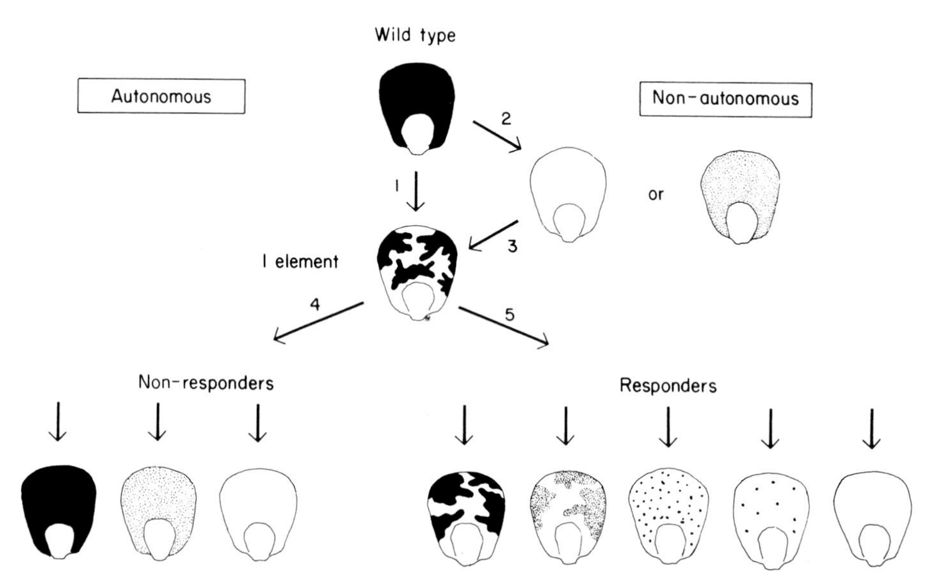

Fig. 1. Mutable loci in maize. A mutable locus in maize can be generated by insertion of an autonomous element (1) or by insertion of a non-autonomous element at a gene locus (2). (The example given is the generation of an unstable mutation in a gene that conditions anthocyanin in the aleurone of the kernel.) When the element is conjoined with the gene locus, gene action is usually totally or partially suppressed. Excision of the element from the gene can restore gene action which appears as a sector of colored aleurone tissue on a colorless or near colorless aleurone background. The non-autonomous mutation is unstable only when an autonomous element is also present in the cell (3). In the absence of transactivation by the autonomous element, a non-autonomous mutation is genetically stable. Premeiotic or meiotic instability of a mutable locus can give germinal mutations that are subsequently stable and can give a wide range of stable levels of gene expression (4). A second type of genetic event can heritably alter the pattern of variegation (5). These events, termed *changes of state,* are mutations of the autonomous or non-autonomous element which still are under controlling element regulation

C. Two-Element Systems in Maize

Early studies of unstable mutations in maize established the basis for two-unit interactive systems (Rhoades, 1936, 1938; McClintock, 1947, 1950a, 1952, 1957). Basically, controlling element systems can comprise two physically distinct entities. One element, referred to as the *autonomous* element (McClintock, 1950a; Fedoroff, 1983) is capable of autonomous transposition, excision, and genetic change. The second element responds by trans-activation, requiring helper functions from the autonomous element for instability and is referred to as a *non-autonomous* element. The non-autonomous element is detected by genetic instability only in the presence of the functional autonomous element elsewhere in the genome. In the absence of trans-activation, non-autonomous mutations often behave as stable alleles showing no somatic or germinal instability (Fig. 2

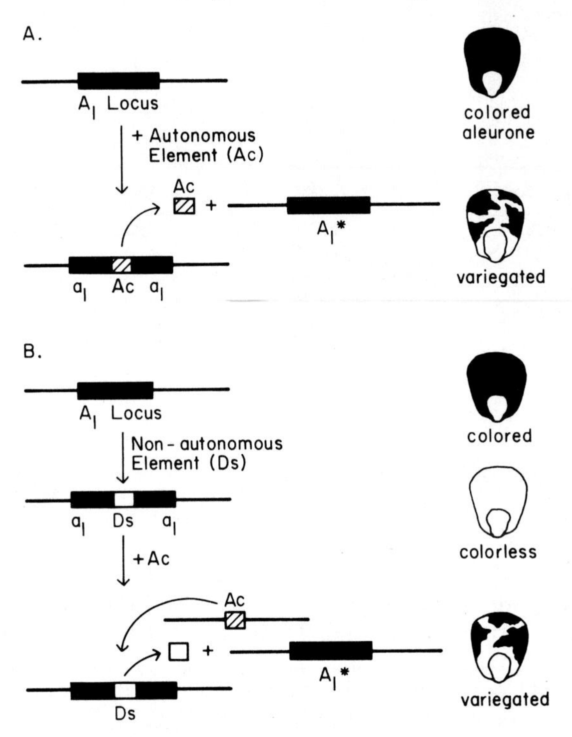

Fig. 2. Two-element controlling system in maize

A. Autonomous unstable mutations are caused by the insertion of an autonomous controlling element at a structural gene. The example shown is the insertion of an autonomous element in a gene which conditions anthocyanin production *(A 1)* in the aleurone of the kernel. If *A 1* is functional, a kernel will have a colored aleurone. When the autonomous element is conjoined with the structural gene, pigmentation capacity is reduced or lost. This results in a colorless aleurone. However, because the autonomous element is inherently unstable, somatic instability will result in restoration of the structural gene. This will result in a variegated aleurone phenotype as shown. The restored *A 1* gene may be modified *(A 1*)* due to imprecise excision of the controlling element.

B. Non-autonomous mutations are unstable only in the presence of the appropriate autonomous element. In the absence of transactivation by an autonomous element, the non-autonomous mutation is genetically stable. In this example, the non-autonomous element conjoined with the *A 1* gene inhibits the pigmentation capacity of the gene and results in a colorless aleurone. When the autonomous element is introduced, the non-autonomous allele becomes unstable and results in a variegated aleurone phenotype.

Fig. 3. Genetic crosses with autonomous and non-autonomous mutations. **Ear 1:** Progeny containing all necessary genes required for anthocyanin pigmentation of the aleurone cells in the kernels. **Ear 2:** An autonomous mutation of an anthocyanin gene. The female parent was homozygous for the autonomous unstable allele and the male parent was homozygous for a stable recessive tester allele. Notice all kernels are variegated, except germinal mutations to fully colored or colorless kernels, since the autonomous element is inherently unstable at the gene locus. **Ear 3:** A non-autonomous mutation of the same gene locus shows that only one-half of the progeny are variegated. The female parent was homozygous for the non-autonomous mutation and heterozygous for an unlinked autonomous element. The male parent was homozygous for a stable recessive tester allele. While each kernel contains the non-autonomous mutation, only one-half contain the autonomous element due to meiotic segregation. **Ear 4:** Cross between a female parent homozygous for a non-autonomous mutation of an anthocyanin gene locus without a transacting autonomous element. The male parent was homozygous for a stable recessive allele of the same locus. Control of the gene by the non-autonomous element in the absence of an autonomous element results in a stable inhibition of gene expression as shown by all non-variegated, colorless kernels

Kernels (from left to right) represent: **1)** kernel with functional anthocyanin synthesis in the aleurone; **2)** a mutable anthocyanin locus caused by insertion of an autonomous element; **3)** a mutable anthocyanin locus caused by a non-autonomous element in the presence of the unlinked autonomous element; **4)** same as kernel **3** except the autonomous element is absent; **5)** a premeiotic germinal mutation of the autonomous mutable allele that restores anthocyanin gene action and 3). Autonomous and non-autonomous controlling elements have also been referred to as the *regulatory* and *receptor* components, respectively (Fincham and Sastry, 1974). An unstable mutation can be caused by insertion of either an autonomous or non-autonomous element and there is genetic and molecular data which shows that some non-autonomous elements can be directly derived from autonomous elements by mutational events (McClintock, 1955, 1956b, 1962; Fedoroff *et al.*, 1983; Pohlman *et al.*, 1984). This evidence supports the idea that certain types of non-autonomous components are defective versions of functional transposable elements. The best studied 2-element system in maize is the *Ac/Ds* controlling element family. *Ac* represents the autonomous component and is capable of trans-activating an unlinked non-autonomous *Ds* element.

D. Controlling Element Families

The interaction between autonomous and non-autonomous elements provides the basis for genetic classification of controlling elements into groups or related families (Peterson, 1980). A non-autonomous allele is a sensitive genetic marker that *reports* on the presence of its corresponding autonomous controlling element. An allele under control of a non-autonomous element is stable unless the appropriate component capable of trans-activation is present elsewhere in the genome. In the presence of the autonomous component, the non-autonomous allele is unstable (e. g. variegated). This interaction has been shown to be highly specific: a given non-autonomous allele can respond to one controlling element system but not other systems. For instance, an autonomous *Ac* element can transactivate *Ds* elements but has no detectable effect on autonomous or non-autonomous elements in other maize controlling element families such as *Spm* or *Dt* elements. Using this genetic test, Peterson and co-workers (Peterson, 1981; Friedemann and Peterson, 1982) have established systematic relationships and classifications for at least seven families of controlling elements in maize. A newly isolated non-autonomous mutation can be crossed with tester stocks containing active autonomous elements to determine the genetic basis of an unstable mutation; or if a plant is suspected of harboring an autonomous element, each responding allele can be crossed with the plant to test for its presence by trans-activation of the appropriate non-autonomous allele. A positive response, instability of the non-autonomous allele, would be subjected to a subsequent genetic segregation test to confirm the specificity of the interaction (reviewed by Peterson, 1981). For instance, a newly isolated non-autonomous *a 1* allele can be identified as an insertion of the controlling element *Ds* if it responds (assayed by somatic instability of *A 1* gene action) when crossed to a tester stock containing the autonomous *Ac* element.

The two-element systems in maize provide powerful genetic analysis of controlling element activity as well as a means to detect the presence of controlling element activity in the maize genome as discussed below. The ability to identify a particular controlling element at a locus by genetic means can also be instrumental for gene cloning if a molecular probe is available for that specific element. This approach has been employed by Fedoroff *et al.* (1984) to obtain genomic clones of the *bronze* locus with *Ac* probes. The general strategy shown in Figure 4 is to obtain insertion of an *Ac* at a gene locus of interest. The mutation is defined as an *Ac* insertion by genetic criteria: i) it is an autonomous mutation which shows a characteristic dosage pattern of *Ac*-induced mutations; and ii) the unstable mutation is capable of trans-activating non-autonomous *Ds*-induced mutations. Using *Ac* DNA probes, genomic clones are obtained and analyzed for the presence of *Ac*. DNA adjacent to *Ac* can then be used as a probe to obtain genomic clones of the wild-type gene.

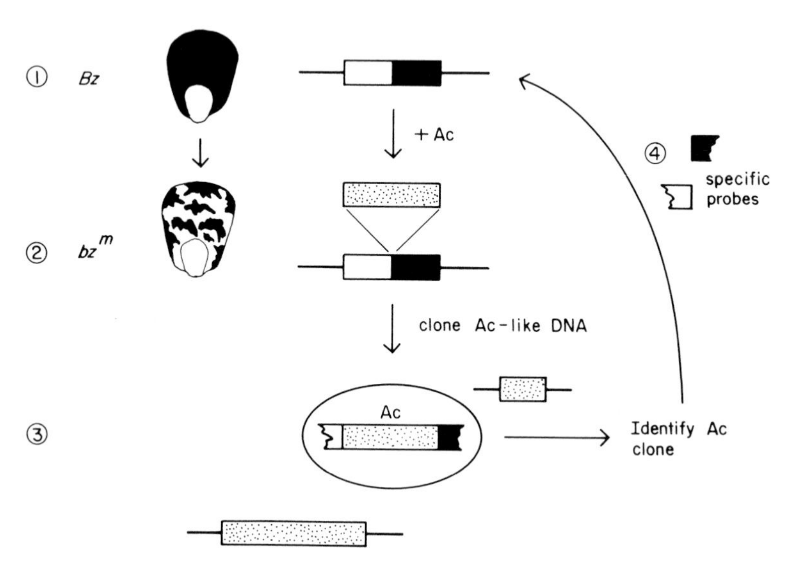

Fig. 4. Gene cloning strategy using *Ac* tagging. **1)** The gene of interest is identified and **2)** mutagenized by insertion of *Ac*. The *Ac* insertion mutation is identified by the ability of *Ac* to transactivate *Ds* elements and by a characteristic dosage effect of *Ac* induced mutations (see text for details). Internal molecular probes of *Ac* hybridize to about 10 homologous sequences in the maize genome. **3)** These DNAs are cloned and screen for the canonical *Ac* element by restriction mapping. **4)** Once the *Ac* element is identified among the genomic clones, subclones of adjacent DNA sequences are used as probes to identify and clone genomic sequences which correspond to the gene of interest. This strategy has successfully been used to clone the *bronze* locus of maize (Fedoroff *et al.*, 1984)

III. Induction of Controlling Element Activity

Studies concerning the origin and nature of controlling elements have shown that genetically undetectable elements are activated after the maize genome has undergone chromosome breakage (McClintock, 1950b; Doerschug, 1973). McClintock's earlier studies on the behavior of broken chromosomes induced by ionizing radiation led to the discovery of transposable elements. An elegant historical perspective of these investigations was presented in the Nobel Lecture by McClintock (1984). We review this work to emphasize the interrelationship between chromosome breakage and the induction of genetic instability caused by transposable elements.

A. Behavior of Broken Chromosomes

Stadler (1928, 1930) was the first to use X-irradiation in plants to recover recessive mutations. Extensive cytological examination of plants carrying Stadler's X-ray induced mutations were conducted by McClintock (1932, 1938a). Maize was particularly suitable for these studies since chromosomes are visible and individually recognized during the pachytene stage of meiosis during male gametogenesis. These experiments revealed that most recessive phenotypes induced by X-rays were due to deficiencies or rearrangement of a chromosomal segment carrying the dominant allele. Other types of complex chromosomal rearrangements such as inversions and translocations were also found in these progeny. McClintock (1932) concluded that gross rearrangments induced by X-irradiation were caused by broken ends of chromosomes, with chromatids fusing pairwise; any broken end would fuse with another broken end regardless of their respective positions in the chromosome complement.

In addition to stable recessive alleles, several progeny of X-irradiated pollen exhibited bizzare variegated phenotypes (McClintock, 1932, 1938a). Cytological examination of variegated plants revealed the presence of ring chromosomes (McClintock, 1932, 1938a). Thus, the variegation could be explained if the dominant marker carried by the ring was lost during mitosis uncovering the recessive phenotype carried by the normal homologue. This instability was shown to be caused by dicentric ring chromosome formation produced by sister strand exchanges which occasionally occurred during ring chromosome replication (Fig. 5) (McClintock, 1938). Mechanical rupture of the ring chromosome was caused by each centromere of the dicentric ring migrating to opposite poles during mitotic anaphase. Most importantly, the two newly broken ends fused, forming a new ring chromosome that indicated when two broken ends are present in a cell they have a tendency to fuse. Depending on the position of the break, deficiencies or duplications may occur and mitotic descendants may differ in their genetic composition, resulting in mutant sectors.

A linear dicentric chromosomes and an acentric chromosome fragment can be formed by crossing over between a normal chromosome and a homologue carrying an inverted DNA segment (Fig. 6). During meiotic ana-

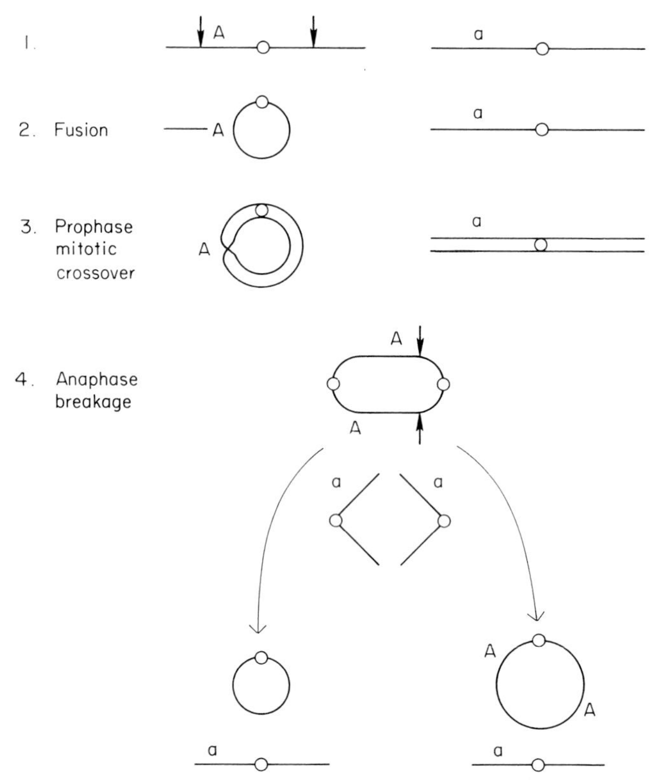

Fig. 5. Instability of a ring chromosome. **1)** Two chromosomal breaks (depicted by arrows) and subsequent fusion could result in **2)** the formation of a centric ring chromosome carrying a dominant marker (e. g. *A 1*) and an acentric fragment. (The normal linear homologue in this example carries the recessive *(a 1)* marker. **3)** A sister strand exchange in the ring chromosome results in dicentic ring formation, shown at anaphase. **4)** Uneven breakage of the double chromosome bridge and fusion of the broken ends results in a smaller *A 1*-deficient ring chromosome and a larger *A 1*-duplicated ring chromosome. The daughter cell carrying the *A 1*-deficient ring gives rise to a recessive *a 1* sector of tissue

phase, each centromere of the dicentric is pulled to opposite poles. Breakage of the chromatid bridge follows, passing a broken chromosome to each telophase nucleus. The trinucleate maize pollen grain develops from two mitoses following meiosis; during these subsequent mitotic divisions of the haploid microsporocytes, McClintock (1938b) observed an unexpected dicentric formation of an original ruptured dicentric chromosome and concluded that replication of the broken chromosome caused chromatid fusion at the previous anaphase breakpoint leading to a dicentric chromosome by the joining of the broken sister chromatids. This cycle continued during the second mitotic division of the microsporocyte and introduced a single broken chromosome into each pollen grain nuclei. This is referred to as the *chromatid* type of breakage-fusion-bridge (BFB) cycle. It was subsequently shown by genetic analysis and marker loss, that

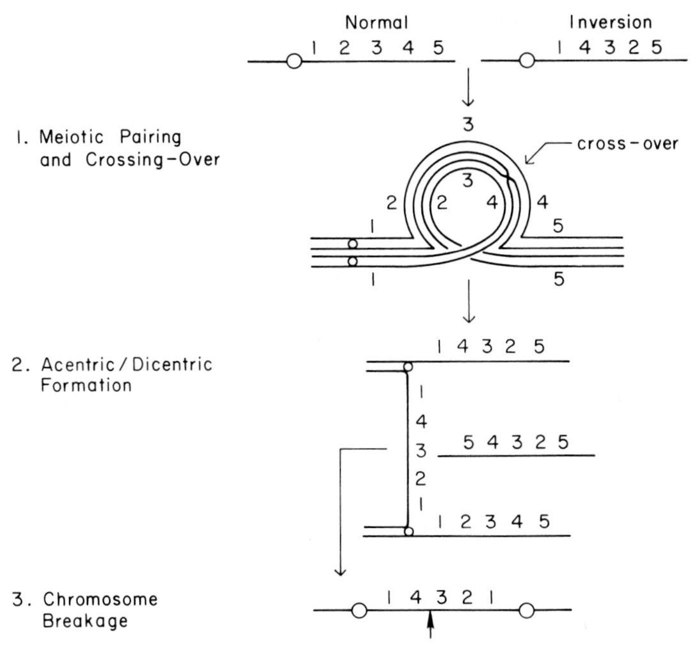

Fig. 6. Dicentric chromosome formation by crossing-over between a normal homologue and a paracentric inversion. **1)** Meiotic pairing between a normal chromosome and paracentric inversion can lead to dicentric formation (2) if crossing over occurs within the inverted region. **3)** During anaphase I the centromeres of the dicentric are pulled to opposite poles resulting in chromosome rupture (solid arrow). The acentric fragment is usually lost during subsequent nuclear divisions. The resulting broken chromosomes will usually carry deficiencies that lead to non-functional gametes. The non-recombinant chromatids result in the passage of the normal and inversion chromosomes to the next generation

this cycle will continue in successive mitoses during endosperm development when a single ruptured chromosome is introduced into the proendosperm nucleus (McClintock, 1941).

Most pollen grains carrying large chromosomal deficiencies are unable to function during fertilization. To follow the fate of broken chromosomes during mitosis after fertilization, it was necessary to construct a broken chromosome that would contain a full complement of genes and therefore produce functional pollen (McClintock, 1941). Plants carrying both an inverted duplication of chromosome 9S and a chromosome 9 deficiency were constructed. Crossing over between the 9S deficiency and the 9S duplication during meiosis produces a dicentric chromosome (see Fig. 7A). Chromosome breakage during meiotic anaphase can occur within the duplicated region resulting in a broken chromosome 9 with a full complement of genes and a terminal duplication. Subsequently, mitotic division during pollen maturation initiated the chromatid BFB cycle (Fig. 7B). If the dicentric chromosome breaks give rise to a chromosome 9 without a deficiency, functional pollen will be produced. Since the 9S deficiency is not pollen transmissable and the 9S duplication is poorly transmitted through the pollen, the net result is that a large fraction of the fertilizing pollen will have a newly ruptured chromosome 9.

These studies demonstrated that when a single broken chromosome is passed to the primary endosperm nucleus, the chromatid type of BFB cycle will continue throughout the development of the endosperm tissue (McClintock, 1941). However, a broken chromosome when delivered to the zygote nucleus was not found to undergo this BFB cycle but rather to *stabilize* or *heal* no longer undergoing fusion and dicentric formation (McClintock, 1941). McClintock (1984) has suggested the healing process is formation of a telomere on the broken end of a chromosome entering the zygote. This process must be repressed in the cells of the endosperm and during the mitotic divisions of haploid cells forming the male and female gametophytes since chromatid fusion and dicentric formation will continue in these cell lineages.

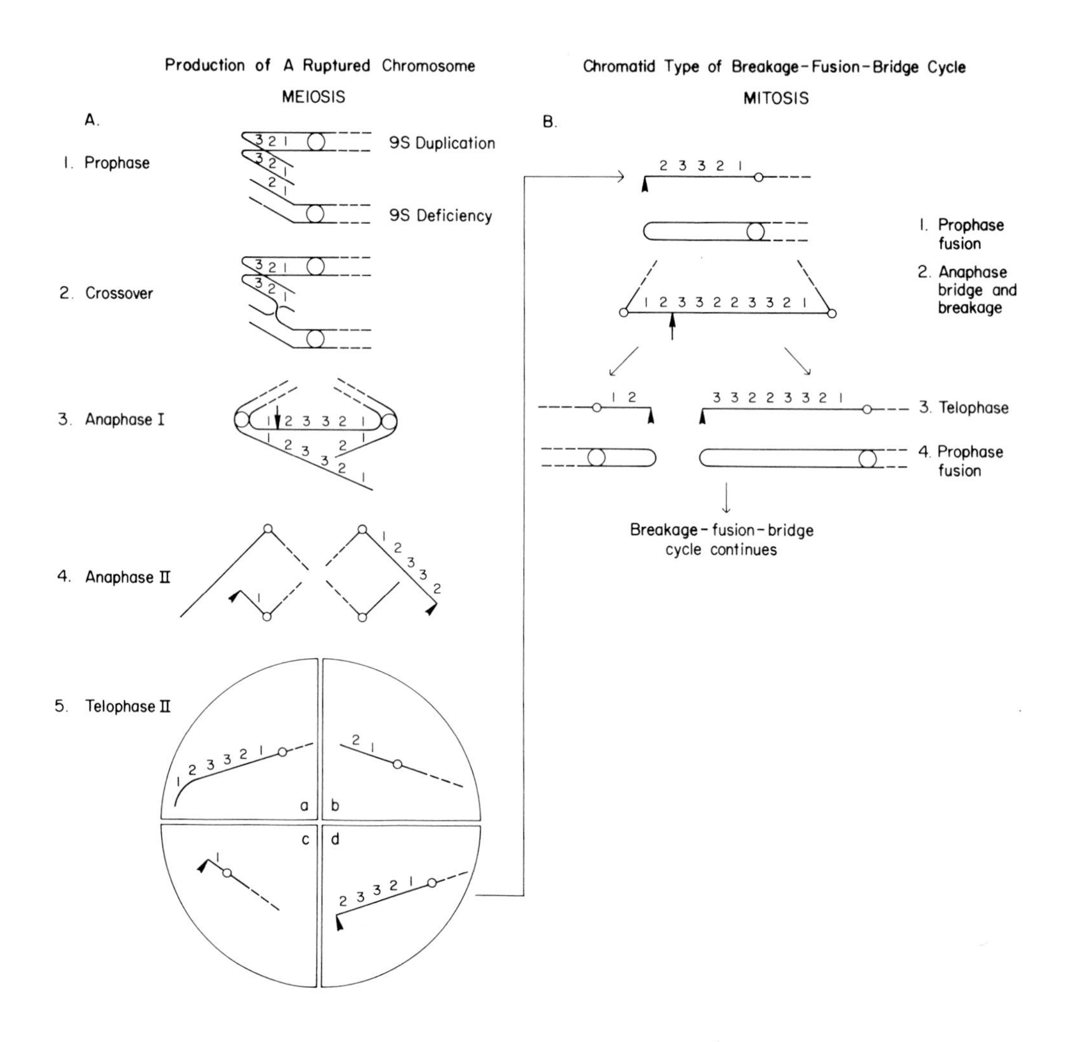

Previous studies on ring chromosome behavior (McClintock, 1932, 1938 a) indicated that when two broken ends of a chromosome enter a cell in sporophytic tissue, the broken ends would fuse. Accordingly, when a broken 9 chromosome was introduced through both the male and female gametophyte, the broken ends fused with each other to form a dicentric chromosome and initiated the *chromosomal* type of BFB cycle (Fig. 8) which can occur in either endosperm or sporophyte tissue (McClintock, 1942, 1944). In the sporophyte tissue, the chromosomal BFB cycle has drastic consequences on cell viability that can prevent the progression of normal tissue development. During plant development, manifestations of continuing BFB cycles are: failure of normal shoot development, tissue sectors defective in chlorophyll or normal morphological growth patterns, and seedling lethality (McClintock, 1950 b, 1951 a). However, the cycle may eventually subside and from these cells tissue will develop into functional shoots. This enabled McClintock (1950 b, 1951 a) to examine the cytological and genetic consequences of the BFB cycle in microsporophyte cells obtained from tassels of BFB plants. Offspring of these plants uncovered a variety of unstable mutations that were primarily responsible for detecting the presence of transposable elements in plants.

Fig. 7. Production of a functional broken chromosome 9

A. Broken chromosome 9 formation following meiotic crossing-over between a 9 S duplication and a 9 S deficiency

1) One type of synaptic configuration between a terminal inverted 9 S duplication of the short arm and a terminal 9 S deficiency. (Only the short arm is illustrated.) The open circle represents the centromere. **2)** One crossover type involving the duplicated and deficient chromosome arms results in a chromatid bridge at the first meiotic anaphase. **3)** Breakage of the bridge during Anaphase I will occur if crossing over occurs between the duplicated and deficient chromosome arms (position depicted by an arrow). If the cross-over occurs between the sister chromatids of the duplicated arm (configuration not shown), then bridge formation will occur during Anaphase II. **4)** The broken chromosomes produced during Anaphase I will segregate during Anaphase II into separate telophase nuclei. **5)** This produces two spores of the tetrad with a broken chromosome 9. In this example, one duplicated and one deleted chromosome, each with a newly ruptured end of the short arm (represented by an arrow head) is shown in nuclei c and d of the resulting tetrad

B. Chromatid type of breakage-fusion-bridge cycle during mitotic divisions of the haploid gametophyte

1) Terminal fusion of sister chromatids occurs at the point of previous meiotic breakage during prophase in the development of the microsporocyte or megasporocyte. **2)** Chromosome bridge and breakage occurs when centromeres are pulled to opposite poles during anaphase. The new break can occur anywhere between the two centromeres (e. g. at the position of the arrow). **3)** The two broken chromosomes segregate into sister telophase nuclei. **4)** This cycle can continue during subsequent mitosis during gametophyte development. If the broken chromosome is delivered to the endosperm, the chromatid BFB cycles will continue with subsequent losses and duplications of chromosome segments involved in the BFB cycle. However, a single broken chromosome delivered to the zygyote will subsequently heal without dicentric formation

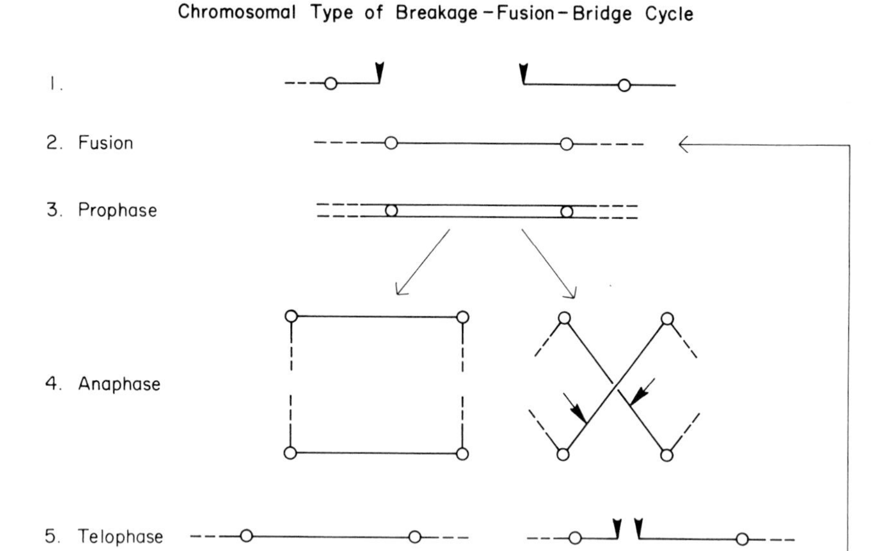

Fig. 8. Chromosomal type of breakage-fusion-bridge cycle. **1)** *Chromosomal* type of breakage-fusion-bridge cycle can be initiated in sporophytic tissue if both male and female gametes donate a newly ruptured chromosome. **2)** At prophase of the first zygote division, each pair of sister chromatids fuse at the ruptured end. **3)** Bridge formation and chromosome breakage, anywhere between the centromeres, at anaphase results in (4) each sister telophase nucleus receiving two newly ruptured chromosomes (only one nucleus is illustrated). **5)** The two ruptured ends fuse, forming a linear dicentric chromosome. **6)** In the next prophase following replication the two sister chromatids are again dicentric. **7)** Two possible configurations of centromere migration during anaphase are illustrated. If the centromeres of the same dicentric chromosome migrate to opposite poles (rightward configuration), two chromatid bridges form and the cycle continues as illustrated from steps 4 to 7. When the centromeres of the same dicentric chromosome migrate to the same pole, no bridge forms and the dicentric chromosome is passed to the daughter cell intact, and the cycle continues as illustrated from steps 5 to 7 (adapted from McClintock, 1951 a)

B. Unorthodox Type of Chromosome Rearrangements in BFB Plants

Viable plants were recovered that had undergone chromosome 9 BFB cycles (McClintock, 1950b, 1951 a). Cytological examination of microsporocytes showed expected gross rearrangements of this region. Since the break during mitotic anaphase could occur anywhere in the fused short arms of chromosome 9, the observed rearrangements included duplications in direct and reverse order, and deficiencies of the 9 S region. Unexpected, but quite prevalent, were rearrangements involving other non-homologous chromosomes. The majority of these changes appeared to be non-random, involving rearrangements between heterochromatic knobs, centromeres, and the nucleolus organizer on chromosome 6. The unexpected types of rearrangements included: (i) fusion of the centromere on chromosome 9 with the centromere of another chromosome not initially involved in the BFB cycle; (ii) knob-to-knob fusions; (iii) isochromosome and pseudo-isochromosome formation; (iv) ring chromosome formation; and (v) translocated chromosomal segments. More complex types of rearrangements involving three or more chromosomes were also detected. Many examples of these unorthodox BFB-induced chromosomal rearrangements have been reviewed by McClintock (1978).

These studies clearly demonstrated that chromosome restructuring occurs as a consequence of BFB cycles in plants. These alterations can spread to chromosomes not directly involved in the initial cycle. The non-random nature of these rearrangements suggests that the action of specific components in the genome may be responsible for generating these new and heritable modifications to the plant genome. At the molecular level, the mechanisms responsible for this action have not been examined. Genetic evidence indicates at least two factors may be involved in mediating non-random chromosome restructuring: (i) the mobile controlling elements that are capable of causing chromosomal rearrangements such as the *Ds* elements as discussed below; and (ii) a chromosomal factor identified by McClintock (1978) that has not been shown to be transposable but can instigate the same general types of chromosomal rearrangements observed in the microsporocytes of BFB plants. This factor has been named the *X* component and it is carried by a small fragment chromosome composed of a centromere, heterochromatin, and part of the chromosome complement of 9S. When a plant contains this fragment chromosome, many unorthodox types of chromosomal rearrangements may occur which appear non-random in nature, in the same manner as non-random chromosomal rearrangements occur in BFB plants. The presence of the fragment chromosome can be easily monitored in plants and the kernel since fragment chromosomes carrying the *X* component have been constructed with a number of genetic markers affecting kernel phenotypes *(I, C, Sh, Bz, W)* (McClintock, personal communication). Variegated kernel phenotypes caused by fragment chromosome loss or rearrangements during endosperm development are strikingly patterned which suggests the *X* component may mediate these events in a regulated fashion (McClintock, 1978).

C. Burst of Mutability Following Chromosome Breakage

Certainly, the cytological consequences of BFB cycles in plants suggest specific components of the plant genome are activated which mediate gross chromosomal modifications. A second type of mutational event was uncovered in the self-pollinated progeny of plants that had undergone the chromosomal type of BFB cycle (McClintock, 1950b, 1951a, 1954). These events were first recognized as somatic sectors in the plant and kernel containing unstable patterns of gene expression and subsequently shown to be due to the action of transposable elements. This mutability caused by controlling element action differed from the somatic variegation associated with ring chromosomes in the following manner: (i) element mediated somatic events usually gave rise to mutations from recessive to dominant at a characteristic time and apparent frequency during plant development; and (ii) somatic instability involved either instability of a specific gene or the programmed loss of an entire chromosome arm from a particular chromosomal location. From the progeny of plants that had undergone BFB cycles, McClintock (1945, 1951, 1954) recovered over 40 different mutable loci including the identification of *Ac/Ds* elements.

D. Examples of Controlling Element Activation by BFB Cycles

A test of the hypothesis that BFB cycles can activate quiescent controlling elements in the maize genome would be to instigate BFB cycles in a genetically marked stock capable of detecting such activity. Studies by McClintock (1950a, 1950b, 1951b) and Doerschug (1973) were designed as such, to test for *Dotted (Dt)* activity following the induction of BFB cycles. *Dt* was originally described by Rhoades (1938) as a factor responsible for destabilizing particular recessive *a 1* alleles *(a 1-dt)* to *A 1* during plant and endosperm development. These alleles, therefore, can report on the presence of *Dt* activity by somatically mutating from a colorless to colored phenotype during plant and kernel development when transactivated by *Dt*.

Plants heterozygous for the 9 S duplication and 9 S deficiency described above and homozygous for an *a 1-dt* allele were crossed as males to plants homozygous for *a 1-dt* (McClintock, 1950a, 1950b, 1951b). As discussed above, plants carrying the chromosome 9 rearrangements would introduce a chromosome 9 with a newly broken end into a high proportion of their offspring which would subsequently initiate the chromatid BFB cycle in developing endosperm tissue (Fig. 7). From these crosses, McClintock (1950a, 1950b, 1951b) was able to detect *Dt* activity as somatic sectors of *a 1-dt* instability in the aleurone of endosperms in which the chromatid BFB cycle was active. Since the *a 1* allele is completely stable in the absence of *Dt* activity and the progenitor strains showed no history of *Dt* activity prior to the induction of BFB cycles, it was concluded that these cycles, directly or indirectly, contributed to the *de novo* induction of *Dt* activity. These experiments were expanded to include an offspring population of over 93,000 kernels with 177 (0.12%) showing somatic mutations *a 1* to *A 1* during kernel development (McClintock, 1951b). Except for a single kernel, all mutations were confined to sectors of the kernel suggesting that the induction of *Dt* activity occurred during somatic development of the endosperm during the BFB cycle. The single exception represented a case with *A1* spots distributed uniformly throughout the aleurone. Since the chromosome breakage cycle was initiated following meiotic anaphase during pollen maturation, this case may be an example of *Dt* induction before fertilization. However, the plant grown from this kernel was shown not to carry an active *Dt* element which suggested that the induction of *Dt* took place following the mitotic division separating the proembryo and proendosperm sperm nuclei. Control experiments were performed by intercrossing plants homozygous for the *a1-dt* allele both with a normal chromosome 9 constitution. Only a single *A1* dot was detected on one kernel in over 21,000 progeny.

These experiments were repeated by Doerschug (1973) using the same stocks and strategy of McClintock (1950a, 1950b, 1951b) with one important exception. The *a1-dt* allele used in Doerschug's studies *[al-m1(cache)]* to detect *Dt* activity, previously isolated by Nuffer (1961), is extremely *sensitive* to *Dt* action. This allele responds very early in endosperm development if *Dt* is present which provides an early detection system for *de novo Dt* induction and reduces the likelihood that the induction of *Dt* activity late in endosperm development would go undetected. Doerschug (1973) found that 0.16% of the offspring of plants carrying the chromosome 9 rearrangements mated as males to plants homozygous for the *a1-m1(cache)* mutation exhibited somatic *A1* sectors. Moreover, two kernels exhibiting uniform *Dt* action in the endosperm carried *Dt* elements in the corresponding embryo. The ability to recover *Dt* elements in the cell lineage of both endosperm and embryo suggests that activation of *Dt* took place before the mitotic division of the spore that separates the proendosperm and proembryo nuclei during pollen maturation. This is just one cell cycle from the induction of the physical break of the dicentric chromosome 9 during meiotic anaphase.

The frequency and speed at which *Dt* is induced is remarkable when considering the experimental design of these studies which is such that the activity of only one controlling element system *(Dt)* will be detected. Since there is no selection, only a screen, for *Dt* activity, it is reasonable to assume that, if detection was possible, many other controlling element activities would be induced following chromosome breakage. If these results can be extrapolated to the minimum of at least seven independent controlling element systems of maize (Peterson, 1981), the potential instability associated with transposable elements following chromosome breakage is extraordinary. *Dt* and *Spm* activity has been reported following ultraviolet or X-ray treatment of pollen (Neuffer, 1966). Both agents are known to cause chromosomal damage, including chromosome breakage. Chromosome breakage may also be indirectly responsible for the origin of genetic instability and mutations mediated by transposable elements in plants challenged with other forms of stress such as systemic virus infection (Sprague and McKinney, 1966; Mottinger *et al.*, 1984a, 1984b; Dellaporta *et al.*, 1984) and the variability associated with *in vitro* regeneration of plants from somatic cells, a process known to induce chromosomal rearrangements that are remarkably similar to the unorthodox types of rearrangements induced by chromosomal BFB cycles in maize (Benzion, 1984; Green *et al.*, 1977; McCoy *et al.*, 1982).

IV. Biology of *Ac/Ds* Elements

A. Chromosome Breakage at Ds

Evidence that transposable elements were responsible for the types of instability uncovered in BFB plants first came from the observation of chromosome breaks at a specific location, termed *Dissociation (Ds),* on chromosome 9 (McClintock, 1948, 1949, 1950 a, 1950 b, 1951 a) in the selfed progeny of plants undergoing the BFB cycle. The breakpoint, first located proximal to the *waxy* locus, was recognized by simultaneous loss of markers distal to the breakpoint *(C Sh Bz Wx)* at a characteristic time and frequency during endosperm development. The consequence of *Ds*-induced breakage was formation of an acentric fragment composed of the two sister chromatids from the *Ds* locus to the end of the short arm of chromosome 9 and the complementary dicentric chromosome from the *Ds* locus to the centromere including the long arms of chromosome 9. Genetic linkage tests determined the initial location of *Ds* as shown in Fig. 9. This genetic location, referred to as the *standard* location, corresponds with the physical location of chromosome breakage in microsporocytes containing *Ds* (McClintock, 1946, 1947). It was soon learned that a dominant factor, termed *Activator (Ac),* was also necessary for *Ds*-induced breakage and transposition to occur (McClintock, 1948, 1949).

Subsequently, many cases were isolated in which *Ds*-type breaks changed position on chromosome 9 to a position distal on the same chromosome or to other chromosomes. Intrachromosomal transpositions of *Ds,* as detected by the change in the position of chromosome breakage, were recovered between all of the loci shown in Fig. 9. Their appearance is immediately revealed by a change in the pattern of marker loss. For instance, transposition of *Ds* from a position distal to the *Sh* locus and proximal to the *I* locus will result in colorless kernel with a sector showing loss of *I* (colored aleurone sectors) with internal subsectors showing BFB instability patterns for markers proximal to *I,* because of dicentric chromosome formation between chromatids at the position of the breakpoint (Fig. 10). In many cases of *Ds* transposition, the appearence of *Ds* at a new location was associated with loss of *Ds*-activity at the former location. It was also recognized that transposition of *Ds* can induce

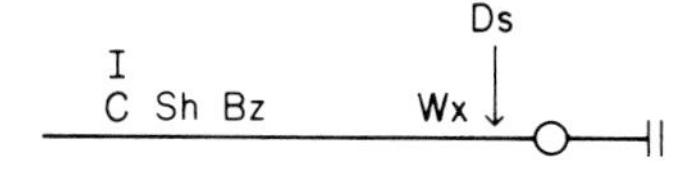

Fig. 9. Standard location of *Ds* on chromosome 9

other types of chromosomal alterations (McClintock, 1947, 1948, 1951 b). These included internal deficiencies, duplications or inversion of segments within chromosome 9, ring chromosome formation, and translocations to other chromosomes, always with one breakpoint at the *Ds* location (McClintock, 1948, 1949, 1951 b).

B. The Ac/Ds Family of Transposable Controlling Elements

The *Ac/Ds* family of transposable controlling elements, first investigated by McClintock (1946, 1947, 1951 a) represents one of several two-unit interactive transposable element systems of maize. (A recent review of the genetics of the *Ac/Ds* system has been provided by Fedoroff, 1983). The two units functionally represent the autonomous *Activator (Ac)* and non-autonomous *Dissociation (Ds)* component. The term *Dissociation* refers to the ability of certain *Ds* elements to catalyze chromosome dissociation but this feature is not a universal characteristic of all *Ds* elements. The autonomous *Ac* element is capable of transposition, excision, and mutational change while the non-autonomous *Ds* element is genetically unstable only in the presence of *Ac* activity. In the absence of *Ac, Ds* elements are genetically inactive. A *Ds* element resident at a gene locus is detected by *Ac*-induced gene instability; in the absence of *Ac, Ds*-induced mutations often behave as stable recessive alleles. Some non-autonomous components, such as *Ds* elements, can be derived from autonomous elements by mutational change (McClintock, 1955, 1956, 1962, 1963; Peterson, 1961, 1968; Fedoroff *et al.*, 1983; Pohlman *et al.*, 1984) which supports the idea that certain types of non-autonomous elements are defective transposons.

The autonomous element *Modulator* was independently described by Brink and Nilan (1952) but subsequently shown to be genetically identical to *Ac* (Barclay and Brink, 1954); that is, both *Ac* and *Mp* elements were capable of *Ds* transactivation and both showed a characteristic dosage response (see below). Since *Ac* and *Mp* elements are genetically indistinguishable, for clarity, we will refer to both elements in this review as *Ac* elements.

C. Mutator Function of Ac

Genetic characterization of *Ac* elements has shown that *Ac* is capable of transposition and causing unstable mutations when inserted at a locus. Because *Ac* elements are autonomous, these mutations are inherently unstable. *Ac* elements are also genetically defined by their capacity to trans-activate *Ds* elements. This can be manifested as *Ds*-induced chromosome breakage, somatic and germinal instability of *Ds*-induced mutations, and transposition of *Ds* elements. It is only necessary that the *Ac* element be present elsewhere in the genome and active for *Ds* trans-activation. This suggests that *Ac* must possess a function that can interact with both *Ac* and *Ds* elements either directly or indirectly through cellular factors. The function required for this instability, a putative transposase factor, has been termed the *mutator* component of *Ac* (McClintock, 1954).

There is genetic evidence that the *mutator* function of *Ac* is affected by the number of active *Ac* elements in the genome. When multiple *Ac* elements are present, there is a change in the pattern of somatic and germinal instability of both *Ac* and *Ds* elements. There is a delay in the timing and a reduction in the apparent frequency of *Ac*-mediated genetic events including chromosome breakage and somatic instability of *Ds* elements as the number of active *Ac* elements increases. This effect is especially evident in the triploid endosperm tissue that is comprised of two maternal and one paternal genomes. Fig. 10 shows that when a single *Ac* is brought into the endosperm through the male gametophyte (1 dose of *Ac*), somatic insta-

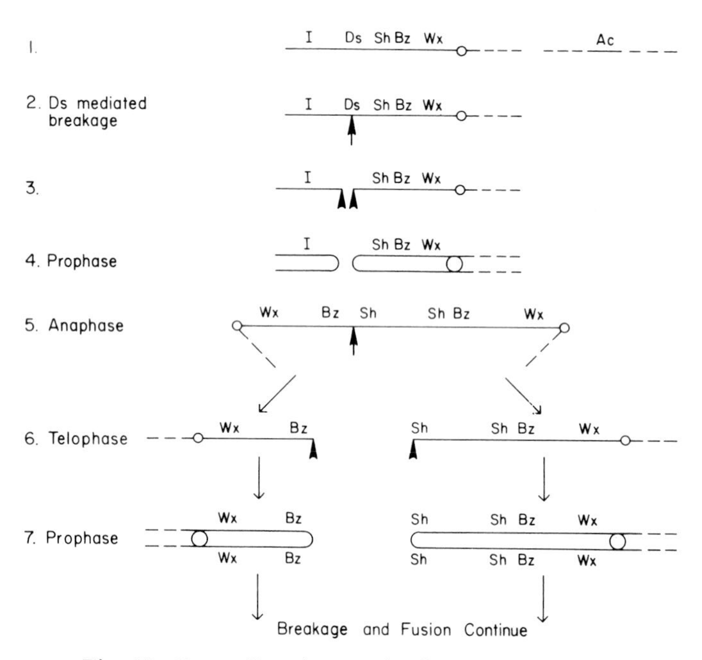

Fig. 10. *Ds*-mediated acentric-dicentric formation

A. **1)** Certain *Ds* elements are characterized by the ability to "dissociate" chromosomes in the presence of *Ac* elements. These *Ds* elements are detected by the loss of chromosome markers distal to *Ds* (e. g. *I:* a dominant colorless aleurone marker). In the given example, the chromosome 9 homologue (not included in the diagram) carries the recessive markers, c sh bz wx. **2)** *Ac* mediates the dissociation of chromosome 9 at the location of *Ds*. **3)** An acentric fragment distal to the *Ds* break point (carrying the *I* marker) is lost, uncovering *C* (colored aleurone) in the descendents of the cell in which the initial dissociation event occurs. This appears as a sector of colored aleurone tissue on a colorless background. **4)** A dicentric chromosome is formed when sister chromatids fuse at the location of the *Ds* break following chromosome replication. This initiates the chromatid BFB cycle. **5)** The dicentric chromosome is physically broken at a position somewhere between the centromeres during mitotic anaphase. Bridge formation and breakage during mitotic anaphase can give rise to two sister telophase nuclei of unequal genetic constitution as diagrammed. **6)** The chromatid BFB cycle continues in successive divisions as a dicentric chromosome is reformed following chromosome replication. This will continue to generate subsectors of tissue hemizygous for the chromosome 9 S markers carried on the recessive homolog. Phenotypically, this will appear as a variegated kernel with a colorless aleurone and sectors of colored *(C)* aleurone containing subsectors of *bz Wx* and *bz wx*.

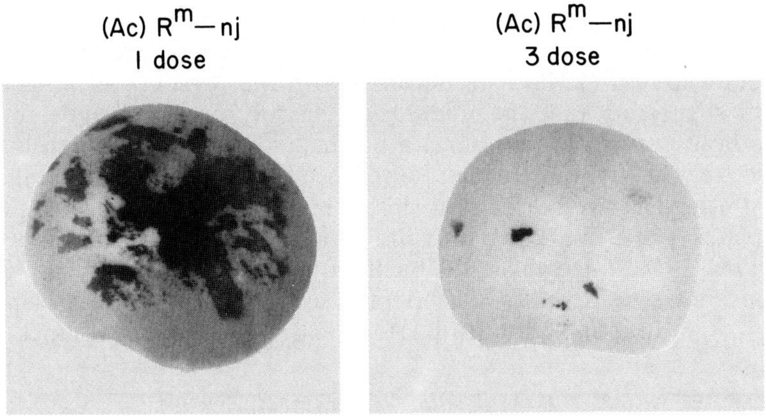

(Ac) R^{m}—nj
1 dose

(Ac) R^{m}—nj
3 dose

B. *Ac* dosage effect can be seen when *Ac* is brought through the male parent into the triploid endosperm *(Ac/-/-)* and through the female parent into the triploid endosperm *(Ac/Ac/Ac)*. The endosperm constitution in kernel A is *r/r/AcR*^{m}*-nj*. [*The AcR*^{m}*-nj* allele is a mutation of the *R-nj* allele caused by insertion of *Ac* (Greenblatt, unpublished).] Somatic mutations from *r* to *R-nj* occur when somatic excision of *Ac* occurs at *AcR*^{m}*-nj*. These somatic events are large and frequent indicating *Ac* transposition early during endosperm development. When the *AcR-nj* mutation is brought into the endosperm through both female and male gametophyte (endosperm constitution *AcR*^{m}*-nj/AcR*^{m}*-nj/AcR*^{m}*-nj*), somatic mutations from *r* to *R-nj* include smaller sectors and apparently less frequently than in kernels containing a single *AcR*^{m}*-nj* gene

bility of *Ac* is developmentally early since the tissue sectors are large. The reciprocal cross brings the same *Ac* element through the female gametophyte, resulting in two doses of *Ac* in the endosperm, and shows a delayed and infrequent pattern of somatic instability.

The repression of somatic and germinal instability of *Ac* and *Ds* elements due to increasing *Ac* dosage appears to be unique to the *Ac/Ds* family. Other known maize transposable element families show either no dosage response or enhanced somatic instability as the number of autonomous elements in the cell increases. For instance, an increase in the apparent frequency of somatic and germinal instability of non-autonomous mutable alleles under the control of the autonomous *Dt* element occurs as the number of *Dt* elements in the cell increases (Rhoades, 1936, 1938; Nuffer, 1955, 1961). The *mutator* component of the *Spm(En)* element does not appear to be sensitive to the dosage of *Spm(En)* elements in the cell (reviewed by Fedoroff, 1983).

Mutational changes in *Ac* or *Ds* elements can give rise to derivatives which have an altered pattern of response to *mutator* activity. For example, a change from a coarse to fine pattern of variegation can result from mutational changes of an *Ac* or *Ds* element. While the phenotypic response of these mutated elements can resemble the delayed patterns of instability apparent with increasing dosages of *Ac*, a mutational change in the element itself is responsible for the phenotype. This is different from the transient change in instability patterns with increasing *Ac* dosage of a particular *Ac* or *Ds* element. These mutational events, termed *changes of state* (McClintock, 1947, 1948), can occur only in the presence of an active *Ac* element suggesting that the *mutator* of *Ac* may be responsible for instigating these events. For instance, a particular state of *Ds* which is highly unstable (e. g. early and frequent chromosome breakage or instability patterns at a locus) can mutate to a state which shows late, infrequent instability patterns in a single step only in the presence of *Ac* (McClintock, 1947, 1948, 1949, 1950b). However, the backmutation from a late to early state of *Ds* appears to be a multiple step process requiring repeated selections through a gradual transition from late to earlier events (McClintock, 1949).

D. Molecular Biology of Ac/Ds

Ac elements have been isolated by molecular cloning techniques (Fedoroff *et al.*, 1983; Behrens *et al.*, 1984; Wessler and Dellaporta, unpublished) and sequenced (Pohlman *et al.*, 1984; Muller-Neumann *et al.*, 1985). To date, all *Ac* elements appear to be the same 4.5 kb element with the conserved structure shown in Fig. 11. This 4.5 kb element has 11 bp imperfect

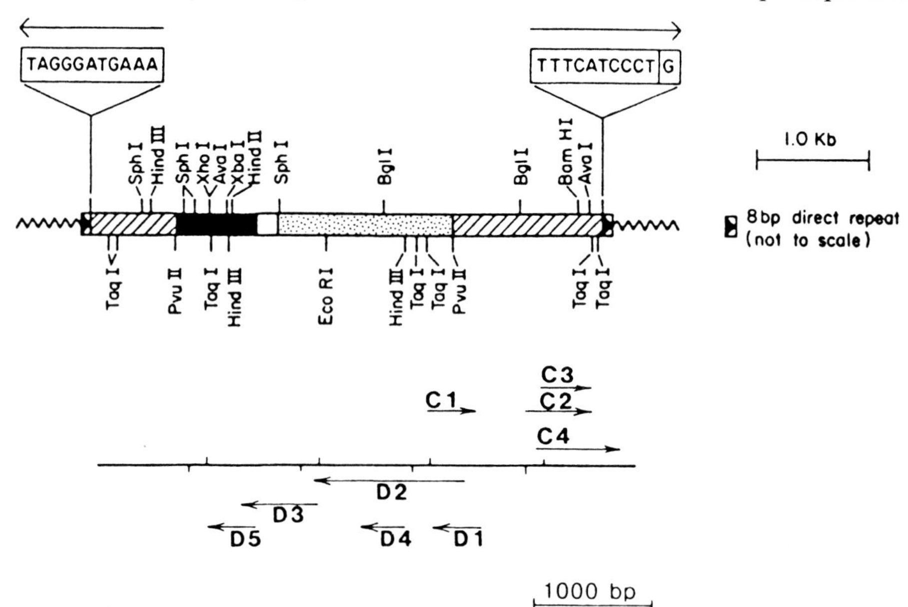

Fig. 11. *Ac* appears to be a canonical 4.5 kb element from comparison of *Ac* elements isolated at *wx-m9* (Fedoroff *et al.*, 1983; Pohlman *et al.*, 1984) *bz-m2 (Fedoroff et al., 1984)*, *wx-m7* (Behrens *et al.*, 1984; Muller-Neumann *et al.*, 1985), and a mutable *R-nj* allele (Dellaporta and Wessler, unpublished results). The *Ac* elements at *wx-m9* and *wx-m7* appear to be identical based on sequence analysis which indicate *Ac* is flanked by 8 bp direct repeats of target site DNA and 11 bp imperfect terminal repeats. The nucleotide sequence data on *wx-m9* indicate several large open reading frames (shown as arrows) are present in *Ac* (J. Messing, personal communication)

terminal inverted repeats and the element is flanked by an 8 bp direct repeat which is present at the target site only once prior to insertion of *Ac* (Pohlman *et al.*, 1984). The direct repeat is thought to occur as a result of staggered cleavage of the target site when *Ac* integrates. Several open reading frames are found in *Ac* that could potentially code for several proteins disregarding the possibility of mRNA splicing between these regions (Pohlman *et al.*, 1984).

Ds elements are less conserved and can broadly be classified in two major categories: those that are structurally related to *Ac* (Type II) (Doring *et al.*, 1984a, 1984b; Fedoroff *et al.*, 1983; Pohlman *et al.*, 1984; Courage-Tebbe *et al.*, 1983) and a conserved subset that are structurally dissimilar to *Ac* sequences except for the inverted termini (Type I) (Sutton *et al.*, 1984; Peacock *et al.*, 1984; Wessler and Dellaporta, unpublished). These two *Ds* types are diagrammed in Fig. 12. Type II *Ds* elements represent those elements directly derived from *Ac* elements (Fedoroff *et al.*, 1983), *Ds* elements found at the *adh 1-2 F11* (Doring *et al.*, 1984b) and *wx-m6* mutations (Fedoroff *et al.*, 1983), and certain *Ds* elements that are characterized by their ability to *dissociate* chromosomes (Courage-Tebbe *et al.*, 1983). The dissociation elements appear to be complex rearrangements of multiple *Ds* sequences that are structurally related to *Ac*. All members of the *Ac/Ds* family are characterized by a similar 11 bp inverted terminal repeat and a 8 bp duplication of target sequences flanking the insertion.

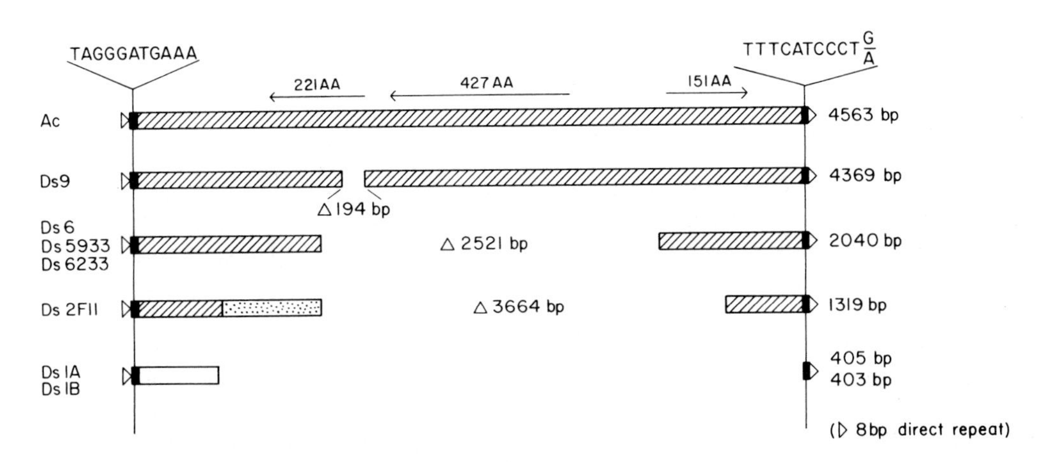

Fig. 12. *Ds* elements. *Ac* represents the 4.5 kb element shown at the top of the figure. *Ds* elements can be directly derived from *Ac* elements as shown by genetic and molecular analysis of a *Ds* element derived from *Ac* at the *wx-m9* mutation. This *Ds* element was shown to be a 194 bp deletion of internal *Ac* DNA (element 2). Other *Ds* elements also appear to represent Ac deletions (element 3 and 4) but others also contain additional sequences that appear less directly related to *Ac* DNA (white box in element 5). The *Ds 1*-like elements are small elements that share little homology to *Ac* DNA except for the terminal regions (element 6)

E. Relationship Between Autonomous and Non-autonomous Components

There is both genetic and molecular evidence that some, but not all, non-autonomous components are directly derived from autonomous elements. Derivatives of the *Ac*-induced mutable alleles *bz-m2* and *wxm-9* have been isolated that behave genetically as *Ds* insertions (McClintock, 1955, 1956, 1962, 1963). Several instances are documented where autonomous *Spm(En)*-induced mutations have given rise to non-autonomous alleles still responsive to trans-activation by unlinked autonomous *Spm(En)* elements (McClintock, 1962; Peterson, 1961, 1968, have been described). Functionally, these events may represent loss of the trans-active *mutator* function(s) required for transposon excision, that would otherwise leave behind a defective version of the element. However, not all non-autonomous elements appear to be directly derived from autonomous elements. Type I *Ds* elements do not share extensive homology with *Ac* elements supporting the idea that defective *Ac* elements represent some but not all *Ds* elements.

Perhaps the clearest molecular evidence for this idea has been provided by Fedoroff *et al.* (1983) and Pohlman *et al.* (1984) that demonstrates the generation of non-autonomous *Ds* from autonomous *Ac* can occur through mutational change of the *Ac* element. Their data show that *Ac* at *wx-m9* differs from the derivative *Dswx-m9* allele by an internal 0.2 kb deletion in the *Ds* element. The structure of *Acwx-m9* and the *Dswx-m9* elements are shown in Fig. 12. The difference between these structures corresponds to a region within an open reading frame of the *Ac* element (Fig. 11) providing further evidence that some non-autonomous components represent loss-of-function derivatives of autonomous elements.

F. Ds Elements That Are Structurally Related to Ac

In addition to *Ds* elements that are known to be direct derivatives of *Ac*, such as *Ds wx-*^{m9}, several other *Ds* elements have been molecularly characterized that share extensive structural homology with *Ac*. These include *Ds* insertions found at the *wx-m6, adh1*-2F11, *sh*-5933, and *sh*-6233 unstable alleles. The latter two mutations represent *Ds* elements that dissociate in the presence of *Ac* and will be considered separately below. The 2 kb *Dswx-m6* element shown in Fig. 12 represents an internal 2.5 kb deletion derivative of the canonical *Ac* element (Fedoroff *et al.*, 1983). This *Ds* structure appears similar to the 2 kb unit of the double 4 kb *Ds* element at *sh*-5933 and *sh*-6233 (Doring *et al.*, 1984a; Geiser *et al.*, 1982; Courage-Tebbe *et al.*, 1983; Weck *et al.*, 1984).

A 1.3 kb *Ds* found in the *adh1*-2F11 allele also hybridizes to *Ac* sequences and possibly represents a modified deletion derivative of the 4.5 kb *Ac* structure with the addition of internal sequences unrelated to *Ac* (Doring *et al.*, 1984; Courage *et al.*, 1984) (Fig. 12). This is one of two *Adh1* mutations recovered from a stock containing both *Ds*-induced *bz2* mutations together with an *Ac* element. The other *Adh1* mutation was the *adh 1*-Fm335 caused by a type I *Ds* element (see below). The *adh 1*-2F11 mutation was initially recovered as a stable *Adh 1* mutant which was subsequently shown to be unstable (high reversion to *Adh+* in pollen) in the presence of *Ac*. The 1.3 kb *Ds* element is found within the transcriptional unit of the *Adh 1* gene. Anaerobically-induced root transcripts contain both an *Adh 1* mRNA that is 1.5 kb larger than the normal *adh 1* transcript and normal size transcripts that appear to lack the *Ds* insertion. The larger message hybridizes to both an *Adh 1* 5′ and 3′ probe and to *Ac* sequences indicating that the *Ds* insertion is most likely contained within the *adh 1*-2F11 mRNA (Döring *et al.*, 1984b).

G. Ds Elements Capable of Chromosome Dissociation

Ds was first identified by McClintock (1945, 1946) as a specific site of chromosome breakage and subsequently shown to be dependent on the presence of *Ac* for instability. Several transpositions of this *Ds* element were identified by their altered pattern of marker loss in endosperm tissue. Two intrachromosomal transposition of *Ds* (*Ds*4864A and *Ds*5245) to a region adjacent to the *shrunken* locus gave high frequencies, of stable and unstable mutations to nearby genes including to the *Sh* locus (McClintock, 1952, 1953). Two of these *sh* mutations (*sh*-6233 and *sh*-5933) derived from these *Ds* stocks are unstable, and so germinal mutations to *Sh* are obtained from plants containing *Ac*. A third mutation *(sh bz-m4)* is a stable *sh* allele both in the absence and presence of *Ac* and mutable for *bronze* only in the presence of *Ac*. This allele has been shown to be deleted for *Sh* DNA (Burr and Burr, 1981; Doring *et al.*, 1981; Chaleff *et al.*, 1981). In all three *sh* mutations, the *Ds* element continues to generate chromosome breaks in the presence of *Ac*.

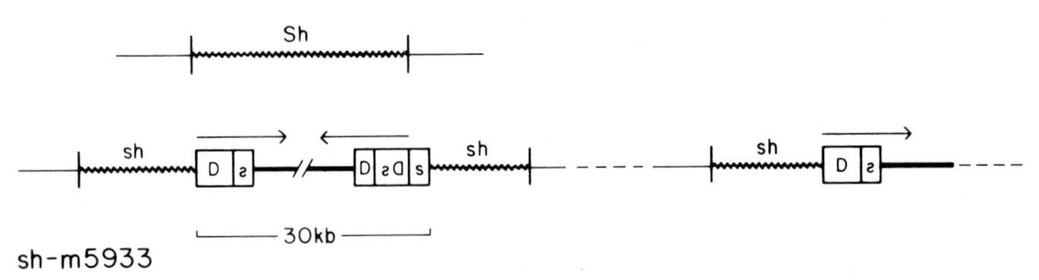

Fig. 13. The structure of *sh-5933* and the composite 4 kb *Ds*. The *sh-5933* mutation has been characterized as a large rearrangement and partial duplication of DNA at *Sh* including the insertion of composite *Ds* elements at the rearrangement breakpoints. The *Sh* gene (1) is interrupted by a 30 kb insertion containing *Ds*-like elements at the endpoints of the rearrangement plus a partial duplication of this rearrangement in the *sh-5933* mutation. The open boxes represent the composite *Ds* element found at the 3′ of the 30 kb insertion and an inverted 3 kb derivative of this element found at the 5′ end. A partial duplication of *Sh* which contains the 3 kb derivative *Ds* sequences and part of the 30 kb insertion is also present but the orientation and position of this duplication in relation to the *Sh* gene is unknown. Several revertants to *Sh* have been analyzed which lose the 30 kb insertion but retain the partial duplication (adapted from Courage *et al.*, 1984)

The structures of *sh*-5933 and *sh*-6233 mutations have been characterized at the molecular level (Döring *et al.*, 1981, 1984; Geiser *et al.*, 1982; Courage-Tebbe *et al.*, 1983; Weck *et al.*, 1984). Allele *sh*-5933 has been shown to contain a 30 kb insertion within the *Sh* gene, coupled with an adjacent partial duplication of *Sh*. This structure is shown in Fig. 13 A. The 3′ terminus of this 30 kb insertion has been sequenced (Döring *et al.*, 1984), revealing the double *Ds* structure shown in Fig. 13 B. This composite structure is built from two 2.04 kb *Ds* elements, one inserted within the other in opposite orientation. Both the inserted and recipient *Ds* have 11 bp inverted repeats flanked by 8 bp duplications. The 5′ terminus of the 30 kb insertion is a 3 kb deletion derivative of the double *Ds* element. The central portion of this 30 kb element is of unknown origin. In addition to the 30 kb insertion, the *sh*-5933 mutation also contains an adjacent duplication that represents a sequence that begins within the central portion of the 30 kb insertion and extends through 5′ *Sh* sequences for more than 25 kb.

Nine germinal mutations of *sh*-5933 to *Sh* have been characterized (Courage-Tebbe *et al.*, 1983). All revertants continue to show *Ds*-induced chromosome breakage in the presence of *Ac*. In all cases the 30 kb insertion is excised restoring an intact sucrose synthase gene that appears to restore *Sh* gene action. However, the duplication that includes part of the 30 kb insertion and upstream sequences is retained in eight of the nine revertants. This could explain continued *Ds*-induced breakage, since the defective 3 kb double *Ds* structure at the 5′ terminus of the insertion was retained. This *Ds* structure may continue to respond to *Ac* causing chromosome dissociation. One *Sh* revertant also contains a 2 kb deletion of the defective 3 kb double *Ds* structure within the duplicated segment. Surprisingly, this revertant has a delayed pattern of *Ac*-activated *Ds*-induced chromosome breakage in the endosperm tissue. Perhaps this deletion is revealing one mechanism that generates the different states of *Ds* elements.

The *sh*-6233 mutation is less complex although the *Ds* insertion is structurally related to the *sh*-5933 mutation (Weck *et al.*, 1984). The unstable *sh*-6233 allele contains only the 4 kb double *Ds* element within the first intron of the *Sh* gene (Weck *et al.*, 1984). One revertant of *sh*-6233 to *Sh* showed excision of the 4 kb double *Ds* element but continues to exhibit *Ac*-mediated chromosome breakage patterns. The possibility that the double *Ds* has transposed nearby could account for continued breakage in this revertant. Analysis of *sh*-6233 suggests that the double *Ds* is itself transposable and is responsible for *Ds*-induced chromosome breakage. A possible mechanism that composite *Ds* elements may mediate chromosome dissociation has been discussed by Döring and Starlinger (1984).

H. Type I Ds Elements

A second class of *Ds* elements, refered to as type I *Ds* (or *Ds 1*), was first characterized at the *adh 1*-Fm 335 mutation by Sutton *et al.* (1984). This mutation, isolated as a low activity *Adh 1* allele that retains the property of anaerobic induction, originated from a strain of maize containing *Ds* at the *bz 2* locus (*bz 2*-m) in the presence of *Ac* (Osterman and Schwartz, 1981). The *adh 1*-Fm335 mutation reverts to *Adh+* at a high frequency in the presence of one dose of *Ac*. This mutation was shown to be caused by a 405 bp *Ds* element inserted in the 5′ transcribed, non-translated region of the *Adh 1* gene (Peacock *et al.*, 1983, 1984; Sachs *et al.*, 1983; Sutton *et al.*, 1984). This *Ds 1* element has the 11 bp terminal inverted repeats characteristic of other type II *Ds* elements and the flanking 8 bp duplication of target site DNA. Between the termini the *Ds 1* sequence is A-T rich and not homologous to *Ac* sequences except for a 26 bp region adjacent to one terminus (Sutton *et al.*, 1984).

The effects of the *Ds*-induced *Adh 1*-Fm335 mutation on *Adh 1* gene activity was shown to result in a sevenfold reduction in enzyme activity (Osterman and Schwartz, 1984). Transcription studies of the *Adh 1*-Fm335 allele indicate that steady state total mRNA levels are decreased by about two orders of magnitude compared to the progenitor allele (Sutton *et al.*, 1984; Peacock *et al.*, 1984). *S 1* mapping experiments indicate that transcription of the *Adh 1*-Fm335 allele begins at the usual site for the *Adh 1* gene and produces a poly A+ mRNA species that co-migrates with the progenitor mRNA on Northern blots. This suggests that if the *DS 1* is initially transcribed in the *Adh 1* RNA, it must subsequently be processed out precisely or imprecisely.

Six *Ac*-induced germinal mutations of the *Adh 1*-Fm335 have been isolated that restore full *Adh 1* enzyme activity levels (Sachs *et al.*, 1983; Peacock *et al.*, 1984). In each event, the *Ds 1* insertion has imprecisely excised from the *Adh 1* gene leaving the 8 bp target site duplications in modified form.

We have subsequently identified a similar *Ds 1*-type element as the insertion in the *wx-m 1* allele (Wessler and Dellaporta, unpublished results). The *wx-m 1* mutation was isolated by McClintock (1948) as a mutation of a single kernel showing an unstable *waxy* phenotype in the presence of *Ac*. The *waxy* locus determines the amylose content of pollen and endosperm tissue (Sprague *et al.*, 1943; Shure *et al.*, 1983). The mutability of *wx-m 1* is expressed as a range of different quantitative levels of *waxy* gene expression. This activity ranges from slightly detectable to wild-type levels of amylose in somatic clonal sectors in the endosperm of kernels containing the *wx-m 1* mutation and an active *Ac* element. Germinal mutations of *wx-m 1* form a series of stable *Wx* alleles which produce an amount of amylose in endosperm tissue that phenotypically corresponds to the different levels of *waxy* expression seen in somatic sectors during endosperm development (McClintock, 1948). The molecular basis for the quantitative differences in *Wx* derivative alleles from the *wx-m 1* mutation appears to be the result of imprecise excision of the *Ds 1* element, leaving behind a modified, but functional, version of the *waxy* gene (Wessler and Dellaporta, unpublished results).

The *Ds 1*-type elements have been shown to represent a family of closely related sequences repeated approximately 40 times in the maize genome (Sutton *et al.*, 1984; Peacock *et al.*, 1984). Several homologous *Ds 1* sequences have been cloned and sequenced were found to be 90—95 % conserved to the 405 bp *Ds adh 1*-Fm 335 element (Peacock *et al.*, 1984). Three of these sequences were shown to contain the 11 bp termini, two of these flanked by 8 bp duplications and the third flanked by a 6 bp direct repeat. These characteristics suggest these elements have been previously mobilized. Remarkably, when the *Ds 1* element is probed to the genome of teosinte, the purported immediate ancestor of maize, and to the genome of *Tripsicum*, a distant relative of maize, multiple homologous fragments are observed (Sachs *et al.*, 1983). Two *Ds 1*-homologous *Tripsicum* sequences have been cloned and sequenced and both are approximately 400 bp, highly homologous to the *Ds 1* element, and have the characteristic 11 bp terminal inverted repeats flanked by 8 bp duplications (Peacock *et al.*, 1984).

These results have some interesting ramifications. It is possible that only the inverted repeat structure is necessary for genetic response to *Ac*. However, the sequence conservation of the repetitive 405 bp element may represent the requirement for additional features necessary for *Ds* activity. Moreover, the *Ds 1* elements at the *Adh 1*-Fm 335 and *wx-m 1* alleles show detectable somatic and germinal instability in the presence of *Ac*. Yet, there exist as many as 40 homologous sequences in the maize genome. The question remains as to the stability of these cryptic *Ds 1* sequences. Because of the high copy number of *Ds 1*-like DNA, it is difficult to determine whether any individual *Ds 1*-like element is unstable. In addition to the multiple *Ds 1*-like sequences, the maize genome contains a significant fraction of *Ac*-like DNA. Surely, the potential for extensive genome restructuring exists if a portion of this DNA is responsive to *Ac*.

Alternatively, many of these *Ds*-like elements may be unable to respond to *Ac* due to DNA, chromatin, or positional differences between authentic *Ds 1* elements at *wx-m 1* and *adh 1*-Fm 335 and their homologous genomic counterparts. The potential mobility of any individual transposable element may be affected by flanking DNA sequences. For instance, independent insertions of *Spm(En)* at a gene locus show differences in controlling element mediated somatic instability patterns which may reflect differences in the position of the *Spm(En)* component (Peterson, 1976, 1977). It is possible that genomic restructuring such as BFB-induced chromosomal rearrangements may change positional controls and contribute to the induction of controlling element instability.

I. Transposition of Ac from a Gene Locus

Several studies indicate that transposition of *Ac* and other maize controlling elements follow rules different from those commonly observed in bacterial transposition (reviewed in Shapiro, 1983): 1. transposition is associated with excision (loss of the element at the donor site) in instances where documentation exists; 2. excisions, while not necessarily precise, can often restore gene activity; 3. the majority of transposition events cover short intrachromosomal distances and may be associated with periods of chromosomal replication.

McClintock's initial discovery of transposition events in maize (McClintock, 1945, 1946, 1950a) were the first studies to note that transposition of *Ds* elements to new chromosomal sites were accompanied by loss of *Ds* activity at the former location. These events were characterized by the recovery of *Ds*-induced chromosome breakage activity at a new location recognized by a change in marker loss patterns in endosperm tissue heterozygous for chromosome 9 kernel markers. Fourteen germinal mutations to *Bz* expression of the unstable *Ac*-induced bronze allele, *bz-m2,* were all associated with loss of *Ac* from the *Bz* locus (McClintock, 1956b). Extensive genetic studies on the somatic movement of *Ac* from the *P* locus are consistent with the idea that transposition often results in the removal of *Ac* from the original site and reinsertion of *Ac* at the recipient site as discussed below.

The *P* locus, located on chromosome 1 of maize, conditions red pigment formation in the somatic pericarp and cob tissues of the ear. Both tissues are maternally derived, and hence the entire pericarp surface of the ear represents somatic diploid tissue that serves as a protective covering during seed maturation. A common cell lineage between the overlying pericarp cells and the female gametophytic tissue in the underlying developing kernel usually means there is a high genetic correspondance between the diploid kernel pericarp tissue and the haploid maternal contribution to endosperm and embryo except for normal meiotic segregation of homologous chromosomes. During sporophyte development, somatic mutations affecting the *P* locus within cell lineages that give rise to pericarp tissue of the ear can be subsequently recovered in one-half (due to normal meiotic segregation) of the underlying offspring if the somatic mutation occurs early enough to be included in germinal cell lineages. The earlier the event, the larger the mutant sector which may sometimes include an area with several underlying kernels. Hence, somatic mutations affecting the *P* locus can sometimes be recovered in offspring for genetic analysis.

Variegated pericarp is an unstable pericarp and cob phenotype conditioned by an allele of the *P* locus *(P-vv)* discovered by Emerson (1914, 1917). The *P-vv* allele has been shown to be a *Modulator (Ac)* element conjoined with a *P-rr* allele which conditions red cob and pericarp. Hence, the *P-vv* mutation can be designated as a *P-rrAc* complex. Plants with the variegated pericarp mutation, heterozygous with colorless pericarp alleles, *(P-rrAc/P-ww or P-rrAc/P-wr)* result in a pericarp phenotype of the ear described as *medium* variegated which comprises numerous red stripes on a colorless pericarp background (Fig. 14). The variegated phenotype is caused by the inhibition of pigment formation when *Ac* is conjoined with the *P* locus *(P-rrAc).* Loss of *Ac* from *P* usually restores full pericarp pigmentation *(P-rr)* and in somatic tissues results in a cell lineage (stripe) of red pericarp cells. The size of the red sector is a function of the timing of *Ac* movement from the *P* locus during the development of the pericarp tissue. The apparent frequency and timing of the variegation in plants heterozygous for *P-rrAc* and a colorless pericarp allele result in an overall ear phenotype called medium variegated. Less frequently, transposition events of *Ac* away from the *P* locus occur relatively early during ear morphogenesis and large sectors of red pericarp tissue covering many kernels may be present on medium variegated ears. About one-half of the offspring from these red pericarp sectors will carry the stable derivative red pericarp allele *(P-rr)* due to normal meiotic segregation of chromosome 1 homologs.

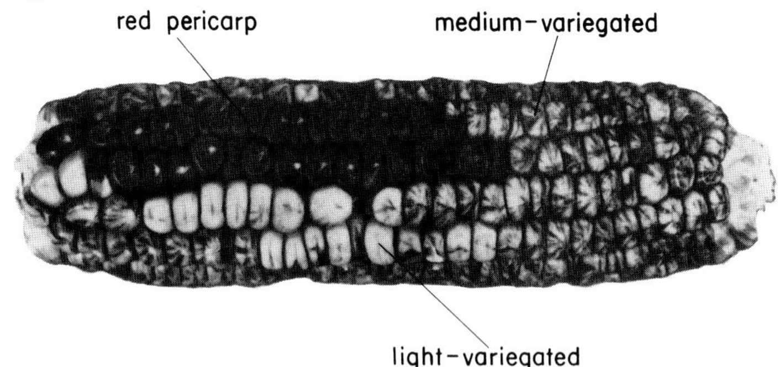

Fig. 14. Red and light-variegated co-twin mutations on medium variegated ear (courtesy of Dr. Irwin Greenblatt)

Surprisingly, more than 80% of red pericarp sectors on medium variegated ears are accompanied by a contiguous sector of *light variegated* pericarp (Fig. 14) (Brink and Nilan, 1952; Van Schaik and Brink, 1959; Greenblatt and Brink, 1962, 1963). These *twinned* mutations are so frequent that they are regarded as the outcome of a *single* transpositional event explained as follows. Genetic tests of the offspring from the light variegated sector were found to differ from sibling medium variegated progeny by possessing an additional *Ac* at some position in the genome other than the *P* locus (Brink and Nilan, 1952; Van Schaik and Brink, 1959). Due to the dosage effects of *Ac* copy number, the second *Ac* element delays the timing and the suppresses the apparent frequency of *Ac* transposition from the *P-rrAc* complex and therefore reduces the size and apparent frequency of striping in the pericarp yielding a light variegated pericarp phenotype. The additional *Ac* element is believed to represent a copy of the *Ac* that transposed (tr-*Ac*) from the *P-rrAc* chromosome during the mutations to red pericarp *(P-rrAc* to *P-rr)* in the contiguous red sector (Brink and Nilan, 1952). (The light variegated sectors may, therefore, be represented by the genotype: *p-rrAc* + tr-*Ac*.) Adjacent red and light variegated pericarp sectors on otherwise medium variegated ears suggests that, during twin formation, transposition of *Ac* occurs from one sister chromatid only after replication of the *P-rrAc* region (Fig. 15) (Brink and Nilan, 1952; Greenblatt and Brink, 1962; Greenblatt, 1968). The next mitotic division results in two daughter cells, one possessing a light variegated phenotype *(P-rrAc* + tr-*Ac)* and the other a red pericarp phenotype *(P-rr),* each cell lineage subsequently forming co-twin pericarp sectors.

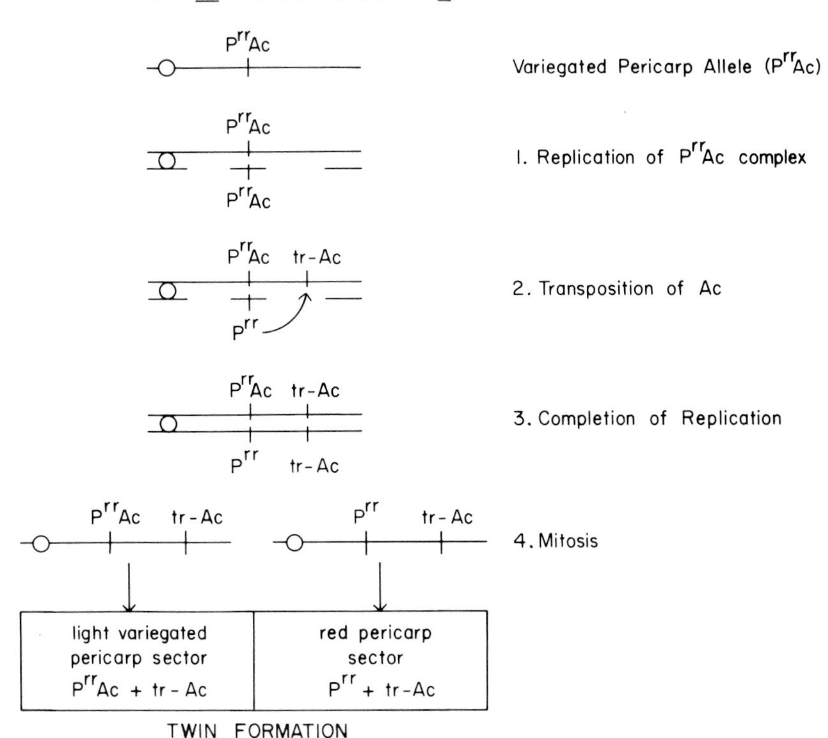

Fig. 15. Model of *Ac* transposition from the *P* locus (adapted from Greenblatt, 1984; used with permission). Analysis of co-twin red and light variegated pericarp mutations on medium variegated ears indicated that the majority of transposition event of *Ac* from the *P* locus are intrachromosomal usually within detectable linkage with the *P* donor site. The model proposed by Greenblatt (1984) to explain co-twin formation is that the *P-rrAc* complex replicates (1) and the *Ac* from one sister chromatid is transposed to an unreplicated recipient site usually on the same chromosome (2). After chromosome replication (3) and mitotic division (4) is complete, the two resulting daughter cells are no longer identical at the *P* locus but both contain a transposed *Ac* at identical positions since *tr-Ac* replication occurs subsequent to transposition. This results in co-twin red *(P-rr + tr-Ac)* and light variegated *(P-rrAc + tr-Ac)* sectors on medium variegated ears *(P-rrAc, no tr-Ac)*

Based on this proposed series of events several questions have been asked. First, what is the evidence that the additional *Ac* element in the light variegated sector arises concurrently with the transposition of *Ac* from *P-rrAc* to yield the *P-rr* sector? Twin spots are physically contiguous tissue sectors suggesting they are daughter lineages from a common cellular event (Brink and Nilan, 1952). Progeny tests show that red and light variegated twin spots can be grouped into two classes depending on the presence or absence of tr-*Ac* in the red sector (Brink, 1955). Since the red pericarp phenotype is unaffected by the presence of *Ac* elsewhere in the genome, detection of tr-*Ac* in this class is based on mating offspring from co-twin red pericarp sectors to a standard *Ds* tester stock. Greenblatt and Brink (1962) have postulated that the class of twin mutations containing tr-*Ac* in the red pericarp sector (about 62% of all light variegated and red pericarp twin mutations) results from a single transposition of *Ac*, after replication at the *P* locus, to a receptor site that replicates along with the new tr-*Ac* (Fig. 15). Cytokinesis would then result in daughter lineages of two types: 1. *P-rrAc* + tr-*Ac* (light variegated pericarp), and 2. *P-rr* + tr-*Ac* (red pericarp containing tr-*Ac* not at *P*).

A test of this hypothesis is the location of the tr-*Ac* in co-twin light variegated and red sectors. If they represent replication of a common unreplicated tr-*Ac* element at the receptor site when *Ac* leaves the *P* locus, then the tr-*Ac* should be located in identical positions in the red and light variegated co-twins. Three point linkage data on the genetic position of tr-*Ac* in relation to the *P* locus in offspring from light variegated and red co-twins confirm that the position of tr-*Ac* in both sectors is indistinguishable when the tr-*Ac* is linked to the *P* locus (Greenblatt and Brink, 1962; Greenblatt, 1984). The experimental design of these studies did not permit mapping events where the tr-*Ac*'s were recombining at random with the *P* locus.

What about untwinned red and light variegated sectors on medium variegated ears? Do these represent alternate modes of *Ac* transposition? Greenblatt (1968, 1974) suggests that these are only visibly untwinned due to loss of the co-twin sector during the three dimensional development of the ear. Since the pericarp tissue is, in a sense, only a two-dimensional surface of the ear, visible twin spots are visible only when both co-twin sectors will eventually emerge on the two-dimensional plane of pericarp tissue. He further suggest that all transposition events of *Ac* from the *P* locus on medium variegated ears result in twinned mutations and that it will depend on the orientation of division planes during the mitosis following *Ac* movement whether or not they are visible.

Evidence for this model of *Ac* transposition is as follows. First, if untwinned red sectors do occur as a result of *Ac* transposition from the *P* locus in medium variegated plants *(P-rrAc/P-ww* or *P-rrAc/P-wr)* by: 1. transposition before replication of the *P-rrAc* complex (red pericarp sector with tr-*Ac*), 2. transposition from the *P* locus prior to chromosome replication but after replication of the receptor site (red pericarp sector with one-half of the red sector cells carrying the tr-*Ac*), or 3. transposition after all replication is complete but prior to cytokinesis (two possibilities: i) potential twin spot with red sector void of tr-*Ac*, or ii) an untwinned red sector with tr-*Ac*), then all potential untwinned red sectors would contain a tr-*Ac* element (Greenblatt, 1968). Since co-twin red sectors contain a tr-*Ac* only about 62% of the time (Greenblatt and Brink, 1962; Greenblatt, 1984), the frequency of tr-*Ac* in apparently untwinned red sectors should exceed the frequency of tr-*Ac* in visibly twinned red sectors if alternate transposition mechanisms do indeed exist. Yet, no detectable differences were found in the offspring of co-twin red sectors when compared to the progeny of visibly untwinned red sectors on medium variegated ears (Greenblatt, 1968). Even if a proportion of apparently untwinned red sectors represented loss of the light variegated co-twin during ear morphogenesis, a detectable difference should be measurable given that untwinned red sectors do occur and that they always would carry the tr-*Ac*.

Secondly, the frequency of red and light variegated types arising from medium variegated parents should be equal if only twinned mutations can occur. Red and light variegated progeny from medium variegated ears represent about 10.2% and 7.4%, repectively, of the total progeny in a backcross progeny of an inbred *P-rrAc/P-wr* line (Greenblatt, 1968). The 16% difference in recoverability of the two types is totally accounted for by recombination between tr-*Ac* and the *P* locus during meiosis in the parental generation. This has the effect of reducing the light variegated types, because meiotic loss of tr-*Ac* from light variegated sectors will yield medium variegated offspring (*P-rrAc* only), while not affecting the recovery of red pericarp types (Greenblatt, 1968). The average frequency of 17.1% recombination between the *P* locus and the tr-*Ac* element in independently occurring red pericarp sectors (Greenblatt, 1966, 1968), can be used to correct light variegated recovery values since the position of tr-*Ac* is apparently identical in the light variegated co-twin sectors. Therefore, the deficit of the light variegated class is totally accounted for by this recombinational loss of tr-*Ac* during meiosis. If recombination between the *P-rrAc* and tr-*Ac* in light variegated sectors is responsible for recovery losses of light variegated offspring, then no losses of the light variegated class is expected in backcrosses of homozygous variegated *(P-rrAc/P-rrAc)* with a colorless pericarp parent since recombination between tr-*Ac* and *P* would not alter the type or number of light variegated or red offspring. Greenblatt (1974) has shown that light variegated and red pericarp types occur at equal frequencies in such a test, supporting his conclusion that untwinned red sectors do occur as a result of *Ac* transposition from the *P* locus. Also, if apparently untwinned sectors are brought about only by loss of one co-twin during ear development, then somatic loss of either co-twin would be expected in equal frequencies. Accordingly, untwinned red sectors were not found to occur more frequently on medium variegated ears (Greenblatt, 1974) suggesting that all transposition events lead to potential twin formation.

If *Ac* transposes from the replicated *P* locus to an unreplicated receptor site, a twinned light variegated : red pericarp mutation will result with the tr-*Ac* in both co-twins. Red co-twin sectors were found to contain tr-*Ac* approximately 62% of the time (Greenblatt and Brink, 1962; Greenblatt, 1984). If there exists a restriction that all transpositions lead to potential twin spots, then the model of *Ac* transposition in Fig. 15 requires that the newly replicated *P-rrAc* does not contribute a tr-*Ac* to a newly replicated receptor site or that the template *P-rrAc* complex does not donate the tr-*Ac* to the template strand of a replicated receptor site. In either situation, an untwinned red sector would result. However, do red co-twin sectors that do not contain a tr-*Ac* element represent transposition of *Ac* to a receptor site that is already replicated? If the tr-*Ac* can integrate in an already replicated receptor site, then the mode of transposition of *Ac*, with the restriction of twinned mutations, would be transposition of the replicated donor *Ac* at *P* to the template strand of the receptor site which is then replicated a second time approximately 62% of the time. An alternate explanation follows.

Greenblatt (1984) has examined 105 independent light variegated : red pericarp twin sectors on medium variegated ears and found 61% of the tr-*Ac* receptor sites are detectably linked with the *P* locus. Hence, the transposition of *Ac* at *P* favors receptor sites within linkage distance to *P* over all other locations combined. These instances of linkage of the *P* locus with tr-*Ac* showed an asymmetry of distribution adjacent to *P* as follows: 1. no tr-*Ac* mapped to a region proximal to *P* for 4 centimorgans, while 23 tr-*Ac* mapped to the corresponding distance distal to the *P* locus; 2. the remaining linked tr-*Ac* elements were equally scattered both proximal and distal to the *P* locus; 3. whether or not the tr-*Ac* element was detected in the offspring of red co-twin sectors did not affect the distribution of receptor sites when tr-*Ac* was linked to the *P* locus.

Greenblatt (1984) has suggested that the asymmetric distribution of linked tr-*Ac* sites reflect the pattern of chromosome replication in the region of the *P* locus. Since *Ac* may transpose only when *P-rrAc* has replicated (Greenblatt and Brink, 1962; Greenblatt, 1966, 1968, 1974) and the receptor site has not begun replication, since the tr-*Ac* is at identical positions when detected in both co-twin sectors, these results can be interpreted as follows (Greenblatt, 1984): the region 4 centomorgans proximal to the *P* locus has completed replication when *P-rrAc* replicates and the corresponding region distal to the *P* locus begins replication when *P-rrAc* has completed replication. More importantly, the same asymmetry of receptor site distribution occuring in both types of twin spots (+/− tr-*Ac* in red co-twin) is consistent with the model of *Ac* transposition and may indicate that only a single mode of transposition of *Ac* is possible (Greenblatt, 1984). That is, *Ac* transposition occurs only from a replicated donor site to a non-replicated receptor site usually over a short intrachromosomal distances. This would imply that about 38% of the time, the tr-*Ac* is present but goes undetected in the red co-twin sector. As discussed below, there is precedence for the idea that *Ac* elements may exist in an inactive state. However, a direct test of this hypothesis must await further molecular and genetic investigation of the tr-*Ac* receptor site in red co-twin sectors with undetected tr-*Ac* activity.

V. Cryptic and Active Forms of *Ac/Ds* Elements

A. General Considerations

A large percentage of a plant genome is comprised of short, interspersed repetitive DNA sequences. In maize, this fraction of the genome consists mainly of repetitive sequences 500—1000 bp interspersed with unique DNA segments with an average length of 2,100 bp (Gupta *et al.*, 1984). A second pattern in maize has been described which consists of middle repetitive sequences interspersed with highly repetitive DNA (Hake and Walbot, 1980). The organization of members of repeat sequence families in plants often show a high degree of homogeneity among related individuals yet their organization may differ substantially between species or even among strains within a species. This suggests a high degree of fluidity of the repetitive elements in the plant genome. Moreover, amplification and displacement of repetitive sequences appear to be fundamental processes in plants (Flavell, 1982; Bedbrook *et al.*, 1980a, 1980b). It has been suggested that the differences in amplification and dispersion patterns of middle repetitive DNA in the genome are generated by DNA transposition mechanisms (Britten and Davidson, 1976; d'Eustachio and Ruddle, 1983) and that this mechanism may be a major factor responsible for diversification among species.

Molecular evidence has been obtained that certain members of this middle repetitive fraction of the maize genome resemble transposable elements (see Flavell, 1982). The *Cin 1* repetitive DNA family of maize has features of transposable elements, such as inverted terminal repeats flanked by direct duplications (Shepherd *et al.*, 1984; Gupta *et al.*, 1984). Moreover, sequences homologous to most active transposable elements in maize are found in all related stocks as dispersed middle repetitive sequences regardless of whether detectable active forms of these elements exist in the genome. The two known exceptions to this dispersion pattern are the *Mu 1* of maize (Bennetzen *et al.*, 1984) and the *copia*-like transposon *Bs 1* of maize (Johns *et al.*, 1985). However, among the dispersed transposon-related sequences in the maize genome dramatic differences in the distribution and copy number exist in closely related maize strains suggesting these sequences may be constantly reshuffled.

Except for DNA homology studies (Fedoroff *et al.*, 1983; Sutton *et al.*, 1984; Dellaporta *et al.*, 1984) the relationship of the known transposable elements of maize to this repetitive fraction of the genome is unknown. These sequences may represent the remnants of once active transposable elements that were destroyed, or they may potentiate the generation of new functional version of transposable elements. In addition, it is unclear what role, if any, repetitive DNA sequences or transposable elements play during normal developmental and evolutionary processes. Are they dispensable sequences or do some members of the population serve an essential role? Is transposition of middle repetitive sequences a normal part of growth and differentiation? Answers to these questions must await further investigation and understanding of transposon function.

Active transposable elements, and other components of the plant genome that respond to certain types of genetic perturbances, are certainly responsible for drastic changes in chromosome structure, genetic variability, and spontaneous mutations (McClintock, 1951a, 1951b, 1978, 1984). In addition, molecular analysis of mutations at the *waxy* locus (S. Wessler, pers. comm.) and *shrunken* locus in maize (Dellaporta *et al.*, 1984) and several loci in *Drosophila* (reviewed by Rubin, 1983) suggests that the majority of spontaneous mutations may be due to chromosomal rearrangements or insertion of moderately repetitive DNA sequences. Hence, these elements may have developed a symbiotic relationship with the genome, capable of activation under the appropriate stimuli. This has led to the suggestion that concerted evolution of these sequences by amplifications and transpositions is partly responsible for diversification in plants.

B. Cryptic Ac-like DNA

It may be possible to activate controlling elements in maize without previous detectable activity through mechanisms such as BFB and agents which damage chromosomes. This suggests mechanisms may exist to change a cryptic sequence to an active element. Therefore, there must exist sequences homologous to controlling elements, but in a quiescent state prior to activation and the activation process may involve either a structural rearrangement, chromatid or methylation alteration, or a positional type of change that subsequently allows mobility and genetic detection of the active form of the element. Although the mechanism of activation has not yet been elucidated, there exist circumstanstial data that pertain to this question. For example, *Ac/Ds* (Fedoroff *et al.*, 1983; Sutton *et al.*, 1984; Peacock *et al.*, 1984; Geiser *et al.*, 1982), the maize transposon *Tz86* (Dellaporta *et al.*, 1984), a receptor element of the *Spm(En)* system (Schwartz-Sommer *et al.*, 1984), and the *Antirrhinum Tam 1* elements all hybridize to middle repetitive DNA sequences.

Even in plants that have no detectable *Ac* activity, there exist sequences homologous to all regions of the active *Ac* element (Fedoroff *et al.*, 1983). The pattern of reiteration of regions within these elements suggests that the internal regions are present in many fewer copies than the highly repetitive ends of the element. Both maize transposons, *Ac* and *Tz86*, have terminal regions present in high copy number but internal regions that are represented in less than 10 copies per genome (Fedoroff *et al.*, 1983; Dellaporta *et al.*, 1984) although no homology exists between these elements. Thus, each transposable element familiy appears to be comprised of a distinct fraction of middle repetitive DNA. The hybridization pattern of the repetitive ends of each element suggests that this DNA is dispersed, rather than tandemly organized, throughout the genome of maize.

C. Differences Between Active and Cryptic Copies of Ac

Since maize elements appear to hybridize to multiple sequences, the question arises as to what distinguishes an active transposable element from these inactive sequences. To address this question, we must first define an active transposable element. As noted above, members of the *Ac/Ds* family share common features such as 11 bp inverted termini and 8 bp target site duplication. Yet the relationship breaks down when internal regions of the element are examined. From the above discussion it appears that elements that are structural derivatives of *Ac* and other poten-

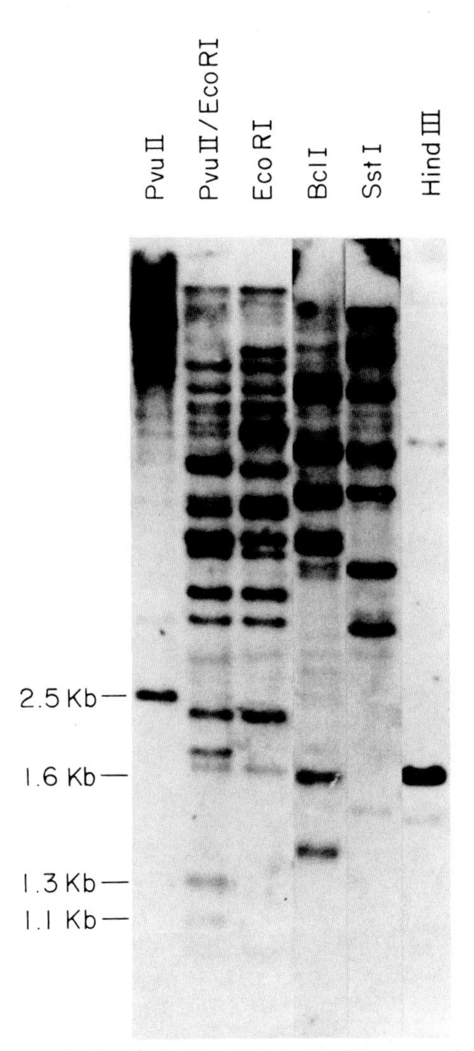

Fig. 16. Genomic blot analysis of *Ac*-like DNA. Maize genomic DNA was digested with the enzymes indicated, subjected to agarose gel electrophoresis, blotted to nitrocellulose filter, and hybridized with an internal 1.6 kb Hind III *Ac* probe. Besides the unique 2.5 kb *Pvu* II fragment characteristic of an active *Ac* element (lane 1), several other sequences which lack genetically detectable *Ac* activity are homologous to this DNA. Most of this *Ac*-like DNA is present in the high molecular weight DNA fraction in the *Pvu* II digest suggesting this DNA may be modified, presumably by methylation. When genomic DNA is digested with enzymes, insensitive to DNA methylation, which do not cut internal to the 4.5 kb *Ac* element, about 8 to 10 fragments are detected (lane 5). A single Eco RI site is present within the 1.6 kb *Hind* III fragment of *Ac*. When genomic DNA is digested with *Hind* III (lane 6) and probed with this 1.6 kb *Hind* III fragment, results suggests that most of the *Ac*-like DNA is structurally conserved for this region. This is confirmed by the Eco RI digest which hybridizes to about twice the number of bands as compared to genomic digests with enzymes which do not cut within this region of *Ac* (lane 3)

tially non-derivative sequences, such as *Ds 1* elements, can respond to *Ac*. The question then becomes if *Ac* is a constant. However, from comparisons of *Ac* elements at three loci, it appears that all active *Ac* elements examined

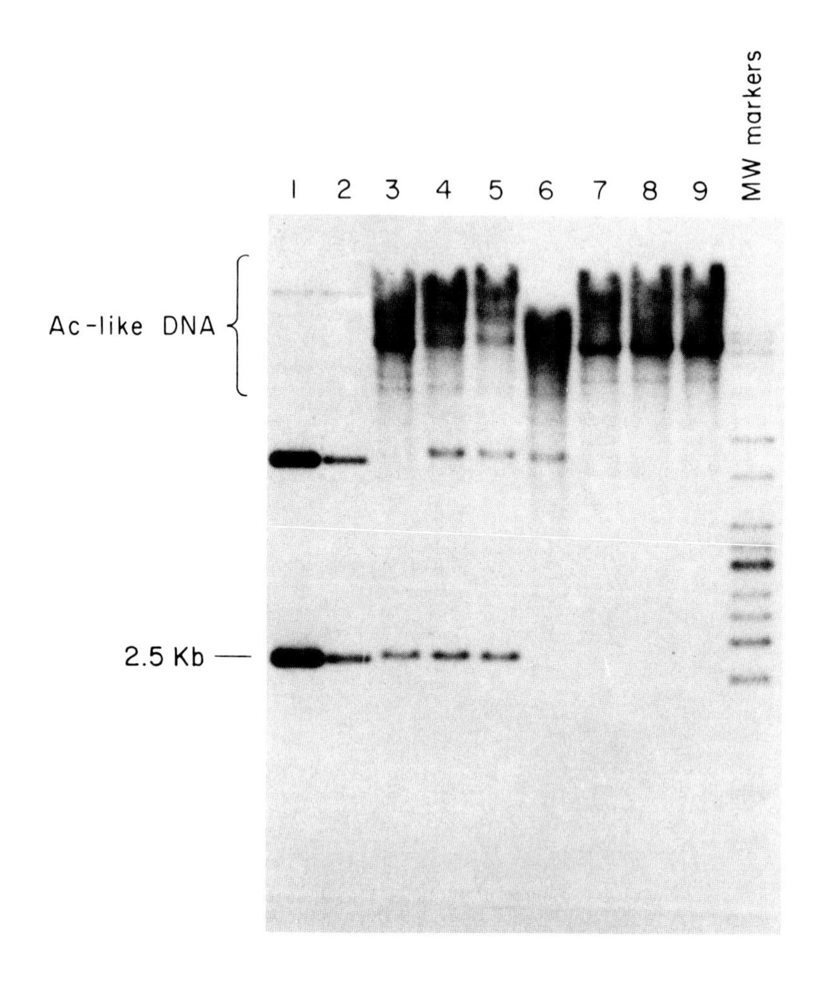

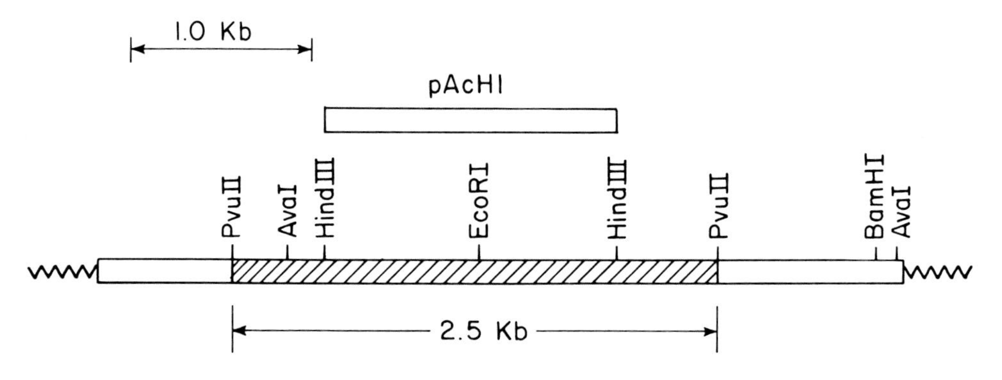

to date are structurally indistinguishable (Fedoroff *et al.*, 1983, 1984; Poulman *et al.*, 1984; Behrens *et al.*, 1984; Muller-Neumann *et al.*, 1984). Sequence comparison of the insertions at *Ac wx-m 9* and *Ac wx-m 7* shows that these elements are identical (Pohlman *et al.*, 1984; J. Messing, pers. comm.; Muller-Neumann *et al.*, 1984). *Ac* elements from unrelated strains of maize such as the *Ac* element found at the *mR-nj* allele (I. Greenblatt, unpublished) also shows a restriction site pattern that is identical to other *Ac* elements (Dellaporta and Wessler, unpublished results). It appears therefore that *Ac* may represent a *canonical* sequence and the homologous, but quiescent, *Ac*-like DNA must differ in some way from the active form. It should therefore be possible to distinguish between the active and cryptic *Ac*-like elements and this distinction may be useful in determining how activation may occur following a genetic perturbance.

Based on structural characteristics, such as restriction site mapping data, the canonical *Ac* element can be distinguished from most cryptic *Ac*-like DNA in the maize genome (Fedoroff *et al.*, 1983, 1984). This situation is not unique to *Ac*. The *Tz 86* element can be structurally distinguished from other homologous sequences based on internal restriction sites (Dellaporta *et al.*, 1984). However, as discussed above, this situation appears to be very different from the one of genomic sequences homologous to the Type I *Ds* family of sequences in maize which appear to represent a highly conserved set of DNA elements repeated about 40 times in the maize genome.

How closely related are these homologous but *cryptic* copies to the known active *Ac* examples? A first indication of the similarity can be obtained from genomic blot analysis of a *Hin*d III digest of DNA from an inbred strain containing a single active *Ac* as shown in Fig. 16. This blot reveals that the 1.6 kb *Hin*d III fragment found in the active *Ac* elements is present in 4—6 single gene equivalents (the very intense band in Fig. 16, lane 1) and therefore must be conserved among some of the cryptic *Ac*-like copies in the genome. Furthermore, the *Eco* RI genomic digest, also shown in Fig. 16 (lane 2), shows twice the number of fragments to that obtained with an enzyme that cuts outside the canonical element (lane 3), indicating that most if not all the cryptic elements with the 1.6 kb *Hin*d III fragment

Fig. 17. Genomic blot analysis of *Ac* elements. Genomic DNAs from plants containing a single active *Ac* element (lanes 1—3) and their siblings which lacks *Ac* were digested with *Pvu* II, an enzyme sensitive to 5-methyl cytosine in the trinucleotide sequence CXG. These samples subject to agarose gel electrophoresis, blotted to nitrocellulose, and hybridized with the radiolabelled 1.3 kb internal *Hind*III fragment of *Ac* (see diagram accompanying figure). This fragment hybridizes to a single gene equivalent of a 2.5 kb *Pvu* II fragment in the DNA of plants containing an *Ac* element. No detectable 2.5 kb Pvu II fragments are seen in DNA of plants lacking *Ac* activity. Notice that all additional hybridization is mainly confined to the high molecular weight DNA fraction at the top of the lanes regardless of whether the plant had *Ac* activity

may contain the *Eco*RI site also found in the 1.6 kb *Hin*dIII fragment of an active *Ac* sequence. We can make the preliminary conclusion based on these results that at least the core region of the active *Ac* restriction map is essentially conserved among some of the cryptic copies.

Probably the most striking feature of the active *Ac* elements identified so far is the presence of an internal 2.5 kb *Pvu*II restriction fragment of *Ac* that is not detectable by genomic blot analysis in strains lacking a genetically active *Ac* element (Dellaporta and Chomet, unpublished). This fragment is identified in Figs. 16 and 17, and detected only in genomic digests of DNA from strains carrying an active *Ac* using the internal 1.6 kb *Hin*dIII fragment of *Ac* as hybridization probe. Plants from kernels exhibiting *Ac* activity show the fragment while sibling plants without *Ac* activity do not. It is clear from Fig. 17 that there are many high molecular weight restriction fragments homologous to the internal *Ac* region in both types of plants.

Why, then, is the 2.5 kb *Pvu*II fragment apparently unique to active copies of *Ac?* There seem to be two possible explanations. Either the cryptic elements have no *Pvu*II sites anywhere near the conserved *Hin*dIII fragment; or, all *Pvu*II sites in the cryptic copies are protected from digestion. Plant DNA, besides containing a significant fraction of 5-methylcytosine (5 mC) in the CG dinucleotide sequence, also contains methylated sites in the trinucleotide sequence CXG (Gruenbaum *et al.*, 1981; reviewed by Vanyushin, 1984). Since *Pvu*II does not cut sites containing methylated C residues (McClelland, 1982; Greenbaum *et al.*, 1981) one possible explanation for the lack of detectable *Ac* homologous *Pvu*II fragments in plants without active *Ac* is the methylation state of the DNA. These results also suggests that active *Ac* elements are undermethylated compared to their inactive genomic counterparts.

D. Cycling Activity of the Mutator Component of Ac

The *mutator* component of the *Ac* element has been shown to undergo changes in the phase of activity: cycling from an active state to an inactive state. The initial observation for this reversible inactivation of *Ac* was made with an *Ac* element found at the *Ac*-induced *wx-m 7* mutation (McClintock, 1964). In the active state, the *Ac* element behaves as other *Ac* elements; that is, the *mutator* component *of Ac* at *wx-m 7* catalyses excision of *Ds* elements in trans and of itself at *wx-m 7.* In the inactive state, no *mutator* activity is genetically detectable. Hence, no instability of the *Ac* at the *waxy* locus or trans-activation of *Ds* elements elsewhere in the genome is observed.

This inactivation of the *mutator* component of *Ac* at *wx-m 7* is evident when the *wx-m 7* mutation is combined with a *Ds*-induced *a 1* mutation *(a 1-m 3)* (Fig. 18). Kernel A *(a 1-m 3/a 1 Acwx-m 7*(active)*/ wx)* shows variegated aleurone pigmentation consisting of deeply colored sectors with a uniformly pale background pigmentation characteristic of this allele in the presence of an active *Ac.* The inactive state of *Ac* at *wx-m 7* is seen in

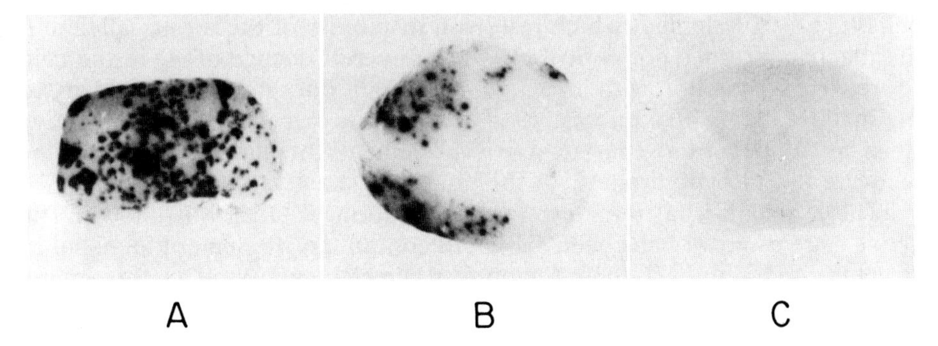

Fig. 18. Phase change in *Ac* activity at the *wx-m 7* allele. The *Ac* element present in the *wx-m 7* allele undergoes changes in the phase of *Ac* activity, cycling between an active and inactive state. The active state of *Ac* at the *wx-m 7* allele can be detected by the ability of *Ac* to transactivate *Ds* elements located elsewhere in the genome. For instance, when the *Acwx-m 7* mutation is combined with the *Ds*-induced *a 1-m 3* mutation, the aleurone of such kernels contain numerous *A 1* sectors representing somatic excision of *Ds* from the *a 1-m 3* allele (**kernel** A). The underlying starchy endosperm (not shown) of these kernels also show somatic *wx* → *Wx* mutations which result from somatic excision of *Ac* from *wx-m 7*. When the *mutator* component of *Ac* cycles to the inactive state, no transactivation of *Ds* occurs. This can be seen as the large sector of colorless aleurone lacking *a 1* → *A 1* mutations in **kernel** B. The underlying starchy endosperm only shows *wx* → *Wx* mutations in the clonal sectors corresponding to the cells containing an active *Ac* element. *Ac* can remain inactive and this state can be recovered germinally. The kernels containing an inactive *Acwx-m 7* allele show no somatic instability of *Ac* at *wx-m 7* nor *Ds* at *a 1-m 3* (**kernel** C)

kernel C. The aleurone has a uniformly pale pigmentation indicating that the *mutator* component of *Ac* is non-functional as a transactivator of *Ds*. The underlying starchy endosperm tissue has a non-clonal, low level of *waxy* gene expression. There is no evidence of clonal *Wx* sectors characteristic of genetic instability of the *Ac* element at *wx-m 7* in the inactive state except where a change of phase has taken place (McClintock, 1964). The basis for the non-clonal background level of *waxy* gene expression is not understood. Phase variation of the inactive state of *Ac* at *wx-m 7* is seen in kernel B with the genetic constitution *a 1-m 3/a 1 Acwx-m 7*(cycling)/*wx*. A large sector of the aleurone shows a uniformly pale anthocyanin pigmentation characteristic of *a 1-m 3* in the absence of mutator function. The corresponding smaller sector of variegated aleurone with underlying somatic instability of the *wx-m 7* mutation indicates an active *Ac* element is present in these cells. Somatic instability of the *wx-m 7* allele occurs only in endosperm tissue corresponding to the variegated portion of the aleurone cells. It appears from this genetic analysis that the mutator component of *Ac* can be reversibly inactivated.

The inactive state of *Ac* can be trans-activated by an active *Ac* element elsewhere in the genome (McClintock, 1964). In this case the inactive *Ac* behaves as a *Ds* element which results in instability of the *wx-m 7* allele but the inactive *Ac* does not contribute to the overall dosage of *Ac* in the cell. Moreover, association with an active *Ac* element does not restore activity to the inactive state (McClintock, 1964). In a plant carrying both the inactive *Acwx-m 7* allele and a separate active *Ac* element, the inactive *Ac* element is recovered in meiotic progeny in the unaltered inactive phase.

The molecular basis for cycling inactivation of *Ac* is not known. The active state of *wx-m 7* has been shown to contain an *Ac* element identical in structure and sequence to the canonical *Ac* element found in the *wx-m 9* mutation (Behrens *et al.,* 1984; Muller-Neumann *et al.,* 1984). We have recently examined the restriction map of the inactive *Ac* element at the *wx-m 7* allele and shown that it too contains an *Ac* element of the same general size and structure as the *Ac* found in the active *wx-m 7* allele with at least one important exception. In the genomic digest the 2.5 kb *Pvu*II restriction fragment described in the previous section as being characteristic of active *Ac* elements is present only in genomic DNA isolated from plants carrying the active state of *wx-m 7.* The 2.5 kb *Pvu*II fragment is undetectable in genomic DNA isolated from plants containing the inactive state of *wx-m 7* (P. Chomet, unpublished results). However, genomic clones of the inactive *Ac* element show that the *Pvu*II sites are still present in the inactive element. This implies that the *Pvu*II sites are protected from digestion in the genomic DNA. Since DNA containing methylated cytosine has been shown to be resistant to *Pvu*II endonuclease digestion (McClelland, 1982; Greenbaum *et al.,* 1981) the lack of detection of the 2.5 kb *Pvu*II fragment in genomic DNA from plants carrying the inactive *wx-m 7* allele may reflect the methylation state of the *Ac* element. Since the inactive *Ac* element has not yet been sequenced, we cannot rule out the possibility that nucleotide differences between the active and inactive *Ac* element are responsible for this change. The differences in methylation patterns may only reflect the phase of activity of *Ac* and not be consequential for inactivation.

The phenomenon of reversible inactivation may not be unique to *Ac.* Several other controlling elements have shown similar alternating cycles in activity such as *Spm(En)* elements (McClintock, 1957, 1958, 1959, 1961, 1962, 1971; reviewed by Fedoroff, 1983) and a *Dt* element (Doerschug, 1968).

VI. Concluding Remarks

A major force in the dynamic flux of the maize genome is the presence of mobile genetic elements capable of causing high rates of spontaneous mutations and gross chromosomal rearrangements. Regardless of whether these elements are genetically detectable, there always appears to exist sequences related to active controlling elements in a quiescent state in the genome. These sequences may be responsible for the generation of active controlling elements following chromosome disruption events. Chromosome breakage appears to be a central mechanism, either directly or indirectly, responsible for this activation signal. The mechanism of activation may involve changes in structure, methylation, or positional changes mediated through chromosome restructuring that occurs following the physical rupture of a chromosome. Once activated, these elements are responsible for heritable genetic instabilities such as non-random chromosome rearrangement and spontaneous unstable mutations. When these elements insert into a structural gene, new patterns of gene regulation are dictated by the controlling element system. Somatic and germinal instability of an element at a gene locus can give rise to altered alleles that differ in quantitative aspects of gene expression. These differences appear to be generated by imprecise excision of a controlling element when excision occurs, modifying the nucleotide sequence of the affected gene and contributing to the polymorphisms in gene expression and genome organization observed in maize progeny. Over extended periods of time, the genetic events mediated by controlling elements may contribute to the widespread genetic diversification in plant populations.

VII. Acknowledgements

We would like to thank our collegues for unpublished information on maize controlling elements with special thanks to Dr. Irwin Greenblatt for his helpful discussion. We are indebted to Dr. Russell Malmberg and Margaret Kelly for their critical reading of the manuscript, and the CSHL Art Department for the line drawings and photographic reproductions.

VIII. References

Bedbrook, J. R., Jones, J., O'Dell, M., Thompson, R. D., Flavell, R. B., 1980a: A molecular description of telomeric heterochromatin in *Secale* species. Cell **19,** 545—560.

Bedbrook, J. R., O'Dell, M., Flavell, R. B., 1980b: Amplification of rearranged repeated DNA sequences in cereal plants. Nature (London) **288,** 133—137.

Behrens, U., Fedoroff, N., Laird, A., Muller-Neumann, M., Starlinger, P., Yoder, J., 1984: Cloning of the *Zea mays* controlling elements *Ac* from the *wx-m 7* allele. Mol. Gen. Genet. **194,** 346—347.

Bennetzen, J. L., Swanson, J., Taylor, W. C., Freeling, M., 1984: DNA insertion in the first intron of maize *Adh*1 affects message levels: Cloning of progenitor and mutant *Adh 1* alleles. Proc. Natl. Acad. Sci., U. S. A. **81** (13), 4125—4128.

Benzion, G., 1984: Genetic and cytogenetic analysis of maize tissue cultures: A cell line pedigree analysis. Ph. D. thesis. Univ. of Minnesota.

Brink, R. A., 1955: Distribution of transposed *Modulator* in red and light variegated twinned mutations from medium variegated pericarp. Maize Genet. Coop. News Letter **29,** 78.

Brink, R. A., and Nilan, R. A., 1952: The relation between light variegated and medium variegated pericarp in maize. Genetics **37,** 519—544.

Britten, R. J., Davidson, E. H., 1976: DNA sequence arrangement and preliminary evidence on its evolution. Fed. Proc. **35** (10), 2151—2157.

Burr, B., Burr, F. A., 1981: Controlling element events at the *Shrunken* locus in maize. Genetics **98,** 143—156.

Chaleff, D., Mauvais, J., McCormick, S., Wessler, S., Fedoroff, N., 1981: Controlling elements in maize. Carnegie Inst. Washington Year Book **80,** 158—174.

Coe, Jr., E. H., Neuffer, M. G., 1977: The genetics of corn. In: *Corn and Corn Improvement.* G. F. Sprague, ed., Amer. Soc. Agronomy, Madison, Wisc. pp. 111—223.

Courage, U., Döring, H. P., Frommer, W. B., Kunze, R., Laird, A., Merckelbach, A., Muller-Neumann, M., Riegel, J., Starlinger, P., Tillmann, E., Weck, E., Werr, W., Yoder, J., 1984: Transposable elements *Ac* and *Ds* at the *Shrunken, Waxy* and *Alcohol Dehydrogenase* loci in Zea mays L. Cold Spring Harbor Symp. Quant. Biol. In press.

Courage-Tebbe, U., Döring, H.-P., Fedoroff, N., Starlinger, P., 1983: The controlling element *Ds* at the *Shrunken* locus in *Zea mays:* Structure of the unstable *sh-m*5933 allele and several revertants. Cell **34,** 383—393.

Dellaporta, S. L., Chomet, P. S., Mottinger, J. P., Wood, J., Yu, S. M., Hicks, J. B., 1984: Endogeneous transposable elements associated with virus infection in maize. Cold Spring Harbor Symp. Quant. Biol. **49,** 321—327.

d'Eustachio, P., Ruddle, F. H., 1983: Somatic cell genetics and gene families. Science **220,** 919—924.

Doerschug, E. B., 1973: Studies of *Dotted,* a regulatory element in maize. I. Induction of Dotted by chromatid breaks. II. Phase variation of Dotted. Theor. Appl. Genet. **43,** 182—189.

Doerschug, E., 1968: Activation cycles of Dt^{TB}. Maize Genet Coop. News. Lett. **42,** 26—28.

Döring, H. P., Tillmann, E., Starlinger, P., 1984a: DNA sequence of the maize transposable element *Dissociation.* Nature **307,** 127—131.

Döring, H. P., Freeling, M., Hake, S., Johns, M. A., Kunze, R., Merckelbach, A., Salamini, F., Starlinger, P., 1984b: A *Ds*-mutation of the *Adh1* gene in *Zea mays* L. Mol. Gen. Genet. **193,** 199—204.

Döring, H. P., Starlinger, P., 1984: Barbara McClintock's controlling elements: Now at the DNA level. Cell **39,** 253—259.

Döring, H. P., Geiser, M., Starlinger, P., 1981: Transposable element *Ds* at the *Shrunken* locus in *Zea mays.* Molec. Gen. Genet. **184,** 377—380.

Emerson, R. A., 1917: Genetical studies on variegated pericarp in maize. Genetics **2,** 1—35.

Emerson, R. A., 1914: The inheritance of a recurring somatic variation in variegated ears of maize. Am. Nat. **48,** 87—115.

Fedoroff, N. V., Furtek, D. B., Nelson, Jr., O. E., 1984: Cloning of the *bronze* locus in maize by a simple and generalizable procedure using the transposable controlling element *Ac.* Proc. Natl. Acad. Sci., U. S. A. **81,** 3825—3829.

Fedoroff, N. V., 1983: Controlling elements in maize. In: Mobile Genetic Elements (ed. J. A. Shapiro). Academic Press, Orlando, Fla., Chp. 1, pp. 1—63.

Fedoroff, N., Wessler, S., Shure, M., 1983: Isolation of the transposable maize controlling elements *Ac* and *Ds.* Cell **35,** 235—242.

Fincham, J. R. S., Sastry, G. R. K., 1974: Controlling elements in maize. Ann. Rev. Genet. **8,** 15—50.

Flavell, R., 1982: In: Genome Evolution. Dover, G. A., Flavell, R. B. eds. London: Academic Press, pp. 301—323.

Freeling, M., 1984: Plant transposable elements and insertion sequences. Ann. Rev. Plant. Physiol. **35,** 271—298.

Friedeman, P., Peterson, P. A., 1982: The *Uq* controlling element system in maize. Mol. Gen. Genet. **187,** 19—29.

Geiser, M., Weck, E., Döring, H. P., Werr, W., Courage-Tebbe, U., Tillmann, E., Starlinger, P., 1982: Genomic clones of a wild type allele and a transposable element-induced mutant allele of the *sucrose synthase* gene of *Zea mays* L. EMBO J. **1,** 1455—1460.

Green, C. E., Phillips, R. L., Wang, A. S., 1977: Cytological analysis of plants regenerated from maize tissue cultures. Maize Genet. Coop. News Lett. **51,** 53—54.

Greenblatt, I. M., 1984: A chromosomal replication pattern deduced from pericarp phenotypes resulting from movements of the transposable element, *Modulator,* in maize. Genetics **108,** 471—485.

Greenblatt, I. M., 1974: Movement of *Modulator* in maize: a test of a hypothesis. Genetics **77,** 671—678.

Greenblatt, I. M., 1968: The mechanisms of *Modulator* transposition in maize. Genetics 58, 585—597.

Greenblatt, I. M., 1966: Transposition and replication of *Modulator* in maize. Genetics **53,** 361—369.

Greenblatt, I. M., Brink, R. A., 1963: Transpositions of *Modulator* in maize into divided and undivided chromosome segments. Nature **197,** 412—413.

Greenblatt, I. M., Brink, R. A., 1962: Twin mutations in medium variegated pericarp maize. Genetics **47,** 489—501.

Greenbaum, Y., Naveh-Many, T., Cedar, H., Razin, A., 1981: Sequence specificity of methylation in higher plant DNA. Nature **292,** 860—862.

Gupta, M., Shepherd, N. S., Bertram, I., Saedler, H., 1984: Repetitive sequences and their organization on genomic clones of *Zea mays.* EMBO J. **3,** 133—139.

Hake, S., Walbot, V., 1980: The genome of *Zea mays,* its organization and homology to related grasses. Chromosoma **79,** 251—270.

Johns, M. A., Mottinger, J., Freeling, M., 1985: A low coy number, *copia*-like transposon in the maize genome. EMBO J., in press.

McClelland, M., 1982: The effect of sequence specific DNA methylation on restriction endonuclease cleavage. Nucleic Acid Res. **9,** 5859—5866.

McClintock, B. 1984: The significance of responses of the genome to challenge. Science **226,** 792—801.

McClintock, B., 1978: Mechanism that rapidly reorganize the genome. Stadler Symp. **10,** 25—48.

McClintock, B., 1971: The contribution of one component of a control system to versatility of gene expression. Carnegie Inst. Washington Year Book **70,** 5—17.

McClintock, B., 1964: Aspects of gene regulation in maize. Carnegie Inst. Washington Year Book **63,** 592—602.

McClintock, B., 1963: Further studies of gene-control systems in maize. Carnegie Inst. Washington Year Book **62,** 486—493.

McClintock, B., 1962: Topographical relations between elements of control systems in maize. Carnegie Inst. Washington Year Book **61,** 448—461.

McClintock, B., 1961: Some parallels between gene control systems in maize and in bacteria. The American Naturalist **95,** 265—277.

McClintock, B., 1959: Genetic and cytological studies of maize. Carnegie Inst. Washington Year Book **58,** 452—456.

McClintock, B., 1958: The *suppressor-mutator* system of control of gene action in maize. Carnegie Inst. Washington Year Book **58,** 452—456.

McClintock, B., 1957: Genetic and cytological studies of maize. Carnegie Inst. Washington Year Book **56,** 393—401.

McClintock, B., 1956a: Controlling elements and the gene. Cold Spring Harbor Symp. Quant. Biol. **21,** 197—216.
McClintock, B., 1956b: Mutation in maize. Carnegie Inst. Washington Year Book **55,** 323—332.
McClintock, B. 1955: Controlled mutation in maize. Carnegie Inst. Washington Year Book **54,** 245—255.
McClintock, B., 1954: Mutations in maize and chromosomal aberrations in *Neurospora.* Carnegie Inst. Washington Year Book **53,** 254—260.
McClintock, B., 1953: Mutation in maize. Carnegie Inst. Washington Year Book **52,** 227—237.
McClintock, B., 1952: Mutable loci in maize. Carnegie Inst. Washington Year Book **51,** 212—219.
McClintock, B., 1951a: Chromosome organization and genetic expression. Cold Spring Harbor Symp. Quant. Biol. **16,** 13—47.
McClintock, B., 1951b: Mutable loci in maize. Carnegie Inst. Washington Year Book. **50,** 174.
McClintock, B., 1950a: Mutable loci in maize. Carnegie Inst. Washington Year Book **49,** 157—167.
McClintock, B., 1950b: The origin and behavior of mutable loci in maize. Proc. Natl. Acad. Sci., U. S. A. **36,** 344—355.
McClintock, B., 1949: Mutable loci in maize. Carnegie Inst. Washington Year Book **48,** 142—154.
McClintock, B., 1948: Mutable loci in maize. Carnegie Inst. Washington Year Book **47,** 155—169.
McClintock, B., 1947: Cytogenetic studies of maize and *Neurospora.* Carnegie Inst. Washington Year Book **46,** 146—152.
McClintock, B., 1946: Maize genetics. Carnegie Inst. Washington Year Book **45,** 176—186.
McClintock, B., 1945: Cytogenetic studies of maize and *Neurospora.* Carnegie Inst. Washington Year Book **44,** 108—112.
McClintock, B., 1944: Maize genetics. Carnegie Inst. Washington Year Book **43,** 127—135.
McClintock, B., 1942: The fusion of broken ends of chromosome following nuclear fusion. Proc. Natl. Acad. Sci., U. S. A. **28,** 458—463.
McClintock, B., 1941: The stability of broken ends of chromosomes in *Zea mays.* Genetics **26,** 234—282.
McClintock, B., 1938a: The production of homozygous deficient tissues with mutant characteristics by means of the aberrant mitotic behavior of ring-shaped chromosomes. Genetics **23,** 315—376.
McClintock, B., 1938b: The fusion of broken ends of sister half-chromatids following chromatid breakage at meiotic anaphases. Univ. Missouri Agr. Exp. Sta. Res. Bull. **290,** 1—48.
McClintock, B., 1932: A correlation of ring shaped chromosomes with variegation in *Zea mays.* Proc. Natl. Acad. Sci., U.S.A. **18,** 677—681.
McCoy, T. J., Phillips, R. L., Rines, H. W., 1982: Cytogenetic variation in tissue culture regenerated plants of *Avena sativa:* High frequency of partial chromosome loss. Can. J. Genet. Cytol. **24,** 37—50.
Mottinger, J. P., Dellaporta, S. L., Keller, P. B., 1984a: Stable and unstable mutations in Aberrant Ratio stock of maize. Genetics **106,** 751—767.
Mottinger, J. P., Johns, M. A., Freeling, M., 1984b: Mutations of the *Adh 1* gene in

maize following infection with barley stripe mosaic virus. Mol. Gen. Genet. **195,** 367—369.
Muller-Neumann, M., Yoder, J., Starlinger, P., 1985: The sequence of the *Ac* element of *Zea mays*. Mol. Gen. Genet. (in press).
Nevers, P., Shepherd, N. S., Saedler, H., 1984: Plant transposable elements. In: Advances in Botanical Research. Academic Press, London, in press.
Nuffer, M. G., 1966: Stability of the suppressor element in two mutator systems at the *A 1* locus in maize. Genetics **53,** 541—549.
Nuffer, M. G., 1961: Mutation studies at the *A 1* locus in maize. 1. A mutable allele controlled by *Dt*. Genetics **46,** 625—640.
Nuffer, M. G., 1955: Dosage effect of multiple *Dt* loci on mutation of *a* in the maize endosperm. Science **121,** 399—400.
Osterman, J. C., Schwartz, D., 1981: Analysis of a controlling element mutation at the *Adh* locus of maize. Genetics **99,** 267—273.
Peacock, W. J., Dennis, E. S., Gerlach, W. L., Sachs, M. M., Schwartz, D., 1984: Insertion and excision of *Ds* controlling elements in maize. Cold Spring Harbor Symp. Quant. Biol. **49,** 347—354.
Peacock, W. J., Dennis, E. S., Gerlach, W. L., Llewellyn, D., Lory, H., Pryor, A. J., Sachs, M. M., Schwartz, D., Sutton, W. D., 1983: Gene transfer in maize: Controlling elements and the alcohol dehydrogenase genes. In: Proc. Miami Winter Symp., Academic Press.
Peterson, P. A., 1981: Instability among the components of a regulatory-element transposon in maize. Cold Spring Harbor Symp. Quant. Biol. **45,** 447—455.
Peterson, P. A., 1977: The position hypothesis for controlling elements in maize. In: DNA insertion elements, plasmids, and episomes. (Ed. A. I. Bukhari *et al.*) Cold Spring Harbor Laboratory, Cold Spring Harbor, N.Y.
Peterson, P. A., 1976: Basis for the diversity of states of controlling elements in maize. Mol. Gen. Genet. **149,** 5—21.
Peterson, P. A., 1968: The origin of an unstable locus in maize. Genetics **59,** 391—398.
Peterson, P. A., 1961: Mutable *a 1* of the *En* system in maize. Genetics **46,** 759—771.
Pohlman, R., Fedoroff, N. V., Messing, J., 1984: The nucleotide sequence of the maize controlling element *Activator*. Cell **37,** 635—643.
Rhoades, M. M., 1938: Effect of the *Dt* gene on the mutability of the *a 1* allele in maize. Genetics **23,** 377—397.
Rhoades, M. M., 1936: The effect of varying gene dosage on aleurone colour in maize. J. Genet. **33,** 347—354.
Rubin, G. M., 1983: Dispersed repetitive DNAs in *Drosophila*. In: *Mobile Genetic Elements*. (Ed. Shapiro, J. A.) Academic Press, Inc. Orlando, Fla., Chapter 8, pp. 329—361.
Sachs, M. M., Peacock, W. J., Dennis, E. S., Gerlach, W. L., 1983: Maize *Ac/Ds* controlling elements: A molecular viewpoint. Maydica **28,** 289—302.
Schwartz-Sommer, Z., Gierl, A., Klosgen, R. B., Wienard, U., Peterson, P. A., Saedler, H., 1984: The *Spm (En)* transposable element controls the excision of a 2-Kb DNA insert at the wx^{m8} allele of *Zea mays*. EMBO J. **3,** 1021—1028.
Shapiro, J. A. (ed.), 1983: *Mobile Genetic Elements*. Academic Press, Orlando, Fla.
Shepherd, N. S., Schwarz-Sommer, Z., Blumberg vel Spalve, J., Gupta, M., Wienand, U., Saedler, H., 1984: Similarities of the *Cin 1* repetitive family of *Zea mays* to eucaryotic transposable elements. Nature **307,** 185—187.
Shure, M., Wessler, S., Fedoroff, N., 1983: Molecular identification and isolation of the *Waxy* locus in maize. Cell **35,** 225—233.

Sprague, G. F., McRinney, H. H., 1966: Aberrant ratio: an anomaly in maize associated with virus infection. Genetics **54,** 1287—1296.
Sprague, G. F., Brimhall, B., Hixon, R. M., 1943: Some effects on the *waxy* gene in corn on the properties of the endosperm starch. J. Am. Soc. Agron. **35,** 817—822.
Stadler, L. J., 1930: Some genetic effects of X-rays in plants. J. Hered. **21,** 3—19.
Stadler, L. J., 1928: Genetic effects of X-rays in maize. Proc. Natl. Acad. Sci., U.S.A. **14,** 69—75.
Sutton, W. D., Gerlach, W. L., Schwartz, D., Peacock, W. J., 1984: Molecular analysis of *Ds* controlling element mutations at the *Adh1* locus of maize. Science **223,** 1265—1268.
Van Schaik, N. W., Brink, R. A., 1959: Transposition of *Modulator,* a component of the variegated pericarp allele in maize. Genetics **44,** 725—738.
Vanyushin, B. F., 1984: Replicative DNA methylation in animals and higher plants. Current Tropics in Micro. and Immun. **108,** 99—114.
Weck, E., Courage, U., Doring, H. P., Fedoroff, N., Starlinger, P., 1984: Analysis of *sh-m*6233, a mutation induced by the transposable element *Ds* in the sucrose synthase gene of *Zea mays.* EMBO J. **3,** 1713—1716.

Chapter 11

Somaclonal Variation: The Myth of Clonal Uniformity

W. R. Scowcroft

CSIRO Division of Plant Industry, Canberra, A.C.T., Australia

Contents

Introduction

The genomes of eukaryotes are in a dynamic state of flux. Nowhere is this more apparent than in higher plants, particularly under *in vitro* culture systems, where the amount of variability generated can be described as staggering. The nature and extent of the variation arising during culture highlights the apparent fragility of plant genomes when the normal course of development is perturbed.

The ability to enhance the level of genetic variability and even the possibility of generating novel genes is exciting to plant biologists in general and provides new experimental options for plant improvement. This article examines the salient features of tissue culture "induced" variability and offers some suggestions as to possible mechanisms which give rise to the variation. Finally, some thoughts will be proffered on the benefits and disbenefits of the phenomenon.

II. *In Vitro* Culture and Genetic Flux

Variability arising from tissue culture has been described at all levels of the tissue culture process. For convenience variability in callus and suspension cultures will be treated separately from that of variation observed among plants regenerated from tissue cultures. The former is usually referred to as tissue culture instability while the latter is described as somaclonal variation. The distinction is arbitrary and the two phenomena are separated here only for clarity and to provide some historicity. Considerably more attention will be devoted to somaclonal variation because the genetic consequences of variability in tissue culture are more readily assessed in regenerated plants and their sexually or asexually propagated progeny.

A. Tissue Culture Instability

Since 1961 reports have appeared which document the inherent instability of tissue culture lines. Instability has been discerned at the karyotypic, morphological, biochemical and molecular levels.

1. Chromosomal Instability

Bayliss (1980) comprehensively reviewed the reported instances of chromosomal variation in tissue cultures. The classes of variant cells included polyploid and aneuploid changes, structural changes in chromosome morphology, and mitotic aberrations including multipolar spindles, lagging chromosomes, fragments and asymetric chromatid separation. In 53 reports where the chromosomal status of callus or suspension cultures was evaluated, Bayliss (1980) records only seven where no chromosomal variants were reported. Subsequently, tissue cultures of at least two of these seven, *Lolium* (Ahloowalia, 1983) and *Medicago sativa* (Reisch and Bingham, 1981) gave rise to plants with extensive chromosomal rearrangements.

As Bayliss (1980) points out chromosomal instability in tissue cultures is clearly the expectation rather than the exception. Polyploid cells can arise from the division of endopolyploid or endoreduplicated cells in the original explant. However, the majority of polyploids or aneuploids arise during culture. For example, callus lines cloned from single cells were shown to contain diploid, tetraploid and aneuploid cells. These could only have arisen during culture subsequent to the single cell cloning procedure.

Structurally aberrant chromosomes can be readily observed even at the gross morphological level. Most commonly these include large and small deletions, as evidenced by loss of satellites, translocations, centric fusions, and the presence of anaphase bridges provide direct evidence of chromosome breakage. At a more refined level of observation specific structural discontinuities have been discerned by Giemsa-banding in cultures of *Vicia faba* (Jelaska *et al.*, 1978), *Crepis capillaris* (Ashmore and Gould, 1981) and *Brachycome dichromosomatica* (Gould, 1982).

There is no clear cause of the enhanced frequency of chromosomal abnormalities in tissue cultures. Bayliss (1980) argues that the increase in chromosomal abnormalities results from disorganized growth of the culture. He further interprets the seeming causal relationship between the auxin 2,4-D and occurrence of chromosomal abnormalities as a result of the stimulation of disorganised growth by 2,4-D rather than a direct effect of the auxin on DNA replication or mitosis.

2. Morphological Changes

A phenomenon frequently encountered is the spontaneous appearance of variation amongst subclones of a parental cell line. Chaturvedi and Mitra (1975) described two subclones of *Citrus grandis* callus grown under identical culture conditions. One would consistently form numerous embryoids while the other formed shoots. Selby and Collins (1976) studied 20 subclones of the cultures of each of three onion varieties. They were found to vary greatly in growth rate (a 3-fold difference), friability, sliminess, pigmentation (from none to deep red), and alliinase activity. Clonal variation in culture morphology and growth rate has often been described as has subclonal differences in pigmentation (see Larkin and Scowcroft, 1981). Changes to auxin habituation and cytokinin habituation (Meins, 1983) have also been frequently observed in cell culture.

3. Biochemical Changes

Tissue cultures of many species produce secondary metabolites, many of which have commercial potential (Staba, 1982). In several instances callus subclones have been described which vary in their ability to produce alkaloids or other secondary metabolites. For example, when 143 colonies of a *Solanum laciniatum* culture were examined individually for the steroidal alkaloid, solasonine, they varied from 0 % to >3 % on a dry weight basis (Zenk, 1978). Similarly, Tabata *et al.* (1978) reported dramatic differences in nicotine production in subclones of *Nicotiana rustica* and *N. tabacum* cultures. *Catharanthus roseus* colonies plated from the one culture were found to differ in their contents of ajmalicine and serpentine (see Zenk, 1978). The content ranged from none to concentrations exceeding that of the parent form. Fujiwara (1982) contains many additional papers which describe variation in secondary metabolite production between callus subclones.

Isoenzyme analysis has been used to a limited extent to detect changes in tissue cultures. Isozyme pattern changes have been observed in cultures of *Phaseolus vulgaris* (Arnison and Boll, 1975), *Hordeum* hybrids (Orton, 1980) and celery (Orton, 1983a). As Orton (1983b) points out these results are equivocal because of known epigenetic and/or developmental effects on isozyme expression.

It is difficult, if not impossible, to interpret variation observed in culture in a genetic mode. Epigenetic events can equally give rise to instability in culture. Non-heritable changes, though of some scientific interest, are trivial with respect to the nature of genomic flux. However, in recent years the genetic analysis of plants regenerated from tissue culture has revealed that extensive genetic changes apparently occur during tissue culture.

B. Somaclonal Variation

It is now firmly established that frequent genetic modifications can occur during the process of tissue and cell culture. Many of these modifications are manifested as heritable mutations among the progeny of regenerated plants. This phenomenon, called somaclonal variation, can be defined as genetic variability generated during tissue culture (Larkin and Scowcroft, 1981). Identifying a variant as a somaclonal mutant requires genetic testing of regenerants by selfing and appropriate crossing for sexually propagated species. For asexually propagated species where meiotic transmission is difficult or impossible, transmission of the trait through at least two successive clonal propagation cycles provides reasonable assurety of a genetic base.

1. Ubiquity of Somaclonal Varation

Table 1 is a comprehensive list of species in which somaclonal variation has been reported to date. Qualitative details for many of these cases are given by Larkin and Scowcroft (1981) and Scowcroft and Larkin (1984).

Table 1. Species Displaying Somaclonal Variation

The explant refers to the plant organ(s) from which the cell lines or *in vitro* culture was initiated. The regeneration mode describes the culture conditions from which plants were regenerated. A single or two word description is admittedly inadequate. Callus, for example can be disorganised as in the case of tobacco, or have organised meristems as is found with most of the monocots. The adventitious shoot mode may involve a more or less discernible callus phase. Androgenesis implies that the somaclones evaluated were either haploids or dihaploids. SC_1, SC_2 — see text, section 4.2.1. nr — not reported

Species	*Explant*	*Regeneration Mode*	*Variant Character in SC_1 or SC_n Plants*	*Transmission*		Reference
				Asex.	*Sex.*	
A. Monocots						
Allium sativa	shoot tips, leaf base	callus	bulb size and shape, clove no., aerial bulbil germination	+		Novak (1980)
Avena sativa	(a) immature embryo, apical meristem	callus	plant ht, heading date, leaf striping, twin culms, awns		+	Cummings *et al.* (1976)
	(b) immature embryo	callus	cytogenetic abnormalities e. g. chromosome loss, interchanges, trisomy, monosomy	nr		McCoy *et al.*, (1982)
Haworthia	flower bud	callus	chromosome no., and behaviour, vigour, leaf, shape, leaf colour, stomata no., esterase isozymes	nr		Ogihara (1981)
Hordeum spp.	(a) immature embryo	callus	plant ht, tillering, albinism		+	Deambrogio & Dale (1980)
	(b) anthers	callus	cytological abnormalities	nr		Mix *et al.* (1978)
	(c) immature ovaries	callus	growth habit and rate, size, head morphology, auricles	nr		Orton (1980)
Lolium spp. hybrid	immature 3 N embryo	callus	leaf size, flower, vigour, survival		+	Ahloowalia (1983)
Oryza sativa	(a) seedling	callus	tiller no., fertile tiller no., panicle length, seed fertility	nr		Henke *et al.* (1978)
	(b) dihaploid seed	callus	chlorophyll content, seed fertility, plant ht, flowering date, grain no., kernel wt.		+	Oono (1981)
	(c) dihaploid immature embryos	callus	height, seed no. and size, panicle size, tiller no., yield, protein	+		Schaeffer (1981) Schaeffer *et al.* (1984)
	(d) mature embryo	callus	heading date, culm length, sterility		+	Fukai (1983)
	(e) seeds	callus	ht, grain wt, chlorophyll mutants		+	Sun *et al.* (1983)

Table 1. (contd.)

Species	*Explant*	*Regeneration Mode*	*Variant Character in SC_1 or SC_n Plants*	*Transmission*		Reference
				Asex.	*Sex.*	
Saccharum officinarum	(a) parenchyma, inflorescence, young leaf	callus	eyespot, Fiji virus, downy mildew diseases	+		Heinz *et al.* (1977)
	(b) meristems, inflorescence, young leaf	callus	auricle length, dewlap shape, top leaf attitude, chromosome no., esterase isozymes, cane yield, sugar yield, stalk no., smut resistance			
	(c) leaf base	callus	eyespot disease, leaf scald	+		Larkin and Scowcroft (1983)
Sorghum bicolor	immature embryos, seedlings	callus	fertility, leaf morphology, growth habit		nr	Gamborg *et al.* (1977)
Triticum aestivum	(a) immature embryo	callus	plant ht, stem thickness, leaf size, spike shape, pollen fertility		nr	Ahloowahlia (1982)
	(b) immature embryo	callus	plant and head morphology, awns, chlorophyll deficiency, gliadins, amylase, grain wt, yield		+	Larkin *et al.* (1984)
	(c) immature embryo	callus	chromosome aberrations		+	Karp and Maddock (1984)
Zea mays	(a) immature embryo	callus	abphyll syndrome, pollen fertility		nr	Green (1977)
	(b) immature embryo	callus	*Dreschslera maydis* race T toxin resistance, male fertility, mtDNA sequence rearrangement		+	Brettell *et al.* (1980); Kemble *et al.* (1981); Umbeck and Gengenbach (1983)
	(c) immature embryo	callus	endosperm and seedling mutants		+	Edallo *et al.* (1981)
	(d) immature embryo	callus	defective kernel, wilting		+	McCoy and Phillips (1982)
	(e) immature embryo	callus	dwarf, leaf morphology, chloroplast mutants		+	Beckert *et al.* (1983)
B. Dicots						
Ananas comosus	syncarp, slip, crown, axillary bud	callus	spine & leaf colour, waxiness, foliage density, leaf width & spines		nr	Wakasa (1979)

Table 1. (contd.)

Species	*Explant*	*Regener-ation Mode*	*Variant Character in SC_1 or SC_n Plants*	*Transmission*		Reference
				Asex.	*Sex.*	
Apium graveolens	petioles	callus	phosphoglucomutase isozyme	nr		Orton (1983b)
Arachis hypogaea	anther	callus	chromosome no.	nr		Bajaj *et al.* (1981)
Begonia × *hiemalis*	leaf	adv. shoot	colour, size, form of flowers & leaves	+		Roest *et al.* (1981)
Brassica napus	(a) anther	andro-genesis	flowering time, glucos-inolates, growth habit		+	Wenzel (1980), Hoffman *et al.* (1982)
	(b) androgenic embryo	callus	restistance/tolerance to *Phoma lingam*		+	Sacristan (1982)
B. oleracea	root crown meristem	adv. shoots	waxiness, stem branching, precocious flower formation	+		Grout and Crisp (1980)
Chry-santhemum	(a) nodes, shoot tip	adv. shoot, callus	flower colour, flower induction temperature	nr		Jung-Heiliger and Horn (1980)
	(b) shoot tip	callus	shoot and leaf mor-phology	nr		Sutter and Langhams (1981)
Frageria ananassa	stolon tips	adv. shoots	vigour, runnering, chlorosis, compact habit, yield	+		Schwartz *et al.* (1981)
Lactuca sativa	(a) cotyledons	callus	leaf wt, length, width, flatness & colour, bud no.		+	Sibi (1976)
	(b) leaf	proto-plasts	cotyledon colour, leaf morphology	nr		Engler and Grogan (1983)
Lycopersicon esculentum	(a) cotyledon	callus	leaf morphology, branching habit		+	Sibi (1981)
	(b) leaf	callus	fruit colour, pedicel, male fertility, growth, chlorophyll		+	Evans and Sharp (1983)
L. peruvianum	anther	callus	self-incompatibility	nr		Sree Ramulu (1982)
Medicago sativa	immature ovaries	suspen-sion/ callus	multifoliate leaves, elongated pertioles, growth habit, primary branch no., plant ht, dry matter yield		+	Reisch and Bingham (1981)
Nicotiana	(a) anther	andro-genesis	crumplead leaf		+	De Paepe *et al.* (1981)
	(b) anther	andro-genesis	amplification AT & GC rich regions	nr		De Paepe *et al.* (1983)
	(c) leaf	proto-plast	ht, flowering time, albino		+	Prat (1983)

Table 1. (contd.)

Species	*Explant*	*Regeneration Mode*	*Variant Character in* SC_1 *or* SC_n *Plants*	*Transmission*		Reference
				Asex.	*Sex.*	
N. glauca	pith	callus	amplification AT & GC rich regions	nr		Durante *et al.* (1983)
N. tabacum	(a) anther	androgenesis	plant ht, stem diameter, leaf size, yield		+	Oinuma and Yoshida (1974)
	(b) anther	androgenesis	yield, grade index, flowering time, plant ht, leaf number, length & width, alkaloids, reducing sugars		+	Burk and Matzinger (1976)
	(c) leaf	protoplast callus, adv. shoots	2 specific leaf colour loci, leaf shape & size		+	Barbier und Dulieu (1981)
	(d) leaf	protoplast	flower & plant morphology, mutator allele		+	Lorz and Scowcroft (1983)
	(e) dihaploid leaf	protoplast	ht, flowering time, leaf morphology		+	Prat (1983)
Pelargonium	stem	callus	plant, leaf & flower morphology, essential oils, fascination, anthocyanin pigmentation, pubescence	+		Skirvin and Janick (1976)
Solanum tuberosum	(a) leaf	protoplast	tuber shape, yield, maturity date, photoperiod, plant morphology, early and late blight resistance, numerous field traits	+		Shepard *et al.* (1980); Secor and Shepard (1981)
	(b) leaf	protoplast	leaf colour & morphology, vigour, ht, anthocyanin pigment	+		Thomas *et al.* (1982)
	(c) leaf	adv. shoots	stem, leaf, flower morphology, skin colour	+		van Harten *et al.* (1981)
	(d) shoots	protoplasts	growth, leaf and stem morphology	nr		Sree Ramulu *et al.* (1983)

Where a number of plants regenerated from cell culture have been observed, somaclonal variation appears to be the general rule. Many earlier reports describing plants regenerated from cell culture failed to mention phenotypic variation with any conviction. This was probably due to the entrenched assumption that clonal uniformity during *in vitro* propagation of plants was maintained. Several authors have made specific reference to the apparent homogeneity among regenerated plants of species such as celery (Williams and Collins, 1976), asparagus and iris (Reuther, 1977), potato (Wenzel *et al.*, 1979), *Pennisetum purpureum* (Haydu and Vasil, 1981) and *Panicum maximum* (Hanna *et al.*, 1984).

Discussion of somaclonal variation in maize, wheat, surgarcane, tomato and potato will serve as examples to cover monocot and dicot species, diploids and polyploids, seed and asexually propagated species. The relevance of this research has largely been couched in term of creating new variability for plant improvement where somaclonal variation is considered a potential benefit. For the purposes of this article, these examples illustrate the reality of the myth of clonal uniformity among plants derived from tissue culture.

2. Maize

The classical genetic, cytogenetic and recently molecular research in maize has enabled a penetrating analysis of somaclonal variation. From only 77 plants regenerated from maize tissue cultures, Edallo *et al.* (1981) were able to identify 17 defective endosperm or seedling mutants which were phenotypically the same as classically identified mutants. In the analysis of somaclone progeny their criterion for classifying a variant as a mutant was that segregation ratios should conform to Mendelian expectations. The somaclones analysed were derived from two different donor genotypes and progeny analysis showed that each somaclone carried an average of one simply inherited mutation. Some plants carried more than one mutation and when more than one plant was regenerated from a callus, they often carried different mutations.

In another analysis of 51 somaclones, 8 segregated for recessive kernel mutations and one segregated for a mutation which caused premature wilting (McCoy and Phillips, 1982). Segregation for some of the mutants did not occur until the second seed generation. This suggests that the male and female flowers on the initial regenerant were genetically different, which is consistent with regenerants arising from more than one progenitor cultured cell. Alternatively, a genetic event induced during the culture phase which gave rise to somaclonal variation may persist after regeneration has occurred. Whatever the cause of such genetic chimaeras, their occurrence at least for germline cells, is rare rather than common.

Maize studies have also provided conclusive evidence that the mitochondrial genome can undergo genetic changes during cell culture. The Texas source of cytoplasmic male sterility is sensitive to the host specific T-toxin elaborated by the causal agent of southern corn leaf blight, *Drechslera maydis.* Normal cytoplasm plants are male fertile and resistant to the toxin. The toxin has been used as a selective agent in tissue cultures of T-cytoplasm maize lines (Gengenbach *et al.*, 1977). Plants regenerated from the selected cell lines were resistant both to the toxin and to infection by the pathogen. Coincidentally, the toxin resistant plants had also reverted to male fertility. Brettell *et al.* (1980) and Umbeck and Gengenbach (1983) have also found that reversion to male-fertility, and toxin resistance may occur without using T-toxin as a selective agent.

The conversion during tissue culture to toxin resistance and male-fertility is maternally inherited and shown to be associated with the mitochondria. A restriction endonuclease analysis of mt-DNA from a number of independently occurring male-fertile, toxin resistant somaclones revealed that in most cases the mitochondrial DNA had undergone a specific change. By comparison with the parent, 15 of 16 converted somaclones had lost a 6.6 kilobase XhoI restriction endonuclease fragment (Gengenbach *et al.*, 1981; Kemble *et al.*, 1982; Umbeck and Gengenbach, 1983).

This maize cell culture research has not fulfilled the original plant breeding objective, namely, the recovery of *D. maydis* T-toxin resistant, male-sterile lines for use in hybrid maize programs. However, it does demonstrate that genetic changes do occur during cell culture and that such alternations may occur at the nucleotide level.

3. Wheat

It has recently become possible to regenerate plants from tissue cultures of wheat with reasonable efficiency. Many cultivars are now amenable to culture, using either immature embryos or young inflorescences as the explant (Maddock *et al.*, 1983; Larkin *et al.*, 1984). In our research, the initial wheat regenerants (SC_1 plants) displayed some phenotypic variation, but it was the analysis of their progeny which revealed the extent of the genetic changes induced during culture (Larkin *et al.*, 1984). The analysis of progeny of 142 regenerants of a double dwarf accession, Yaqui 50 E, through several consecutive seed generations showed segregation for increased height, reduced height, heading date, presence or absence of awns, glume colour, grain colour, and leaf waxiness. Segregation ratios for some, but not all, of the qualitative characters apparently conformed to Mendelian expectations. For the quantitatively varying character, plant height, an offspring parent regression analysis gave a heritability estimate of 0.67.

A preliminary analysis of yield and yield component data from a replicated hill plot experiment involving 256 somaclonal lines derived from three cultivars indicated that somaclonal variation did affect yield parameters. Significant effects, both positive and negative relative to parental controls, were observed for grain number and total grain weight per spike, grain weight, grain yield and harvest index. In the cultivar Millewa, for example, 6 of the 100 lines significantly ($p < 0.05$) outyielded the parent control and 3 lines had significantly higher harvest index. A more extensive field trial is in progress to confirm these differences.

Biochemical analysis on seed of the progeny of some of these somaclones has also revealed extensive variation in the electrophoretic pattern of the gliadin proteins. Particular protein bands were not only lost, but "new" bands were observed and intensity changes for other fractions were also recorded. This contrasts with the striking uniformity of the gliadin patterns among seed of the parent. Some of the variant patterns breed true in subsequent generations while some continue to segregate.

An analysis of the synthesis of the secretory enzyme, α-amylase, induced by gibberellic acid (GA_3) has revealed changes in response both to GA_3 induction and to abscisic acid repression of GA_3 α-amylase induction. For these genetic variants, isoelectric focussing of both α- and β-amylases indicates that some variants have lost and/or gained specific amylase protein fractions. In others the relative concentration of particular fractions has altered relative to the parental control seed.

The results of research on somaclonal variation in wheat can be summarised as follows:

a) Variation was manifested for both morphological and biochemical characters and for traits under simple genetic control (gliadins, grain colour) and polygenic control (height, heading date, yield).
b) A single somaclone could be variant for a number of traits which appeared to assort independently in progeny analysis.
c) Both heterozygous and homozygous mutants were recovered in the primary regenerant. A single somaclone could contain both states at different loci.
d) Aneuploidy did occur but variants were present in euploid plants.
e) Mutations affected characters for which major gene loci are known to be located on all seven homoeologous groups.
f) Somaclonal mutants could be recessive (awns, grain colour), dominant (glume color, awn inhibition) or codominant (gliadins).

4. Tomato

Plants regerated from cultured leaf explants of an inbred variety carried a wide variety of mutants (Evans and Sharp, 1983). These mutants were detected in the segregating progeny of selfed somaclones. Among the progeny of 230 plants, 13 nuclear gene mutations were recovered. These mutants affected growth habit, pedicel jointedness, fruit colour, albinism, and male sterility. Three of the thirteen mutants were dominant and for one of the recessive mutants the original regenerant was homozygous for the mutant. As a control no mutants were found among more than 2,000 plants from seed of the donor parent.

In the wild species, *Lycopersicon peruvianum,* plants regenerated from *in vitro* cultures of anthers carried a surprisingly large number of mutations to new gametophytic incompatibility alleles (S-alleles) as well as S-allele reversions (Sree Ramulu, 1982). Among only 37 anther culture regenerants of one genotype, 16 had S-allele changes. From a second genotype one regenerant out of 16 showed a new S-allele specificity. This result stands in sharp contrast to numerous conventional experiments which have failed to generate any new S-alleles despite mutagenic treatments.

5. Sugarcane

Several independent studies have shown that disease resistant segregants can be recovered in sugarcane presumably as a consequence of somaclonal variation. These have been summarised by Larkin and Scowcroft (1981). They include resistance of Fiji virus disease and downy mildew *(Sclero-*

spora sacchari) and culmicolous smut *(Ustilago scitaminea)*. Following earlier reports of resistance to eyespot disease *(Helminthosporium sacchari)* among sugarcane somaclones in Hawaii (Heinz *et al.*, 1977), an extensive analysis has been carried out on somaclones of Australian sugarcane cultivars (Larkin and Scowcroft, 1983).

Complex callus cultures were initiated from the cultivar, Q 101, (Larkin, 1982) from which plants were regenerated over a period of 6—18 months. Plants were then assayed for their reaction to the host-specific pathotoxin by measuring leachate conductivity of leaf discs exposed to a defined pathotoxin concentration (Larkin and Scowcroft, 1983). A total of 480 plants were assayed. The distribution of toxin reaction was biassed towards resistance and approximately 10 % of the somaclones were judged resistant at least at the toxin screening level. When toxin was included in the callus phase the distribution of toxin reaction was further biassed towards resistance.

A total of 85 somaclones were analysed for the stability of their increased toxin tolerance in each of up to five subsequent vegetative generations. The majority (70 %) maintained a stable reaction to the toxin while only 10 % reverted towards increased sensitivity. The remaining 20 % somatically segregated for resistant and susceptible plants. They may represent somaclones which were mosaic for resistant and susceptible cells. To further characterise stability, six somaclones showing vegetative stability were subjected to a second 6 month tissue culture cycle. Among 60 secondary somaclones assayed, 60 % had similar or enhanced toxin tolerance relative to that of the respective primary somaclone. The remainder were more susceptible to the toxin.

The relatively high level of stability of toxin tolerance through a second culture cycle argues in favour of the notion advanced by Shepard *et al.* (1980) that characters can be "stacked". It seems possible that one can screen for modification of a second characteristic following a second culture cycle and expect that a workable proportion of these will retain the first characteristic recovered from selection and/or screening of somaclones derived from a first culture cycle.

6. Potato

Shepard and his colleagues (Shepard *et al.*, 1980; Secor and Shepard, 1981; Shepard, 1981) have described the extensive variation observed among plants regenerated from cultured protoplasts of the cultivar, Russet Burbank. This old, but particularly valuable cultivar, has been effectively excluded from potato improvement programs because it is sterile. In their initial somaclone (protoclone) experiments some 1700 plants were evaluated for overall morphological variation. In two subsequent growing seasons this population was reduced to a total of 65 clones which generally possessed acceptable vigour, vine and tuber characteristics. Statistically significant differences were found for 22 of the 35 characters and each of the 65 somaclones differed from the parent by at least one trait. The modal class of 15 somaclones differed from Russet Burbank in 4 characters and one plant was distinguished from the parent by no less than 17 traits. Shepard concluded that variation from cell culture would provide enough variability to facilitate the selective improvement of Russet Burbank.

The potato somaclones were also screened for both late and early blight resistance. The parent, Russet Burbank, is highly susceptible to both of these diseases. From among more than 800 plants a considerable range of variation in reaction to late blight *(Phytophthora infestans)* was found. About 2% of the somaclones displayed enhanced resistance which was transmitted through subsequent tuber generations. In a similar fashion several disease resistant somaclones were recovered from a population of 500 plants screened for resistance to early blight *(Alternaria solani).* Five plants were initially identified as resistant on the basis of a toxin assay. Four of these subsequently displayed field resistance.

Somaclonal variation has been confirmed for several other important cultivars of potato such as Maris Bard (Thomas *et al.*, 1982) and the widely grown European variety Bintje (Sree Ramulu *et al.*, 1983). Significantly, somaclonal variation in potato is now being used as a breeding option in the U.K., U.S.A. and Germany.

III. Factors Influencing Somaclonal Variation

Extensive, orthogonal experiments have not yet been conducted to define the relative impact of various components and aspects of tissue culture on somaclonal variation. However, several features of tissue culture are already known to have an influence on the level of somaclonal variation in plants.

A. Sexual Versus Asexual Species

One of the more important consequences of somaclonal variation is the greatly enhanced frequency of chromosomal abnormalities during tissue culture. Many of these chromosome changes will affect totipotency and will not be carried through to regenerated plants, i. e. morphogenesis selectivity eliminates many, though undefined, chromosomal changes. For example, Ogihara (1981) found a significant reduction in aneuploids when comparing chromosome numbers of tissue culture lines of *Haworthia* with plants regenerated from them. However, meiotic analysis showed that many of the regenerated plants still carried chromosome rearrangements.

There is a further barrier which eliminates chromosomal abnormalities in sexually reproducing species, namely gametogenesis and fertilisation. Somaclonal mutations which cause sterility obviously will be eliminated. Thus analysis of tissue culture generated variability is likely to give different interpretations for asexually as against sexually propagated species. This is further compounded by ploidy level. In diploids such as maize (McCoy and Phillips, 1982) and tomato, gross chromosomal abnormalities among somaclone progeny are rare, whereas in polyploid species such as wheat (Karp and Maddock, 1984) and oats (McCoy *et al.*, 1982) they are more frequent.

B. Preexisting Versus Culture Induced Variation

It has been argued that variation in plants regenerated from tissue culture was pre-existing in the cells of the donor explant, either as a somatic mutation or residual heterozygosity. Though neither can be ruled out as contributing some variation, evidence indicates that the great majority of variation occurs during tissue culture.

Barbier and Dulieu's (1980) extensive analysis of variation among plants regenerated from protoplasts of leaf callus showed that about 3 % of regenerated plants carried a mutation at either one of two defined loci affecting chlorophyll synthesis. A genetic analysis of a sample of the altered genotypes indicated that the alleles had either reverted to wild type or had been deleted. From this analysis they could not discriminate between pre-existing and culture induced variation.

Subsequently, Barbier and Dulieu (1983), conducted an experiment in which protoplasts isolated from plants heterozygous at each of two loci affecting chlorophyll development — the a_1 and *yg* loci — were cultured to produce single protoplast derived colonies. Each of more than 1000 colonies were subdivided into four subcolonies from each of which a single plant was regenerated and phenotypically classified with respect to variation at the a_1 or *yg* loci. This analysis allowed them to conclude that 17 % of the original colonies, each of which had been derived from a single protoplast, had become heterogeneous at the a_1 or *yg* locus. While this suggests that the variation did occur during culture, it is also consistent with the notion that single strand lesions were present in the cell of the leaves from which the protoplasts were isolated. Following DNA replication during protoplast culture, mitotic segregation yielded daughter cells in which the lesion became homozygous.

A similar type of experiment in tobacco used the semi-dominant aurea mutant, Sulphur, to discriminate between pre-existing and culture induced variation (Lörz and Scowcroft, 1983). Both homozygotes and the heterozygote can be readily distinguished by phenotype as normal green *(su/su)*, albino *(Su/Su)* and yellow-green *(Su/su)* leaves. In leaves of the *Su/su* heterozygote, single green or albino spots and twin spots are observed as a consequence of somatic genetic events. Protoplasts were isolated from leaves of the heterozygote and cultured to produce 2156 morphogenic colonies each of which was derived from a single protoplast. From observations on the regenerated shoots, colonies were classified as parental where all shoots were *Su/su*, homogeneous variant where all shoots were either *Su/su* or *su/su*, or heterogeneous variant where colonies gave rise to both *Su/su* and *su/su*, or *Su/su* and *Su/Su* shoots. The homogeneous variant colonies represent pre-existing variation, whereas the heterogeneous ones reflected genetic alteration occurring after protoplast isolation. The frequency of this latter class of colony was three times that of the homogeneous variant class, indicating that the occurrence of the somatic genetic event was significantly enhanced during tissue culture.

The genetic analyses of dihaploids produced by microspore culture in both *Nicotiana sylvestris* (De Paepe *et al.,* 1981) and rape (Hoffman *et al.,* 1982) provide convincing proof that variation occurred during the culture phase following microspore culture when spontaneous chromosome doubling occurred. In both rape and tobacco, some dihaploids, which are expected to be completely homozygous, produced progeny which segregated for morphological mutants.

In an extension of this approach, Prat (1983) cultured protoplasts from a tobacco line produced by five consecutive cycles of androgenesis and chromosome doubling and subsequently selfed for two generations. A genetic analysis of plants regenerated from cultured protoplasts of this line showed that some of the regenerants were heterozygous for mutations separately affecting plant height, flowering time and albinism. Additional variation was shown for quantitatively inherited traits such as leaf and flower morphology. The pedigree of the line from which protoplasts were isolated ruled out residual heterozygosity as a source of variation and it can be concluded that the variation arose during the culture phase. Similar experiments with second cycle dihaploids in *N. tabacum* (Brown *et al.,* 1983) confirm that the anther culture process generates more variability than can be accounted for by residual heterozygosity.

C. Genotype

Evidence, though limited, indicates that the genotype of the donor has a significant effect on the extent of variation generated during culture. McCoy *et al.* (1982) following a detailed examination of chromosomal abnormalities among plants regenerated from tissue cultures of two oat cultivars found that one cultivar consistently gave a higher frequence of chromosomal variants. Similarly, in a comparison of chromosome variation in regenerants derived from protoplasts of two potato cultivars, Karp *et al.* (1982) found that while aneuploidy was frequent in both, the nature of the variation differed. Maris Bard regenerants had high chromosome numbers (46—92) and a wide range of aneuploidy while in the cultivar, Fortyfold, a higher proportion of plants had chromosome numbers which were near normal.

The frequency of morphological variants which appeared in the analysis of somaclonal progeny of wheat differed between cultivars. The cultivar Yaqui 50 E had a greater frequency than that of Millewa which in turn produced more somaclonal variants than the cultivar Warigal. In strawberry, where meristem culture is routinely used for clonal propagation, varietal differences occur in the frequency of off-type plants (Schwartz *et al.,* 1981). Among plants derived by adventitious shoot formation from leaf explants in *Begonia* × *hiemalis,* Roest *et al.* (1981) found that in one variety 43% of regenerants were variant (colour, size and form of leaves and flowers) whereas for another variety only 7% were variant. Some varieties of *Pelargonium* are known to be inherently unstable through conventional stem cutting propagation. As expected such cultivars also produce a high frequency of abnormal plants regenerated from callus culture (Skirvin and Janick, 1976).

D. Explant Type and Culture Mode

It can be seen from Table 1 that quite diverse explants can give rise to somaclonal variants. Some explant sources and modes of producing *in vitro* plants, as for example adventitious shoots from petiole or leaf explants, minimise the duration of the callus phase. While such a procedure does reduce the extent of somaclonal variation, it does not obviate it. Barbier and Dulieu (1981) found somaclonal variants not only among plants regenerated from callus and cultured protoplasts but also among plants derived from adventitious shoots on leaf explants.

By inference it was considered that the extensive variation identified in potatoes by Shepard required the use of a protoplast culture system. While this is desirable from the standpoint that individual somaclones have a unique single cell origin, it is not essential for the elaboration of somaclonal variation. Thomas *et al.* (1982) found variation between different plants regenerated from a single callus which itself had been derived from cultured protoplasts of the cultivar Maris Bard. They also produced plants by stem embryogenesis from a different cultivar Majestic and found relatively low levels of variability which could reflect varietal differences as discussed earlier. In contrast, from a different cultivar of potato, Desirée, van Harten *et al.* (1981) found substantial variation for stem, leaf, flower and skin colour variants among plants regenerated by the rapid adventitious shoot technique from rachis, petiole and leaf-disc explants. In the case of plants produced from rachis and petiole explants, not previously exposed to X-rays (as had other explants), 50% of the plants were variant relative to the donor parent.

The mutagenic activity of culture medium and particularly growth regulators have been evaluated directly using the *Tradescantia* stamen hair system (Dolezal and Novak, 1984) which was developed as a sensitive system for somatic mutation resulting from ionizing radiation or chemical mutagenesis. In this system spontaneous mutant events occured at a frequency of 0.5/100 stamens. There is a linear response to N-methyl-N-nitrosourea; at 500 μM NMU there is a 21 fold increase in mutant events. There was no mutagenic effect of culture media containing 5 μM of either 2,4-D, kinetin or indole acetic acid (IAA) or a combination of 5 μM 2,4-D, 10 μM kinetin and 10 μM IAA. In each test the somatic mutation rate was not significantly greater than spontaneous rate. This result provides substantial evidence that tissue culture media is not mutagenic *per se*.

E. Duration of Culture

Prolonged period of tissue culture are known to result in an increased frequency of gross chromosomal aberrations (Bayliss, 1980; Meins, 1983). As expected the frequency of somaclonal variants among plants regenerated from tissue cultures also increases with prolongation of the culture period. A common observation is that tissue cultures lose topipotency with time in culture. The most logical explanation for this is the loss or mutation of genes which are essential for regeneration. Indeed such mutants which block specific steps in the regeneration process are being recovered from tissue culture derived plants.

In oats the frequency of cytogenetically abnormal plants increased dramatically with increased time in culture (McCoy *et al.*, 1982). Abnormalities included chromosome breakage, chromosome loss, interchanges and aneuploids. Plants were regenerated from cultures of two oat cultivars, Lodi and Tippecanoe, after 4, 8, 12, 16 and 20 months in culture. Among Tippecanoe regenerants the frequency of observable chromosome aberrations rose from 49% after 4 months culture to 88% after 20 months. For Lodi over the same time period the change was 12% to 48%. Similar observations have been made among plants regenerated from tissue cultures of different ages of triticale (Armstrong *et al.*, 1983) and wheat (Karp and Maddock, 1984). In sugarcane, plants regerated from tissue culture at various intervals from 6 months to 18 months after culture initiation were assayed for tolerance to the toxin produced by *Helminthosporium sacchari* which causes eyespot disease (Larkin and Scowcroft, 1983). The frequency of toxin tolerant plant predictably increased with duration of tissue culture.

A similar effect of culture duration was also found for tobacco plants regenerated from leaf protoplasts of the sulphur mutant (Lörz and Scowcroft, 1983). Doubling the time in culture from 3 to 6 months more than doubled the frequency of somaclonal mutations at the *Su* locus. However, more extensive experiments by Barbier and Dulieu (1983) gave a result which is at variance with the consensus that culture duration increases the level of variation. Protoplasts were isolated from leaves heterozygous for two loci affecting chloroplast development, and cultured. Twenty five colonies each were subcloned into 4 sectors after each of 3 successive culture periods. The analysis of plants regenerated from the subclones enabled Barbier and Dulieu (1983) to conclude that variation occurred only during the first culture period. Additional variants at either of the two loci were not observed in the second or third culture phases.

IV. Origin of Tissue Culture Instability and Somaclonal Variation

An understanding of the mechanisms which give rise to tissue culture instability and somaclonal variation is necessary for several practical reasons. First, where somaclonal variation is to be utilised to generate variability for plant improvement it could be desirable to enhance its occurrence. Equally important where uniformity of tissue culture plants is essential, as in rapid micropropagation or for *in vitro* germplasm conservation, it is necessary to be able to control the mechanisms which generate variation.

Apart from chromosomal aberrations there is insufficient evidence to support, let alone favour, any one of several possible mechanisms to account for somaclonal variation. In addition to DNA amplification and transposable elements which will be considered here primarily because they reflect the direction of current research, other possibilities include, somatic rearrangement of genes particularly those in multigene families, somatic crossing over, sister chromatid exchange, altered nucleotide methylation patterns, perturbation of DNA replication by altered nucleotide pools, and silencing or activation of genes by mutations in associated non-coding regions.

A. Chromosomal Aberrations

It has been known for some time that chromosomal abnormalities occur at a high frequency in tissue cultures. Amont regenerated plants and their progeny the incidence of gross chromosomal aberrations is greatly reduced relative to cells in tissue cultures. In the analysis of regenerants, chromosomal aberrations which preclude morphogenesis will not be observed. Where somaclonal progeny are analysed those aberrations which cause sterility will also be eliminated. As pointed out earlier the requirement of morphogenesis and fertility is a selection process which eliminates many chromosomal abnormalities and strongly favours euploidy.

In an extensive meiotic analysis of plants regenerated from two cultivars of oats, McCoy *et al.* (1982) found that the common abnormality was breakage and loss of chromosomal segments. Among more than 300 regenerants for each of two cultivars, approximately 20 % of the plants carried one or more interchanges and/or breakage events as evidence by the occurrence of heteromorphic pairs at diakinesis. Only about 6% of regenerated plants were aneuploid. This analysis also indicated that the frequency of chromosomal rearrangement increased with time in culture and that there was a cultivar effect on the frequency of aberrations among the regenerants.

In wheat, karyotype abnormalities have also been identified in regenerated plants (Karp and Maddock, 1984). Amont 192 plants regenerated from four different wheat cultivars, 29% were aneuploid. A meiotic analysis of regenerants showed that interchanges were particularly common. In our own cytogenetic analysis of wheat somaclones we have found that aneuploidy is infrequent, as would be expected since our analysis was done on progeny of fertile somaclones (unpublished data). However, among these progeny we have observed both both ring and open quadrivalents which presumably result from translocation events. These interchanges are not necessarily associated with morphological mutants.

A causal relationship between gross chromosomal aberration and genetic changes in somaclones is incompletely resolved. Orton (1983 a) sought a correlation between chromosome loss and genetic changes at the phosphoglucomutase (PGM) locus in celery callus cultures and regenerated plants. A causal relationship could not be established despite the fact that chromosome loss and instability at the Pgm-2 locus was observed. The two manifestations of somaclonal variation appeared to be independent. However, since chromosome rearrangements can only be seen cytologically if they are a certain size or type, it may only be our discontinuous powers of resolution of the variation which make chromosome rearrangements and single gene mutations appear to result from separate mechanisms. Specific gene effects can result from chromosomal interchanges because of gene inactivation, position effects and altered developmental timing. Chromosomal abnormalities are part of the spectrum of somaclonal variation and they, as well as other classes of somaclonal variants, may have a more fundamental biological basis.

B. DNA Amplification

Plant genomes are large with haploid genome sizes in the range of 10^9—10^{10} nucleotide paris. Though variable from one species to another, about 60% of the genome of a "typical" plant is composed of repeated DNA. Most of the repeated sequence DNA is low or moderately repetitive DNA with a reiteration frequency per sequence of less than 10^3. Some highly repetitive DNA (10^5—10^6 copies) also occurs in most plants and can comprise up to 20% of the total DNA as in maize (Sorensen, 1984). Repetitive DNA is of two basic types. The repeat units can be clustered as arrays of tandem repeats with specific genome location, or the repeat family can be interpersed throughout the genome, usually in different repeat unit permutations.

One class of genes which is highly repeated in plants is that which codes for ribosomal RNA (rDNA). The 26, 18 and 5S rDNA genes in plants, as in animals, are transcribed as a single unit. Typically this major transcriptional unit ranges from 7.8 to 12.7 kilobase pairs. The reiteration frequency is highly variable from species to species. The number of copies can be as few as 1300 per nucleus or as many as 31,900. Sorensen (1984) reflects a consensus that the reiteration frequency of rDNA in higher plants is dynamic.

Both qualitative and quantitative changes in DNA of tissue cultures and regenerated plants have been documented. A kinetic analysis of DNA extracted from pith callus cultures (Durante *et al.,* 1983) indicated that there was differential replication of both GC and AT rich sequences during the early phases of dedifferentiation. They interpret these results as a primary step in the cascade leading to dedifferentiation.

In an extensive analysis of dihaploid plants of *N. sylvestris* derived by consecutive androgenesis, De Paepe *et al.* (1982) identified both heritable quantitative and qualitative changes in the nuclear DNA. An increasing proportion of a 1.703 g/cm^3 satellite signalled an increase in GC rich sequences or alternatively an increase in the degree of methylation. Thermal denaturation also indicated an increase in AT rich sequences. An analysis of zero-time binding sequences following S1 nuclease treatment showed that there was a progressive increase of inverted repeat sequences in consecutive dihaploid plants. This was particularly prevalent in DNA sequences 200—400 nucleotides long.

It is reasonable to conclude therefore, that DNA sequence amplification, or deamplification, could be one of the mechanisms responsible for somaclonal variation. Such amplification could lead either to increased synthesis of a specific gene product or to perturbations in developmental timing of gene activity if the repeated sequences function in new chromosomal locations.

C. Transposable Elements

The penetrating genetic analysis of unstable mutations in maize enabled Barbara McClintock to postulate the existence of genetic elements which transpose from one location in the genome to another (McClintock, 1951). Transposition occurs in both somatic and germline cells. Recent molecular and genetic analysis has shown the occurrence of transposable elements in eukaryotes to be the likely rule rather than the exception (Shapiro and Cordell, 1982). In plants, mutants at the bronze, shrunken, waxy and alcohol dehydrogenase loci of maize and the chalcone synthase locus of *Antirrhinum* have been found which contain identifiable transposable elements (see Flavell, 1984). Mutant phenotypes can result either from insertion of the element into the structural gene or into the regulator gene sequences.

By virtue of their movement, transposable elements can inactivate structural genes, alter gene regulation, possibly reactivate silenced genes, and can generate duplications and deficiencies. Though not understood, genomic and developmental shock can induce the transposition of mobile elements. The breakage-fusion-bridge cycle used by McClintock (1951) to generate unstable patterns of mitotic segregation for a chromosome segment is a classical example. In Drosophila strains which have been destabilized by hybrid dysgenesis, "transposition bursts" occur in which many transposition events occur (Gerasimova *et al.,* 1984). The transposition bursts were interpreted as occurring in pre-meiotic cells of the germ line.

Concrete evidence has yet to be presented that transposition events are a cause of somaclonal variation in plants. Several examples however do provide circumstantial evidence of tissue culture events which are analogous to transposition. In Shepard's analysis of somaclonal variation in potatoe, somaclones were identified which simultaneously differed from the parent cultivar in up to 17 characters. Among the 65 somaclones, each one was distinguishable from the parent by an average of 7 characters. Such coincident variation could be explained by transposition burst. Similarly we have observed a concurrence of variation among wheat somaclones (Larkin *et al.,* 1984). Several somaclone families were simultaneously variant for many characters. For example, progeny derived from a single somaclone segregated for height, maturity date, presence or absence of awns, grain colour and had altered gliadin protein fractionation patterns.

Simultaneous independent mutants have also been reported in tobacco and lucerne. Chaleff and Keil (1981) utilised selection in cell culture to produce several independent herbicide (picloram) resistant plants. The resistance was inherited as either a dominant or semi-dominant gene. Subsequently they found that three of the five picloram resistant genotypes, unlike the parental plants, were also resistant to hydroxyurea. This resistance was inherited as a single dominant mutation and surprisingly, in two of the three cases, assorted independently from picloram resistance. In alfalfa, Reisch and Bingham (1981) observed a far greater amount of variation among plants regenerated from cell lines previously subjected to selection for resistance to ethionine than among plants regenerated from unselected callus. The variation was reflected in leaf and growth habit, plant height, shoot length, number of primary branches and field plot yield. The authors attributed the variation to impaired DNA replication and/or repair as a consequence of ethionine induced transethylation of nucleic acids. It is equally likely, in both this and the tobacco cases, that plants selected for a particular mutation would also carry mutants at other loci if multiple transposition had occurred.

Unstable mutants have also been reported among plants regenerated from cell and tissue culture. In section 3.2 the results of the effect of culture on the allelic state of the *Su* locus in tobacco were discussed (Lörz and Scowcroft, 1983). Plants regenerated and grown to maturity carried other independent mutants which affected leaf, flower and whole plant morphology and in many plants meiotic segregation was aberrant. One type of somaclonal mutant, for which there were two independent observations among 120 regenerants, affected the frequency of both single and twin spots on the leaves of *Su/su* heterozygotes. The frequency of leaf spotting relative to non-culture derived plants was elevated up to 500 fold. This "Su mutator" somaclonal variant is inherited as a dominant mutant which appears to be lethal in the homozygous condition. The mutator is unstable and can be spontaneously lost either in somatic tissue or during meiosis. In alfalfa, Groose and Bingham (1984) recovered a white flowered somaclone from a purple-flowered parent. Chromosome loss did not account for this variant. Following a second culture cycle of the white variant, some regenerants had reverted to purple flower type. Both the original mutant and the revertant are heritable. Reversion rate among culture derived plants was as high as 50% and the instability of the mutant allele was itself heritable.

D. Somaclonal Variation — Analysis and Understanding

As for any other type of genetic variation, many experimental approaches are employed to gain greater insight into the nature of somaclonal variation. Analytical techniques as discussed by Orton (1983b), Rivin *et al.* (1983) and Sorenson (1984) are now being applied at the whole plant, cell, biochemical and now molecular level. The increasing availability of cloned DNA sequences provides powerful tools to dissect the genetic changes which occur during tissue culture. Probes for genes which code for a known function will enable the characterisation of somaclonal mutants at specific loci. Non-specific unique and reiterated DNA sequences probes will be essential to determine the extent of genomic rearrangements which occur during tissue culture. Conventionally, the mapping of mutants depends on the ability to visualise morphological or biochemical changes for a particular phenotype. Isozyme polymorphisms have aided the mapping of new mutants in many species. Now the identification of restriction endonuclease fragment polymorphisms provides plant genetics with a powerful new tool to characterize and localise new genetic changes. Such molecular maps will permit even more precision to characterize genomic flux during tissue culture.

V. Benefits and Disbenefits of Somaclonal Variation

The impact of somaclonal variation is seen primarily in the plant breeding mode. One of the major potential benefits of somaclonal variation is the creation of additional genetic variability in co-adapted, agronomically useful cultivars, without the need to resort to hybridization. This takes on added appeal if *in vitro* selection is possible or if rapid plant screening procedures are available. The use of *in vitro* selection to enrich for mutants of potential agronomic importance has been discussed in detail by Chaleff (1981), Scowcroft *et al.* (1983), Maliga (1984), Negrutiu *et al.* (1984), Scowcroft and Ryan (1984). Characteristics for which somaclonal mutants can be enriched during *in vitro* culture include resistance to disease pathotoxins, herbicides, tolerance to environmental and edaphic stresses such as heat and aluminium tolerance. Somaclonal variation for other characters for which there is no adequately defined *in vitro* response, such as seed protein quality, photosynthetic efficiency and yield, can be of benefit provided effective whole plant screening protocols are available.

A serious disbenefit of somaclonal variation occurs in operations which require clonal uniformity as in the horticulture and forestry industries where tissue culture is employed for rapid propagation of elite genotypes. Similarly, the genetic integrity of germplasm conserved by *in vitro* techniques is at risk because of somaclonal variation. Where clonal uniformity is desired an understanding of the mechanisms which cause variation might help establish protocols to reduce its incidence. On current evidence clonal uniformity is more likely to be enhanced if tissue culture micropropagation avoids the use of callus. The incidence of somaclonal variation appears to be reduced where meristem culture or somatic embryogenesis is utilised for propagation (Scowcroft, 1984).

A new option to introgress alien genes from wild relatives, into crop species has emerged from the observation that chromosome rearrangements occur in plants regenerated from tissue culture (see Section 4.1.). The occurrence of chromosomal rearrangements during the tissue culture of interspecific hybrids and synthetic crop species, such as triticale, is particularly compelling. Armstrong *et al.* (1983) found that karyotypes of triticale were quite unstable during culture. Aneuploidy was frequent and among 51 plants regenerated from 6 month old cultures more than half the plants possessed altered chromosomes. These included telocentrics, acrocentrics, deletions and euchromatic additions. C-banding enabled identification of rye-wheat recombinant chromosomes.

In ryegrass, 400 plants have been derived from callus initiated from a single immature triploid embryo (Ahloowalia, 1983). The triploid embryo was from a cross between diploid *Lolium multiflorum* and tetraploid *L. perenne.* The regenerated plants showed wide variation in leaf shape and size, floral development, vigour and survival. Interestingly some plants displayed novel phenotypes. Cytological analysis revealed frequent multivalent formation which is characteristic of reciprocal translocations. In some plants penta- and hexavalent associations were observed which signifies that more than one reciprocal translocation had occurred during culture.

Larkin and Scowcroft (1981), McCoy and Phillips (1982) and Ahloowalia (1983) recognized that the enhanced level of chromosome exchange in tissue culture might be employed to effect alien gene transfer. In crops such as wheat, wide crossing programs have produced alien chromosome addition lines following hybridization with relatives such as *Aegilops, Agropyron, Elymus, Haynaldia* and *Hordeum.* These alien species carry genes for resistance to wheat disease pathogens and tolerance to evironmental stress such as heat, salinity and metal toxicity. Research is now being initiated to utilise tissue culture in an attempt to enhance the rate of transfer of specific alien genes from these wheat alien addition lines. This could emerge as one of the most salutary consequences of somaclonal variation for plant improvement.

VI. Conclusions

Chromosomal, morphological and biochemical variation has been known to occur in tissue cultures virtually since the earliest days of *in vitro* plant research. In contrast, and somewhat anachronically, it is only in the past few years that the expectation of clonal uniformity among plants regenerated from tissue culture has been shown to be erroneous. In a complete turn about, clonal uniformity is now recognized as the exception rather than the rule. Why such a palingenetic myth should have been retained for so long is unclear.

Doubtless the recent ready acceptance of somaclonal variation as a real biological phenomenon is due largely to the overwhelming body of genetic evidence now available. Acceptance has been aided however, by other evidence that eukaryote genomes are in a far greater state of genetic flux than previously considered possible. It has yet to be conclusively demonstrated that genetic events such as induction of transposition, specific gene sequence amplication, gene conversion, somatic recombination and activation of silenced gene are mechanisms which contribute to somaclonal variation. The increasing availability and use of molecular probes will provide some of the experimental tools necessary to understand the relative contribution, if any, of such mechanisms to somaclonal variation.

While still limited in scale somaclonal variation is beginning to impact on plant breeding. Variation from tissue culture, with or without *in vitro* selection, is being utilised for disease resistance or stress tolerance breeding in sugarcane, potato, rice, wheat, tomato, tobacco, alfalfa and oilseed rape. Tissue culture provides a tremendous fillip to broaden the germplasm base for genetic improvement of crop species. Embryo culture to rescue interspecific sexual hybrids and somatic hybridization enables even more disparate genomes to be brought together. Somaclonal variation may provide the only feasible opportunity for exchange of genetic information between disparate genomes in such wide crosses.

VII. Acknowledgements

Many of the thoughts and ideas expressed in this article result from discussion with colleagues in the Division of Plant Industry. In particular I wish to acknowledge Philip Larkin, Sarah Ryan and Richard Brettel for their continuing contribution to the evolution of knowledge about somaclonal variation and tissue culture.

VIII. References

Ahloowalia, B. S., 1983: Spectrum of variation in somaclones of triploid ryegrass. Crop Sci. **23,** 1141—1147.

Armstrong, K. C., Nakamura, C., Keller, W. A., 1983: Karyotype instability in tissue culture regenerants of tritical (*X Triticosecale* Wittmack) cv. 'Welsh' from 6-month-old callus cultures. Z. Pflanzenzüchtg. **91,** 233—245.

Arnison, P. G., Boll, W. G., 1975: Isoenzymes in cell cultures of bush bean (*Phaseolus vulgaris* cv. Contender): Isoenzymatic differences between stock suspension cultures derived from a single seedling. Can. J. Bot. **53,** 261—271.

Ashmore, S. E., Gould, A. R., 1981: Karyotype evolution in a tumour-derived plant tissue culture analysed by Giemsa C-banding. Protoplasma **106,** 297—308.

Bajaj, Y. P. S., Ram, A. K., Labana, K. S., Singh, H., 1981: Regeneration of genetically variable plants from the anther derived callus of *Arachis hypogea* and *Arachis villosa*. Plant Sci. Lett. **23,** 35—49.

Barbier, M., Dulieu, H. L., 1980: Effects génétiques observés sur des plantes de Tabac régénérees à partin de cotylédons par culture *in vitro*. Ann. Amelior. Plantes. **30,** 321—344.

Barbier, M., Dulieu, H., 1983: Early occurrence of genetic variants in protoplast cultures. Plant Sci. Lett. **29,** 201—206.

Bayliss, M. W., 1980: Chromosomal variation in plant tissue culture. Int. Rev. Cytol. **11 A,** 113—143.

Beckert, M., Polacsek, M., Caenen, M., 1983: Etude de la variabilité génétique obtenue chez le mais après callogenese et régénération de la plantes *in vitro*. Agronomie **3,** 9—10.

Brettel, R. I. S., Thomas, E., Ingram, D. S., 1980: Reversion of Texas male-sterile cytoplasm in culture to give fertile, T-toxin resistant plants. Theor. Appl. Genet. **58,** 55—58.

Brown, J. S., Wernsman, E. A., Schnell, R. J., 1983: Effect of a second cycle of anther culture on flue-cured lines of tobacco. Crop Sci. **23,** 729—733.

Chaleff, R. S., 1981: Genetics of Higher Plants — Applications of Cell Culture. Cambridge: Cambridge Univ. Press.

Chaleff, R. S., Keil, K. L., 1981: Genetic and physiological variability among cultured cells and regenerated plants of *Nicotiana tabacum.* Molec. Gen. Genet. **181,** 254—258.

Chaturvedi, H. C., Mitra, G. C., 1975: A shift in morphogenetic pattern in Citrus callus tissue during prolonged culture. Ann. Bot. **39,** 683—687.

Cummings, D. P., Green, C. E., Stuthman, D. D., 1976: Callus induction and plant regeneration in oats. Crop Sci. **16,** 465—470.

De Paepe, R., Bleton, D., Gnangbe, F., 1981: Basis and extent of genetic variability among doubled haploid plants obtained by pollen culture in *Nicotiana sylvestris.* Theor. Appl. Genet. **59,** 177—184.

De Paepe, R., Prat, D., Huquet, T., 1982: Heritable nuclear DNA changes in doubled haploid plants obtained by pollen culture of *Nicotiana sylvestris.* Plant Sci. Lett. **28,** 11—28.

Deambrogio, E., Dale, P. J., 1980: Effect of 2,4-D on the frequency of regenerated plants in barley and on genetic variability between them. Cereal Res. Commun. **8,** 417—423.

Dolézel, J., Novàk, F. J., 1984: Effect of plant tissue culture media on the frequency of somatic mutations in *Tradescantia stamen* hairs. Z. Pflanzenphysiol. **144,** 51—58.

Durante, D., Geri, C., Grisvard, J., Guille, E., Parenti, R., Buiatti, M., 1983: Variation in DNA complexity in *Nicotiana glauca* tissue cultures. I. Pith tissue dedifferentiation. Protoplasma **114,** 114—118.

Edallo, S., Zucchinali, C., Perzenzin, M., Salamini, F., 1981: Chromosomal variation and frequency of spontaneous mutation associated with *in vitro* culture and plant regeneration in maize. Maydica **26,** 39—56.

Engler, D. E., Grogan, R. G., 1983: Isolation, culture and regeneration of lettuce leaf mesophyll protoplasts. Plant Sci. Lett. **28,** 223—229.

Evans, D. A., Sharp, W. R., 1983: Single gene mutations in tomato plants regenerated from tissue culture. Science **221,** 949—951.

Flavel, R. B., 1984: Transposable elements. Plant Mol. Cell. Biol. **1,** 207—219.

Fujiwara, A., 1982: Plant Tissue Culture 1982. Tokyo: Japn. Assoc. Plant Tissue Culture.

Fukai, K., 1983: Sequential occurrence of mutations in a growing rice callus. Theor. Appl. Genet **65,** 225—230.

Gamborg, O. L., Shyluck, J. P., Brar, D. S., Constabel, F., 1977: Morphogenesis and plant regeneration from callus of immature embryos of sorghum. Plant Sci. Lett. **10,** 67—74.

Gengenbach, B. G., Green, C. E., Donovan, C. M., 1977: Inheritance of selected pathotoxin resistance in maize plants regenerated from cell cultures. Proc. Natl. Acad. Sci., U.S.A. **74,** 5113—5117.

Gengenbach, B. G., Connelly, J. A., Pring, D. R., Conde, M. F., 1981: Mitochondrial DNA variation in maize plant regenerated during tissue culture selection. Theor. Appl. Genet. **59,** 161—167.

Gerasimova, T. I., Mizrokhi, L. J., Georgiev, G. P., 1984: Transposition bursts in genetically unstable *Drosophila melanogaster.* Nature **309,** 714—716.

Gould, A. R., 1982: Chromosome instability in plant tissue cultures studied with banding techniques. In: Fujiwara, A. (ed.), Plant Tissue Culture 1982. pp. 431—432. Tokyo: Jap. Assoc. Plant Tiss. Culture.

Green, C. E., 1977: Prospects of crop improvement in the field of cell culture. Hort. Sci. **12**, 7—10.

Groose, R. W., Bingham, E. T., 1984: Variation in plants regenerated from tissue culture of tetraploid alfalfa heterozygous for several traits. Crop Sci. (in press).

Grout, B. W., Crisp, E. T., 1980: The origin and nature of shoots propagated from cauliflower roots. J. Hort. Sci. **55**, 65—70.

Hanna, W. W., Lu, C., Vasil, I. K., 1984: Uniformity of plants regenerated from somatic embryos of *Panicum maximum* Jacq. (Guinea grass). Theor. Appl. Genet. **67**, 155—159.

Haydu, Z., Vasil, I. K., 1981: Somatic embryogenesis and plant regeneration from leaf tissues and anthers of *Pennisetum purpureum* Schum. Theor. Appl. Genet. **59**, 269—273.

Heinz, D. J., Krishnamurthi, M., Nickell, L. G., Maretzki, A., 1977: Cell, tissue and organ culture in sugarcane improvement. In: Reinert, J., Bajaj, Y. P. S. (eds.), Applied and Fundamental Aspects of Plant Cell, Tissue and Organ Culture. pp. 3—17. Berlin: Springer-Verlag.

Henke, R. R., Mansur, M. A., Constantin, M. J., 1978: Organogenesis and plantlet formation from organ and seedling-derived calli of rice *(Oryza sativa)*. Physiol. Plant **44**, 11—14.

Hoffman, F., Thomas, E., Wenzel, G., 1982: Anther culture as a breeding tool in rape. II. Progeny analyses of androgenetic lines and induced mutants from haploid cultures. Theor. Appl. Genet. **61**, 225—232.

Jelaska, S., Papes, D., Pevalek, B., Devide, Z., 1978: Developmental and karyological studies of *Vicia faba* callus cultures. pp. 101. Calgary: Fourth Intl. Cong. Plant Tissue Cell. Cult.

Jung-Heilinger, H., Horn, W., 1980: Variation nach mutagener Behandlung von Stecklingen und *In vitro*-Kulturen bei *Chrysanthemum*. Z. Pflanzenzüchtg. **85**, 185—199.

Karp, A., Nelson, R. S., Thomas, E., Bright, S. W. J., 1982: Chromosome variation in protoplast derived potato plants. Theor. Appl. Genet. **63**, 265—272.

Karp, A., Maddock, S. E., 1984: Chromosome variation in wheat plants regenerated from cultured immature embryos. Theor. Appl. Genet. **67**, 249—255.

Kemble, R. J., Flavell, R. B., Brettell, R. I. S., 1982: Mitochondrial DNA analysis of fertile and sterile maize plants from tissue culture with the Texas male sterile cytoplasm. Theor. Appl. Genet. **62**, 213—217.

Larkin, P. J., 1982: Sugarcane tissue and protoplast culture. Plant Cell. Tiss. Org. Cult. **1**, 149—164.

Larkin, P. J., Ryan, S. A., Brettell, R. I. S., Scowcroft, W. R., 1984: Heritable somaclonal variation in wheat. Theor. Appl. Genet. **67**, 443—455.

Larkin, P. J., Scowcroft, W. R., 1981: Somaclonal variation — a novel source of variability from cell culture for plant improvement. Theor. Appl. Genet. **60**, 197—214.

Larkin, P. J., Scowcroft, W. R., 1983: Somaclonal variation and crop improvement. In: Kosuge, T., Meredith, C. P., Hollaender, A. (eds.), Genetic Engineering of Plants. pp. 289—314. New York: Plenum.

Larkin, P. J., Scowcroft, W. R., 1983: Somaclonal variation and eyespot toxin tolerance in sugarcane. Plant Cell. Tiss. Organ. Cult **2**, 111—121.

Liu, M.-C., 1981: *In vitro* methods applied to sugar cane improvement. In: Thorpe, T. A. (ed.), Plant Tissue Culture. pp. 299—323. New York: Academic Press.

Lörz, H., Scowcroft, W. R., 1983: Variability among plants and their progeny regenerated from protoplasts of Su/su heterozygotes of *Nicotiana tabacum*. Theor. Appl. Genet. **66**, 67—75.

Maddock, S. E., Lancaster, V. A., Risiott, R., Franklin, J., 1983: Plant regeneration from cultured immature embryos and inflorescences of 25 cultures of wheat *(Triticum aestivum).* Exp. Bot. **34,** 915—926.

Maliga, P., 1984: Isolation and characterization of mutants in plant cell culture. Ann. Rev. Plant. Physiol. **35,** 519—542.

McClintock, B., 1951: Chromosome organization and genic expression. Cold Spr. Harb. Quant. Biol. **16,** 13—47.

McCoy, T. J., Phillips, R. L., 1982: Chromosome stability in maize *(Zea mays)* tissue cultures and sectoring in some regenerated plants. Can. J. Genet. Cytol. **24,** 559—565.

McCoy, T. J., Phillips, R. L., Rines, H. W., 1982: Cytogenetic analysis of plants regenerated from oat *(Avena sativa)* tissue cultures: high frequency of partial chromosome loss. Can. J. Genet. Cytol. **24,** 37—50.

Meins, F., 1983: Heritable variation in plant cell culture. Ann. Rev. Plant Physiol. **34,** 327—346.

Mix, G., Wilson, H. M., Foroughi-Wehr, B., 1978: The cytological status of plants of *Hordeum vulgare* L. regenerated from microspore callus. Z. Pflanzenzüchtg. **80,** 89—99.

Negrutiu, I., Jacobs, M., Caboche, M., 1984: Advances in somatic cell genetics of higher plants — the protoplast approach in basic studies on mutagenesis and isolation of biochemical mutants. Theor. Appl. Genet. **67,** 289—304.

Novàk, F. J., 1980: Phenotype and cytological status of plants regenerated from callus cultures of *Allium sativum* L. Z. Pflanzenzüchtg. **84,** 250—260.

Ogihara, Y., 1981: Tissue culture in Haworthia. Part 4: Genetic characterization of plants regenerated from callus. Theor. Appl. Genet. **60,** 353—363.

Oinuma, T., Yoshida, T., 1974: Genetic variation among doubled haploid lines of burley tobacco varieties. Jap. J. Breed. **24,** 211—216.

Oono, K., 1981: *In vitro* methods applied to rice. In: Thorpe, T. A. (ed.), Plant Tissue Culture. pp. 273—298. New York: Academic Press.

Orton, T. J., 1980: Chromosomal variability in tissue cultures and regenerated plants of *Hordeum.* Theor. Appl. Genet. **56,** 101—112.

Orton, T. J., 1983a: Spontaneous electrophoretic and chromosomal variability in callus cultures and regenerated plants of celery. Theor. Appl. Genet. **67,** 17—24.

Orton, T. J., 1983b: Experimental approaches to the study of somaclonal variation. Plant Mol. Biol. Rep. **1,** 67—76.

Prat, D., 1983: Genetic variability induced in *Nicotiana sylvestris* by protoplast culture. Theor. Appl. Genet. **64,** 223—230.

Reisch, B., Bingham, E. T., 1981: Plants from ethionine-resistant alfalfa tissue cultures: variation in growth and morphological characteristics. Crop Sci. **21,** 783—788.

Reuther, G., 1977: Adventitious organ formation and somatic embryogenesis in callus of Asparagus and Iris and its potential applications. Acta Hort. **78,** 217—224.

Roest, S., Van Berkel, M., Bokelmann, G., Broertjes, C., 1981: The use of an *in vitro* adventitious bud technique for mutation breeding of *Begonia* × *hiemalis.* Euphytica **30,** 381—388.

Sacristan, M. D., 1982: Resistance to *Phoma lingam* of plants regenerated from selected cell and embryogenic culture of haploid *Brassica napus.* Theor. Appl. Genet. **61,** 193—200.

Schaeffer, G. W., 1982: Recovery of heritable variability in anther-derived doubled-haploid rice. Crop Sci. **22,** 1160—1164.

Schaeffer, G. W., Sharpe, F. T., Cregan, P. B., 1984: Variation for improved protein and yield from rice anther culture. Theor. Appl. Genet. **67,** 383—389.

Schwartz, H. J., Galletta, G. J., Zimmerman, R. H., 1981: Field performance and phenotypic stability of tissue culture-propagated strawberries. J. Amer. Soc. Hort. Sci. **106,** 667—673.

Scowcroft, W. R., 1984: Genetic variability in tissue culture: Impact on germplasm conservation and utilization. IBPGR Report/84/152. Rome IBPGR Secretariat.

Scowcroft, W. R., Larkin, P. J., 1982: Somaclonal variation: a new option for plant improvement. In: Vasil, I. K., Scowcroft, W. R., Frey, K. J. (eds.), Plant Improvement and Somatic Cell Genetics. pp. 159—178. New York: Academic Press.

Scowcroft, W. R., Larkin, P. J., Brettell, R. I. S., 1983: Genetic variation from tissue culture. In: Helgeson, J. P., Deverall, B. J. (eds.), Use of Tissue Culture and Protoplasts in Plant Pathology. pp. 139—162. Sydney: Academic Press.

Scowcroft, W. R., Ryan, S. A., 1984: Tissue culture and plant breeding. In: Yeoman, M. M. (ed.), Plant Cell Culture Technology, in press. Oxford: Blackwell Scientific.

Secor, G. A., Shepard, J. F., 1981: Variability of protoplast-derived potato clones. Crop Sci. **21,** 102—105.

Selby, C., Collins, H. A., 1976: Clonal variation in growth and flavour production in tissue cultures of *Allium cepa* L. Ann. Bot. **40,** 911—918.

Shapiro, J. A., Cordell, B., 1982: Eukaryotic mobile and repeated genetic elements. Biol. Cell. **43,** 31—54.

Shepard, J. F., 1981: Protoplasts as sources of disease resistance in plants. Ann. Rev. Phytopathol. **19,** 145—166.

Shepard, J. F., Bidney, D., Shahin, E., 1980: Potato protoplasts in crop improvement. Science **208,** 17—24.

Sibi, M., 1976: La notion de programme génétique chez les végétaux supérieurs. II. Aspect expérimental: obtention de variants par culture de tissus *in vitro* sur *Lactuca sativa* L. Apparition de vigeur chez les croisements. Ann. Amélior. Plantes. **26,** 523—547.

Sibi, M., 1982: Heritable epigenic variations from *in vitro* tissue culture of *Lycopersicon esculentum* (var. Monalbo). In: Earle, E. D., Demarly, Y. (eds.), Variability in Plants Regenerated from Tissue Culture. pp. 228—244. New York: Praeger.

Skirvin, R. M., Janick, J., 1976: Tissue culture-induced variation in scented *Pelargonium* spp. J. Amer. Soc. Hort. Sci. **101,** 281—290.

Sorenson, J. C., 1984: The structure and expression of nuclear genes in higher plants. Adv. Genet. **22,** 109—144.

Sree Ramulu, K., 1982: Genetic instability at the S-locus of *Lycopersicon peruvianum* plants regenerated from *in vitro* culture of anthers: generation of new S-allele reversions. Heredity **49,** 319—330.

Sree Ramulu, K., Dijkuis, P., Roest, S., 1983: Phenotypic variation and ploidy level of plants regenerated from protoplasts of tetraploid potato (*Solanum tuberosum* L. cv. 'Bintje'). Theor. Appl. Genet. **65,** 329—338.

Staba, E. J., 1982: Production of useful compounds from plant tissue cultures. In: Fujiwara, A. (ed.), Plant Tissue Culture 1982. pp. 25—30. Tokyo: Jap. Ass. Plant. Tissue Culture.

Sun, Z., Zhao, C., Zheng, K., Qi, X., Fu, Y., 1983: Somaclonal genetics of rice, *Oryza sativa* L. Theor. Appl. Genet. **67,** 67—73.

Sutter, E., Langhans, R. W. L., 1981: Abnormalities in *Chrysantheum* regenerated from long term cultures. Ann. Bot. **48,** 559—568.

Tabata, M., Ogino, T., Yoshioka, K., Yoshikawa, M., Hiraoka, N., 1978: Selection of lines with higher yield of secondary products. In: Thorpe, T. A. (ed.), Frontiers of Plant Tissue Culture 1978. pp. 213—222. Calgary: Int. Ass. Plant Tiss. Culture.

Thomas, E., Bright, S. W. J., Franklin, J., Lancaster, V., Miflin, B. J., 1982: Variation amongst protoplast-derived potato plants (*Solanum tuberosum* cv. 'Maris Bard'). Theor. Appl. Genet. **62,** 65—68.

Umbeck, P. F., Gengenbach, B. G., 1983: Reversion of male-sterile T-cytoplasm maize to male fertility in tissue culture. Crop Sci. **23,** 584—588.

Van Harten, A. M., Bouter, H., Broertjes, C., 1981: *In vitro* adventitious bud techniques for vegetative propagation and mutation breeding of potato (*Solanum tuberosum* L.). Euphytica **30,** 1—8.

Wakasa, K., 1979: Variation in the plant differentiated from the tissue culture of pineapple. Jap. J. Breed. **29,** 13—22.

Wenzel, G., 1980: Recent progress in microspore culture of crop plants. In: Davies, D. R., Hopwood, D. A. (eds.), The Plant Genome. pp. 185—196. Norwich: John Innes Charity.

Wenzel, G., Schieder, O., Przewozny, T., Sopory, S. K., Melchers, G., 1979: Comparison of single cell culture derived *Solanum tuberosum* L. plants and a model for their application in plant breeding programs. Theor. Appl. Genet. **55,** 49—55.

Wernicke, W., Potrykus, I., Thomas, E., 1982: Morphogenesis from cultured leaf tissue of *Sorghum bicolour* — the morphogenetic pathways. Protoplasma **111,** 53—62.

Williams, L., Collins, H. A., 1976: Growth and cytology of celery plants derived from tissue cultures. Ann. Bot. **40,** 333—338.

Zenk, M. H., 1978: The impact of plant tissue culture on industry. In: Thorpe, T. A. (ed.), Frontiers of Plant Tissue Culture 1978. pp. 1—13. Calgary: Int. Assoc. Plant. Tiss. Culture.

Subject Index

A pagenumber with * refers to a whole chapter